PHYSICS for ARCHITECTS

Yehuda Salu

Howard University

Disclaimer: This publication is designed to provide accurate information about the subject matter that it covers. It is presented in good faith, and although the author and publisher have made every reasonable effort to represent every information accurately, they do not warrant or assume any liability for its accuracy, completeness, or suitability for any purpose. This book is published with the understanding that neither the publisher nor the author are attempting to render any engineering or other professional services. If such services are required, the reader/user should apply his/her own professional knowledge, or seek appropriate professional advice.

Cover: A collage of The Gateway Arch, Saint Louis. Based on an original photograph from 1965, courtesy National Park Service, Jefferson National Expansion Memorial.

ISBN-13: 978-1463708115
ISBN-10: 1463708114

TABLE OF CONTENTS

PREFACE TO THE FIRST EDITION

Students of most undergraduate architecture programs in the United States are required to take an introductory physics course. There are three good reasons for that requirement. First, architects have to understand the fundamentals of physics as they apply to processes taking place in buildings and in structures. Second, as part of general education, physics broadens our understanding of the physical world around us. Third, since physics is an exact science that relies on mathematics, solving physics problems enhances the analytical and scientific thinking skills of the student. "Physics-for-Architects" was written specifically for architecture students, aiming to satisfy those three basic requirements.

The specific details of an introductory physics course may vary from one architecture program to another. Different programs may have different overall concentrations, which may affect the relative weight of their physics component. Within any given architecture program, physics topics may be distributed between introductory physics and other professional courses. Also, different programs may put different emphasis on the mathematical aspects of physics, sometimes referred to as the difficulty level of the physics course. What is common to all architecture programs is that the total time allotted to introductory physics is between one and two semesters. "Physics-for-Architects" has been designed so that it could be used in a wide variety of undergraduate architecture programs, according to their specific needs.

Time is an important factor when compiling any curriculum and a textbook to support it. Class time as well as the time available for students' homework is limited. Therefore, the topics that should be included in the physics curriculum and the depth of each topic have to be selected very judiciously. Topics have to be prioritized, and eventually some topics will have to be dropped out altogether. In "Physics-for-Architects", the fundamental principles of physics and topics that have direct bearing on architecture receive the highest priority. Other topics are secondary. For example, Newton's Laws are fundamental, and they are discussed in detail both theoretically and in numerous applications. On the other hand, some topics of modern physics had to be considered secondary, and they are interwoven at a qualitative level within other chapters. Some general concepts, such as series and parallel connections, apply to several topics. Since architects encounter them mainly in situations of heat flow, they are introduced here in that context. Most physics textbooks introduce those concepts in relation to electrical circuits of resistors and capacitors. The latter has no relevance to architects, and is not presented here.

The math pre-requisites to all the topics in the book are high-school-level algebra and trigonometry. An intensive review, only of the mathematical concepts and techniques that are needed for the course, is provided as an introductory chapter. Detailed solved examples are embedded in the text, and problems of various levels of mathematical difficulty are provided at the end of each chapter. That should allow teachers to tailor the math level of each topic to the specific needs of their program and their students. Teachers may opt to teach some topics at the highest level of mathematics, while other topics could be addressed with lesser mathematical rigor, or even only qualitatively.

It is common to distinguish between physics-type problems and engineering-type problems. Physics-type problems are those that can be solved by using physics formulas. Engineering-type problems rely not only on closed physics formulas, but also on practical approximations, simplifications, ad hock tables, and the likes. Most physics textbooks limit their scope to physics-type problems. However, many physics related problems that architects encounter are of the engineering-type. Most architecture programs teach that kind of problems in specialized courses, distinct from introductory physics. That creates an unnecessary dichotomy between physics and its applications. One of the tenets of "Physics-for-Architects" is to bridge that gap and to demonstrate to architecture students the relevance of physics. Many engineering-type problems are included in the text. The transition from axiomatic, physics-type approach to practical, engineering-type approach in analyzing situations and solving problems is illustrated throughout the text.

It should be noted, though, that "Physics-for-Architects" is a physics textbook, and it is not intended to be a replacement for the professional architecture books. Engineering-type discussions and problems are brought here as an introductory material, for illustrative purposes. Whenever a real situation has to be addressed, it is recommended that the reader seek professional advice and consult the professional literature and building codes.

Although "physics-for-Architects" has been designed for architecture students, "physics is physics". The physics principles, concepts, formulas and ideas brought here are the same as those presented in other general purpose, introductory physics textbooks. The main difference is in the examples used to introduce and to explain the physics. Those examples are related to buildings and structures and to their use by humans. Readers from other disciplines too may find the book interesting and effective in their study of physics.

Acknowledgements
This book evolved from my notes for the course Physics for Architecture that I have taught at Howard University for more than a decade. I am indebted to all my students for their comments, suggestions, and insights that they have shared with me during those years. Especially I am grateful for their patience and understanding during the entire process of trial and error that we had to go through in order to test and finally select the material for the book. The partial support of the National Science Foundation, through grant number DUE-0087360, is gratefully appreciated. Special thanks for permitting the use of their materials are given to: the American Society of Heating, Refrigerating, and Air-conditioning Engineers; Arthur Hsu; Boston Properties; English Heritage; John Wiley and Sons; Bureau of Reclamation, U.S. Department of the Interior; U.S. Geological Survey; and Washington University.

PREFACE TO THE SECOND EDITION

In the second edition, the envelope of the presentations was pushed a bit further. Topics that are seldom mentioned in general-purpose introductory physics textbook, but that have relevance to architects and builders, were included, as long as they could be thoroughly explained by basic algebra and trigonometry. One example is a section about the catenary and stone arches. In addition, explanations of basic concepts were expanded, some problems were modified and some new ones were added, and humbling errors were corrected thanks to feedback that I received from my students and from instructors that have been using the book. My special thanks are to William Moore from the University of Miami, Peter Persans from Rensselaer Polytechnic Institute, Al Glodowski from Illinois Institute of Technology, Melissa Carubia from Boston Architectural College, Zackery Belanger from Kirkegaard Associates, and Gary Lythe and Greg Kopp from the Boundary Layer Wind Tunnel Lab for their insightful comments. Last but not least, Jennifer Clark, archivist, US National Park Service, for her support in getting the picture for the book's cover.

Yehuda Salu, October 2008

1. MATHEMATICAL REVIEW

1.1 Algebra

1.1.1 Equations and formulas

A formula is a relationship between variables that are represented by letters. A formula has a left-hand side (LHS) expression, an equal sign, and a right hand side (RHS) expression. Quite often, the LHS consists of only one variable. For example, $A = a \cdot b$ is a formula that relates the area of a rectangle, represented by A, to its length, represented by a, and its width, represented by b. When numbers are substituted for the variables of the RHS, the value of the corresponding LHS variable can be found. The use of formulas is very common in physics. The variables in physics formulas represent a variety of physics entities, and the formulas express in concise ways the relationships between those variables.

Sometimes, we know the values of some variables that appear on both sides of a formula, and we want to know the values of the unknown variable or variables. For example, we may know that the area of a certain rectangle is 15 square meters, its length is 5 m, and we want to know its width. If we substitute these values in the formula for the area of a rectangle, we get $15 = 5 \cdot b$. This is an equation for the unknown b. We can solve it and find b=3, meaning that the width of that rectangle is 3 meters long. Solving equations is a main staple of physics. In this course, we will encounter a variety of equations, so a quick review of methods of solving equations is in place.

There are many kinds of equations, and there are specific techniques for solving each kind. The equations that we will encounter the most belong to the class of what is called "linear algebraic equations" (LAE). In LAE's, the LHS and the RHS contain combinations of numbers, operators, and letters. The operators are + (add), - (subtract), $\times$ or $\cdot$ (multiply), and / (divide). The letters stand for variables. An unknown is a variable that we want to find its value. In a LAE, unknowns are multiplied by numbers but not by themselves or by other unknowns. For example $2x+3y=x/5$ is a LAE with two unknown (x and y). The equation $2xy+3x=x/5$ is not a LAE because of the term xy (the unknown x multiplies the unknown y). The equation $2x+3y=x^2/5$ is also not a LAE because of the term x^2.

The basic idea used in solving equations is that if we start with an equation A=B, and if we modify A and B in the same way, the modified A will remain equal to the modified B. For example, if A=2+2 and B=1+3 we start with 2+2=1+3. If we add 1 to both sides, we'll have 2+2+1=1+3+1, which is true. If we divide both sides by 2, the equality will hold (2+2)/2=(1+3)/2. If we take the square root of both sides, the equality will hold $\sqrt{2+2} = \sqrt{1+3}$, and so on.

In order to solve a LAE with one unknown, we operate the same way on the LHS and on the RHS. We choose the operations such that finally only the unknown remains in one side and only numbers remain on the other side of the equation. For example, if the equation that we want to solve is: 2x+3=4, our goal is to get x=some number. The 2 and the 3 are in our way. We can get rid of the 3 in the LHS by subtracting 3 from both sides: 2x+3-3=4-3. This gives us 2x=1. Now only the 2 is in our way. We get rid of it by dividing both sides by 2: 2x/2=1/2, which gives us x=1/2.

When the LHS of an equation consists only of one variable, and the RHS contains other variables and numbers, we say that this is a formula for the variable of the LHS. For example, $V=I\cdot R$ is a formula for V. By applying the same operation on both side, this formula may become a formula for I: I=V/R, and a formula for R: R=V/I. We say that in the first case the original formula was solved for I, and in the second case it was solved for R. All three formulas express the same idea. When it comes to memorizing formulas, if you memorize one formula you should be able to derive the others.

We will encounter situations where it is needed to solve sets of two equations with two unknowns. There are two main methods of doing that: the method of substitution and the method of Gauss' elimination.

Let's first see in an example how the method of substitution works:
The two equations are:
(1) x+y=5
(2) 2x-y=1

Step 1. We choose equation (1). We want to modify it to x=something. The unknown y is in our way, so we get rid of it by subtracting y from both sides. We get x+y-y=5-y, which simplifies to x=5-y.
Step 2. We substitute the expression for x in equation (2). We get 2(5-y)-y=1. This is an equation with one unknown y.
Step 3. We solve 2(5-y)-y=1. We want to bring it to the form y=something. The 2, the parenthesis, and the 5 are in our way. We first get rid of the parenthesis: 10-2y-y=1, or 10-3y=1. Now we get rid of the 10: 10-3y-10=1-10, or –3y=-9. We divide both sides by (-3) to get: -3y/(-3)=-9/(-3), or y=3. This is the first half of the solution.
Step 4. We substitute y=3 in equation (1): x+3=5
Step 5. We solve x+3=5 to get x=2. This is the second half of the solution.

In general terms, the method of substitution has the following steps (corresponding to the last example):

Choose one equation (say the first), and express one of the unknowns in it with the other unknown and the numbers.

Substitute that expressed unknown in the other equation (the second), which now becomes an equation with one unknown.

Solve that equation (the modified second equation). This is the first part of the solution.

Substitute the value of the unknown that you have found in the previous step in the other equation (the first), which now becomes an equation with one unknown.

Solve the equation of the previous step. This is the second part of the solution.

The second method of solving sets of LAE's is by Gauss' elimination. The following is an example that illustrates the method of Gauss' Elimination.

The original equations are:

(1) $2X-3=-3Y$

(2) $4Y=5-3X$

First we bring these equations to the standard format:

(1') $2X+3Y=3$

(2') $3X+4Y=5$

We want to eliminate X from equations (1') and (2'). The coefficient of X in (1') is 2, and the coefficient of X in (2') is 3. If we multiply both sides of (1') by 3, and both sides of (2') by 2 we get:

(1'') $6X+9Y=9$

(2'') $6X+8Y=10$

We now subtract the LHS of (2'') from the LHS of (1'') and the RHS of (2'') from the RHS of (1'') to get

(1'')-(2''): $6X-6X+9Y-8Y=9-10$ or $Y=-1$.

This is half of the solution. We substitute it in (1) (we may also substitute it in (2) to get:

$2X-3=-3(-1)$ or $2X-3=3$. --> $2X=6$. -->$X=3$, which is the second half of the solution.

In general terms, the following steps are taken when Gauss' elimination is used:

First, the two equations are manipulated to the pre-elimination form: The coefficient of one of the unknowns (say the x) is the same in both equations, all the unknowns are in the LHS's, and only numbers appear in the RHS's:

(1) $mX+nY=p$

(2) $mX+kY=q$

m, n, p, q, and k are numbers, and X and Y are the unknowns. (These correspond to (1'') and (2'') in the last example).

In the elimination step we subtract from the LHS of (1) the LHS of (2), and from the RHS of (1) we subtract the RHS of (2). This is justified because we subtract the same amount from the LHS and the RHS of (1).

We get:

nY-kY=p-q, or

(n-k)Y=p-q

This corresponds to (1'')-(2'') from the last example.

(The word elimination in the name of the method indicates that one of the unknowns (X in this case) was eliminated in the process (mX-mX=0).

(n-k) is a number and so is p-q. We get Y=(p-q)/(n-k), which is the first half of the solution. We substitute this value of Y in either (1) or (2), and get one equation with one unknown, whose solution is the second half of the solution of the set (1) and (2).

1.1.2 Understanding formulas

Although formulas enable us to calculate the precise relationships between variables, quite often it is possible to get a general idea about such relationships without actually doing any calculation. Consider for example the formula

$$C = P \cdot N$$

Where C = total cost of apples bought, P = price of 1 pounds of apples, and N = number of pounds bought.

By using this formula wc can calculate the total cost of any amounts of apples that we buy, for any price. But the formula tells us also that the more apples we buy, the more we pay, or in other words, the total cost is proportional to the number of pounds bought. We can see it from the formula, because the total cost is equal to the product of a constant number P (the price of one pound) and a variable N (number of pounds bought). The property of a product of two numbers is that if one is held constant and the second is increased, the outcome increases. If the second number is decreased, the outcome decreases.

So, in general, if a variable appears as a part of a product in the right hand side of a formula, we know that the variable on the left-hand-side is proportional to that right-hand-side variable.

Consider as a second example the formula

$$S = \frac{G}{N}$$

Where S is the share of each winning ticket in a lottery, G is the grand prize, and N is the number of winning tickets. (According to the rules of this lottery, there may be a number of winning tickets, and the grand prize is divided equally between them). We can use this formula to calculate exactly the share of each winning ticket (if there is at least one) for any number of winning tickets and for any grand prize. But this formula tells us also that the more winning tickets there are, the less is the share of each ticket. We can see it in the formula, because the variable on the left-hand-side (S) is equal to a fraction. When the numerator of a fraction (G) is held constant and the denominator (N) is increased, the value of the fraction (S) decreases. When the denominator decreases, the value of the fraction as a whole increases. We say that S is inversely proportional to N.

In a formula, if the right-hand-side is expressed as a fraction, the value of the variable in the left-hand-side is inversely proportional to the variable in the denominator of the right-hand-side. (Inversely proportional is the same as proportional to the inverse. An inverse of the number N is 1/N.)

From this formula we can see also that S is proportional to G, meaning that if G increases S increases, and if G decreases S decrease. This is because S is equal to a fraction whose numerator is G. By increasing the numerator we increase the value of the fraction, and by decreasing the numerator the value of the fraction decreases.

To summarize: in a formula, if the right-hand-side is expressed as a fraction, the value of the variable in the left-hand-side is proportional to the variable in the numerator of the right-hand-side, and inversely proportional to the variable in the denominator.

There are formulas in which the left-hand-side is proportional to the square of a variable on the right-hand-side. For example, the formula for the area A of a circle of radius r is $A = \pi \cdot r^2$. In this case, the area is proportional to the square of the radius. Let's compare this formula with the formula $L = 2\pi \cdot r$, which gives the length L of the circumference of a circle whose radius is r. The circumference is proportional to r. If r is doubled, the circumference is doubled, if r is increased by a factor of 10, the circumference is increased by a factor of 10. However, if r is doubled, the area A is increased by a factor for $2^2=4$, and if r is increased by a factor of 10, the area A is increased by a factor of $10^2=100$. Certain formulas enable us to evaluate relative changes in their left-hand-side variable, without actually going through the whole calculation.

When we have a formula, we may be able to tell which variables are proportional to which. However, if we just know that two variables are proportional to each other, we cannot express this relationship by a formula that has an equal sign. For example, if we want to express the fact that the cost C of apples is proportional the number of pounds N that we buy, we can write

$C \propto N$, where $\propto$ is the proportionality symbol.

In order to "upgrade" a proportionality relationship to an equality, we have to multiply one of the variables by a constant. (In this example, we can multiply N by P, the price of one pound of apples). Such a constant is called a proportionality factor, or a proportionality constant. (In this example, P serves as a proportionality constant).

PROBLEMS

Expressing relations by formulas

1. How many days (d) are there in w weeks? (answer: d=7w)
2. How many hours (h) are there in d days?
3. How many cents (c) are d dollars?
4. How many dollars (d) are c cents?
5. How many ounces (O) are p pounds?
6. How many cents (c) are d dimes and q quarters?
7. How many days (d) are w weeks and m months?
8. How many yards (y) are f feet and i inches?
9. What is the average score (S) of a student who got x points in the first test, 75 in the second, y in the third, and 98 in the fourth?
10. How many feet (F) have c cows, p pigs, and d ducks?

Solve the following equations of one unknown:

11. $2x+121=543$
12. $x-120=-40.5$
13. $-x=13$
14. $46-x=46$
15. $88=2x-12$
16. $4x=150$
17. $x \cdot \frac{3}{2} = 9$
18. $\frac{x}{5} = 9$
19. $\frac{x \cdot 5}{2} = \frac{25}{3}$
20., $\frac{5}{x} = 9$
21. $\frac{9}{7} = \frac{3}{x}$
22. $4-5x=2$
23. $2\frac{3}{7}x = 11$
24. $4\frac{2}{5}x + 1\frac{2}{9} = 13$
25. $7+11=4+2x$
26. $13+4x=-2x+9$
27. $4x+9=12x$
28. $-3x=x-2$
29. $\frac{x+4}{2} = 15$
30. $\frac{2x-9}{6} = \frac{4x+2}{3}$
31. $\frac{2x-9}{6} = \frac{4x+2}{3} + 1$
32. $4(2x+11)=7$
33. $6(2x+5-9+7)=(3-4x+5)(5+2)$
34. $\frac{2}{x+1} = 15$
35. $\frac{1}{4} + \frac{1}{x} = \frac{1}{8}$
36. $\frac{1}{2} + \frac{1}{x} = -\frac{1}{4}$
37. $\frac{x+9}{x+9} = 1$
38. $\frac{x+9}{x+9} = 2$

Manipulating formulas

39. Given $S = v \cdot t$, solve for t; solve for v.

40. Given $S = \frac{1}{2} a t^2$, solve for a; solve for t.

41. Given $X = x_0 + v_0 t + \frac{1}{2} a t^2$, solve for a; solve for v_0.

42. Given $E = m c^2$, solve for m.

43. Given $H = \frac{kA(t_2 - t_1)}{L}$, solve for k; solve for t_2.

44. Given $\frac{P_1 V_1}{T_1} = \frac{P_2 V_2}{T_2}$, solve for V_1; solve for T_2.

45. Given $f_1 = \frac{1}{2L}\sqrt{\frac{S}{\mu}}$, solve for S; solve for μ.

46. Given $\frac{1}{R} = \frac{1}{R_1} + \frac{1}{R_2}$, solve for R.

Solve the following sets of equations with two unknowns:

47. x+y=7
x=3

48. 2x+3y=17
y=3

49. x-y=9
x-2=9-x

50. 7x+2y=15
2x-1=3

51. 2x+3y=9
x=y

52. 4x+6y=18
x-y+3=3

53. -2x-7y=10
3x-5y=x+2y+6

54. x+y=4
x-y=0

55. 2x+3y=7
x+5y=10.5

56. 2x+3y=9
4x+2y=14

57. 2x-3y=-1
5x+5y=10

58. 3x+2=-x+6y
10y=20-10x

Understanding formulas

In the following formulas, you can figure out relationships between variables, even though you may not know (yet) the meaning of these variables.

59. The voltage across a resistor V is given by the formula $V = I \cdot R$, where I is the current through the resistor, and R is the resistance of the resistor.
True or False?
The voltage is proportional to the current.

The voltage is inversely proportional to the resistance.

60. The heat flow Q through a wall is given by the formula: $Q = \frac{k \cdot A \cdot T \cdot t}{d}$, where k is the heat conductivity constant of the wall, A is the area of the wall, T is the temperature difference across the wall, t is the duration of the heat flow, and d is the thickness of the wall.

True or False?
The heat flow is proportional to the area of the wall.
The heat flow is inversely proportional to the duration of the heat flow.
The heat flow is inversely proportional to the thickness of the wall.
The heat flow is inversely proportional to d/k.

61. The distance S covered by a car accelerating from rest in time t is given by $S = 0.5 \cdot a \cdot t^2$, where a is the acceleration of the car. If the time is tripled, what would be the relative change in the distance covered by that car?

62. The electric potential V created at a distance of r from a point source is given by $V = \frac{C}{r}$, and the electric field E is given by $E = \frac{C}{r^2}$, where C is a constant that depends on the source. What would be the relative change in V and in E if the distance r is quadrupled?

63. Using the proportionality symbol, write down the fact that the area A of a triangle is proportional to its height h. What is the proportionality factor needed to upgrade this relationship to an equality?

1.2 Geometry

The term "congruent", which is used frequently in geometry, means identical. For example, two congruent triangles mean two triangles that are identical to each other, but which are found in different locations. It is possible to move around one of those congruent triangles without distorting it, and bring it to an exact match with the other.

1.2.1 Length, area, volume

Lines have length. In order to measure the length of a line, a unit of length has to be used. There are many units of length e.g. meter, foot, yard, mile, etc. When we say that the length of a straight-line-segment is three meters, we mean that three units of length, one meter each, can exactly cover that segment. This coverage will not leave any part of that segment uncovered and will not cover any part of it more than once.

Surfaces have surface-area, or in short area. Similar to lines, areas are measured by units of area. Each unit of length has a corresponding unit of area. For example, the unit of length 'foot' has a corresponding unit of area 'square-foot'. If the unit 'foot' can be realized by a ruler one foot long, the unit 'square-foot' can be realized by a square tile, whose side is one foot long. When we say that the area of a certain rectangle is three square-feet, we mean that that rectangle can be exactly covered by three units of area, each of which is one square foot. No part of the surface is left uncovered or covered more than once.

Three-dimensional objects have volume. Like in the cases of lines and surfaces, units of volume are used to measure the volume of three-dimensional objects. Each unit-length has a corresponding unit of volume e.g. cubic foot is the unit of volume that corresponds to foot. A cubic foot can be realized by a cube whose size is one foot long. When we say that the volume of a certain box is ten cubic-feet, we mean that ten volume-units, one cubic foot each, can exactly feel that entire space of the box. No part of that box will be left out or be filled more than once.

Any of those units of length, area, and volume can be cut into smaller pieces. These smaller pieces can be used to cover, and thus measure, objects that are curved or that cannot be covered by an integer number of the appropriate whole units.

Some geometric objects have formulas for their sizes:

The circumference (length) of a polygon is the sum of the lengths of its sides.
The circumference L of a circle of radius r is $L=2\pi r$.
The area A of a rectangle of length a and width b is $A=a \cdot b$.
The area A of a triangle of base a and height h is $A=1/2\ a \cdot h$
The area A of a circle of radius r is $A=\pi r^2$.
The area A of the surface of a sphere of radius r is $A=4\pi r^2$.
The volume V of a prism or a cylinder of base area B and height h is $V= B \cdot h$
The volume V of a sphere of radius r is $V=4/3\pi r^3$.

Since there are several units for length, area, and volume, it is important to know how to convert from one unit to another. This is done with the aid of conversion relations. Here are some conversion-relations for length:

1 yard= 3 feet.
1m=100cm=3.28ft=39.4in
1ft=30.5cm=12in
1mi=1.61km=5280ft
It should be remembered that these conversion-relations are only for length.

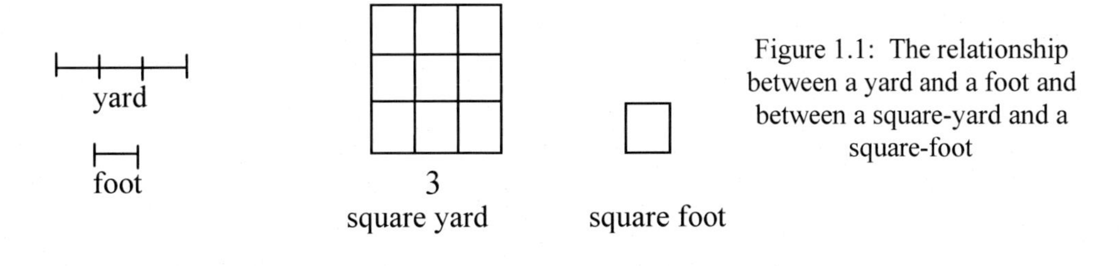

Figure 1.1: The relationship between a yard and a foot and between a square-yard and a square-foot

Conversion-relations for area and volume are derived from these ones, based on the definition of unit-length, unit-area, and unit-volume. To illustrate how this is done, consider the yard and the foot. Figure 1.1 shows the relationship between a yard and a foot and between a square-yard and a square-foot. There are 3x3=9 square-foot in a square yard.

Similarly, there are 3x3x3=27 cubic-foot in a cubic-yard (Figure 1.2)

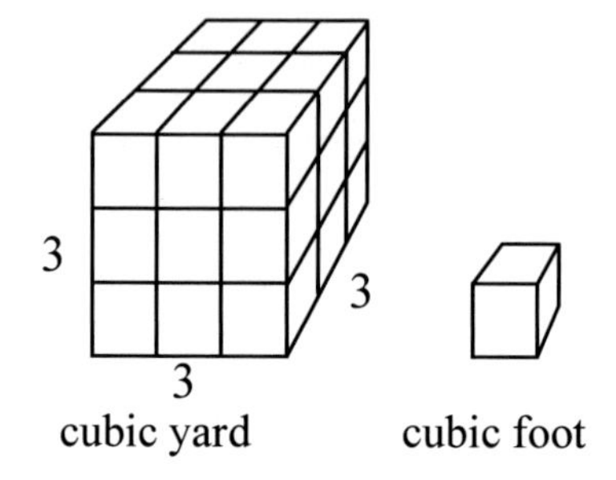

Figure 1.2: The relationship between a cubic yard and cubic foot

1.2.2 Angles

Angles are formed by two rays that start from the same point. The size of an angle can be determined by placing its head at the center of a circle, and then by considering the intersections of its rays with the circle. The size of the angle is proportional to the ratio between the length of the arc that connects those two intersection points and the circumference of the entire circle.

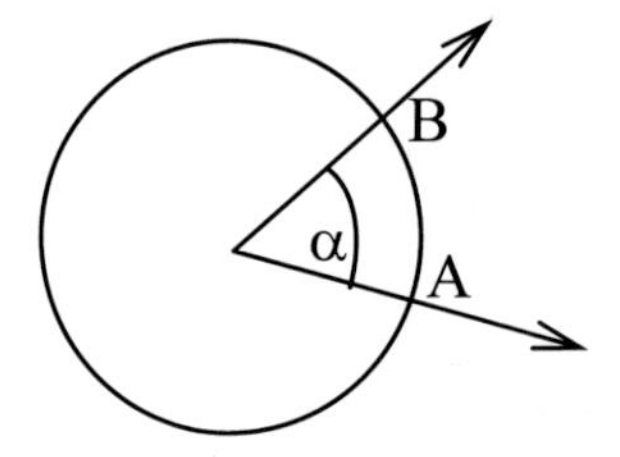

Figure 1.3: The size of the angle α depends on the ratio between the length of the arch that connects points A and B and the circumference of the circle.

When degrees are used to express the size of angles, 360 degrees correspond to the entire circle, 180 degrees to a straight angle (half a circle), 90 degrees to a right angle (a quarter of a circle) and so on. When angles are measured in radians, 2π radians correspond to the entire circle (360 degrees), π radians correspond to a straight line (180 degrees), $\pi /2$ radians correspond to a right angle, and so on. In these expressions, π=3.14159 approximately. The conversion formulas between degrees and radians are:

degrees=180 x radians/π and radians=π x degrees/180.

Based on these formulas, 1 radian=57.2958 degrees (approximately).

The intersection of two lines creates two pairs of vertical angles. In figure 1.4, angles α and β are vertical and angles γ and δ are vertical. Vertical angles are congruent:

$\alpha = \beta$ and $\gamma = \delta$

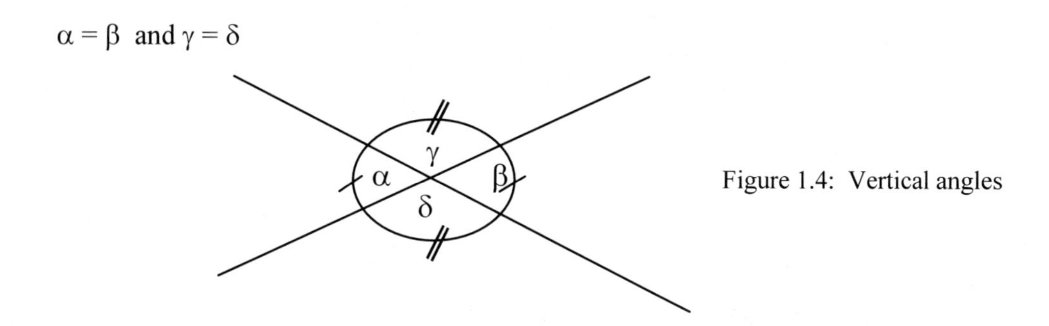

Figure 1.4: Vertical angles

Eight angles are formed when two parallel lines are intersected by a third line, as shown in figure 1.5. The following angles are congruent:

a=d=A=D, and b=c=B=C

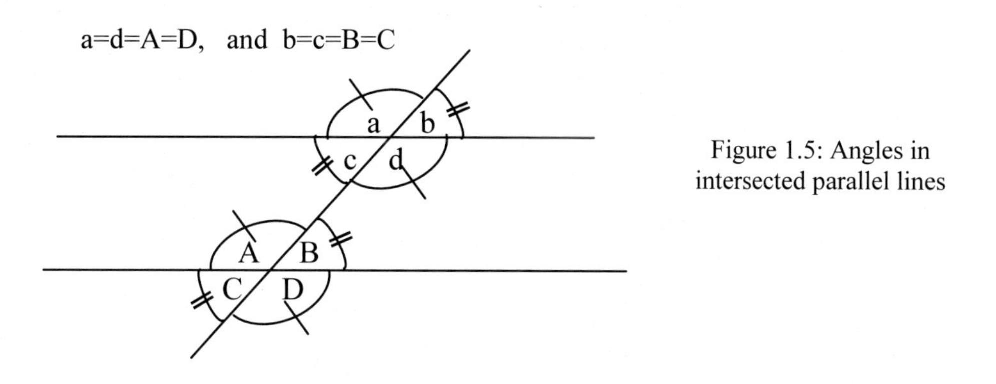

Figure 1.5: Angles in intersected parallel lines

The sum of the sizes of the angles in a triangle is 180 degrees. In an isosceles triangle, the angles adjacent to the congruent sides are congruent. In an equilateral triangle, each angle is 60 degrees.

PROBLEMS

Use drawings when solving problems 1 and 2.

1. A concrete sidewalk 3 feet wide has to be poured around a rectangular 14f by 35ft swimming pool. The sidewalk has to be 4 inches thick. How many cubic feet of concrete are needed?

2. A fence has to be erected around the pool, 2 feet away from the sidewalk. The fence has two gates, each 3 feet wide. How long is the fence?

3. Find the size of the angle α in each of the drawings. ⊏ indicates right angle.

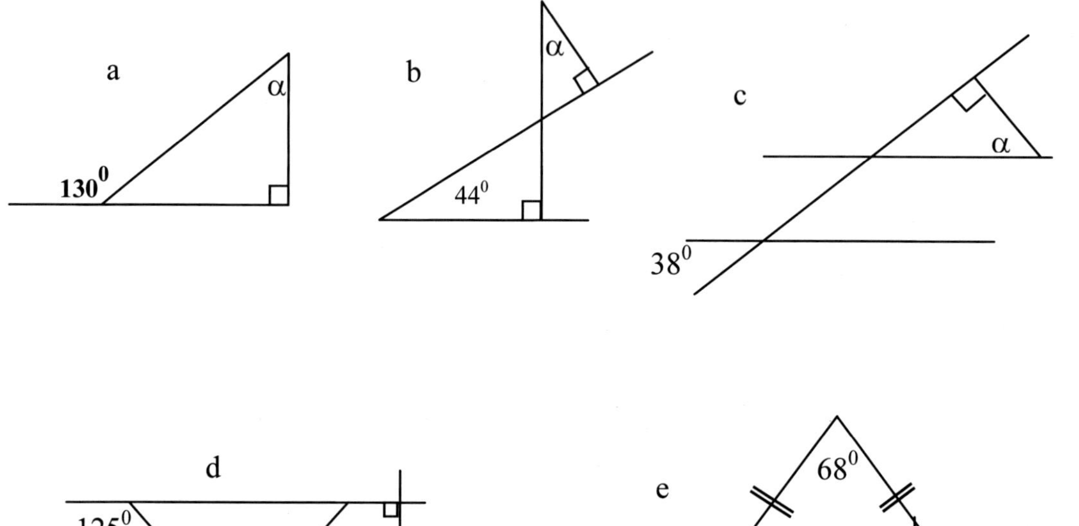

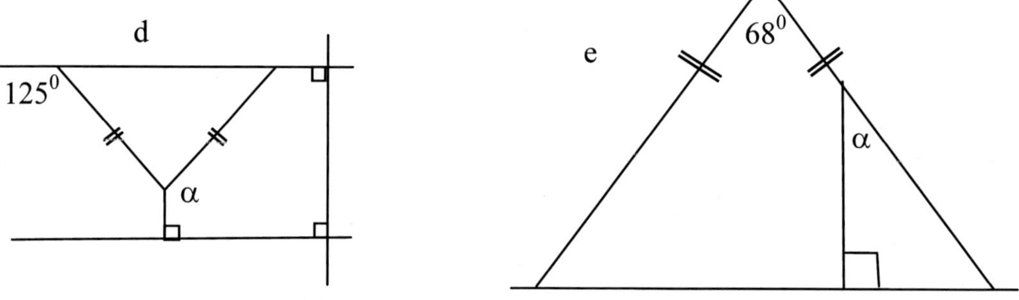

4. A semispherical dome tops the cylindrical main room of the Pantheon in Rome. The apex of the dome is 43 m above the floor, and its internal diameter is also 43 m. Built in 118-128 AD by the emperor Hadrian, the concrete dome has at its center a circular opening (oculus) of a diameter 8.7 m, through which light enters the room and moves around it, as the sun and the moon traverse the sky.
What is the area of the oculus?
What is the surface area of the inside of the dome (assume a smooth, complete semi-sphere)? What is the area of the internal walls? (Assume a complete, smooth cylinder)
What is the volume enclosed by the dome and its supporting cylinder?

Figure 1.6: Left: a model of the Pantheon.

Right: The inside of the Pantheon. The dome and its oculus and a part of the side walls.

1.3 Trigonometry

1.3.1 The trigonometric functions of acute angles

Similar triangles are triangles that have the exact same shape, but different sizes. In similar triangles, the corresponding angles are equal to each other. Figure 1.7 shows two right angle triangles that are similar to each other.

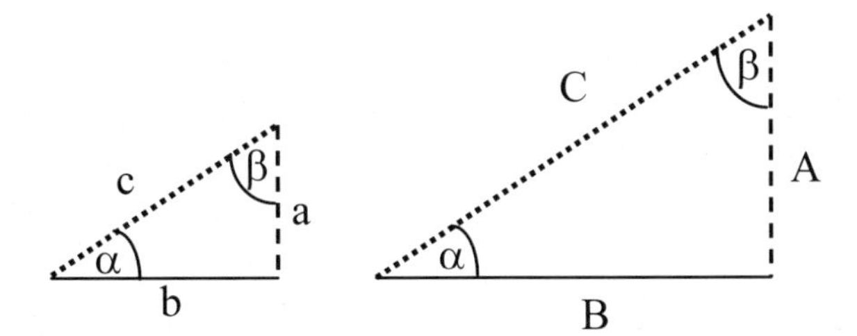

Figure 1.7: Two similar right-angle triangles

The corresponding sides (same line-patterns) are clearly not equal to each other. For example, 'a'–the length of the side opposite the angle α in the left triangle–is smaller than the length of the corresponding side ‘A’ in the right triangle. However, it was found that the corresponding ratios of two sides in similar triangles are equal to each other. So,

$$\frac{a}{c}=\frac{A}{C};\qquad \frac{b}{c}=\frac{B}{C};\quad \text{and}\qquad \frac{a}{b}=\frac{A}{B}. \qquad [1.1]$$

(You can verify for yourself these equalities by measuring the lengths of the sides and calculating their ratios.) Because of this property, each ratio was given a name as follows:

The ratio between the length of the side opposite the angle α and the hypotenuse is called the sinus of α, or in short sin(α) :

$$\sin(\alpha)=\frac{a}{c} \qquad [1.2a]$$

The ratio between the length of the side adjacent to the angle α and the hypotenuse is called the cosine of α, or in short cos(α) :

$$\cos(\alpha)=\frac{b}{c} \qquad [1.2b]$$

The ratio between the length of the side opposite the angle α and the adjacent side is called tangent of α, or in short tan(α) :

$$\tan(\alpha)=\frac{a}{b} \qquad [1.2c]$$

Those three ratios do not depend on the size of the triangle, but they depend on the angle α. The ratios sinus, cosine, and tangent are called ‘trigonometric functions’. Mathematicians found ways for calculating the trigonometric functions for any given angle. Nowadays, many calculators would give you the value of any trigonometric function of any angle that you key-in. For example, if you key-in $\sin(30^0)$, the calculator will display 0.5, which is the ratio between the lengths of the side opposite the 30^0 angle and the hypotenuse.

Since the sides of a right angle triangle are always smaller than the hypotenuse, sin(α) and cos(α) will always be between 0 and 1 for any α between 0^0 and 90^0. When $\alpha=0^0$, the triangle in figure 1.7 collapses into a line. The same thing happens when $\alpha=90^0$. It is helpful to remember the values of the trigonometric functions for these angles:

$\sin(0^0)=0$; $\cos(0^0)=1$; $\sin(90^0)=1$; $\cos(90^0)=0$.

The lengths of the sides of a right angle triangle are related through Pythagorean Theorem stating that

$$a^2 + b^2 = c^2 \quad [1.3a]$$

Using the notations of figure 1.7. This is the same as

$$c = \sqrt{a^2 + b^2} \quad [1.3b]$$

If both sides of equation 1.3a are divided by c^2 we get:

$$\frac{a^2}{c^2} + \frac{b^2}{c^2} = \frac{c^2}{c^2} \quad \text{or} \quad \sin^2(\alpha) + \cos^2(\alpha) = 1 \quad [1.3c]$$

1.3.2 Trigonometric functions of any angle

So far, we discussed the trigonometric functions of acute angles. A right angle triangle cannot have any angle that is greater than 90 degrees. The definition of the trigonometric functions has been expanded to include also angles that are greater than 90 degrees. This is done with the aid of a coordinate system.

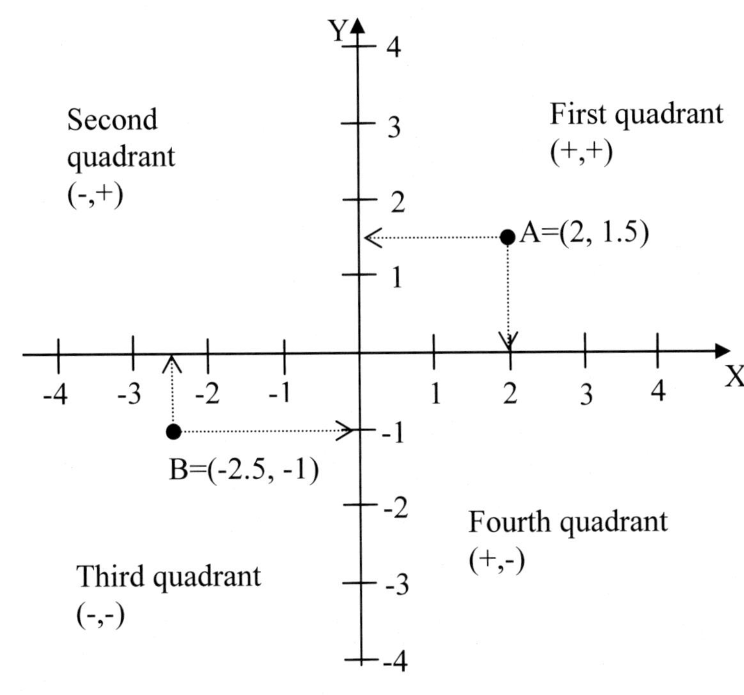

Figure 1.8: An x-y coordinate system, two points A and B, the four quadrants, and the signs of the x and y coordinates in each quadrant.

Figure 1.8 shows a standard x-y coordinate system. It has two perpendicular axes x and y. Their intersection is called the origin. Each axis is divided at the origin to a positive half (marked by an arrowhead) and a negative half. Each point on an axis corresponds to a real number. Positive numbers are sequences along the positive half of the axis, and negative number on the negative side. At the origin, each axis has the value of zero. The figure shows two points: **A** whose coordinates are x=2 and y=1.5, and **B** whose coordinates are x=-2.5 and y=-1.

In order to define the trigonometric functions of any angle, a circle whose radius is one unit and whose center is at the origin of the coordinate system is drawn (figure 1.9). A point **A** is chosen on the circle in the first quadrant. A right angle triangle whose head is the point **A** is then drawn, as shown in the figure. The length of the hypotenuse of this triangle, which is also the radius vector of point **A,** is one. The x-coordinate of **A** plays also the role of the adjacent side b of the angle α. The y-coordinate of point **A** is also the opposite side a of the angle α. Since c=1, cos(α)=(the x-coordinate of **A**), and sin(α)=(the y-coordinate of **A**). As point **A** slides counter clockwise on the unit circle, the angle between its radius vector and the positive half of the x-axis keeps increasing. For example, when it reaches the point marked as **A'**, it makes an angle of α'=225 degrees with the x-axis. The trigonometric functions of this and any other α' are now defined using the x and y coordinates of **A'**, the same as for α: cos(α') =(the x-coordinate of **A'**)=b', which happens to be a negative number. sin(α')=(the y-coordinate of **A'**)=a', which happens to be a negative number. The sign of the radius vector is always positive. In general, when α is in a certain quadrant, the sign of sin(α) is the same as the y coordinate at this quadrant, and the sign of cos(α) is the same as the x-coordinate in that quadrant (see figure 1.8). tan(α) is positive in the first and third quadrants, and negative in the second and fourth quadrants.

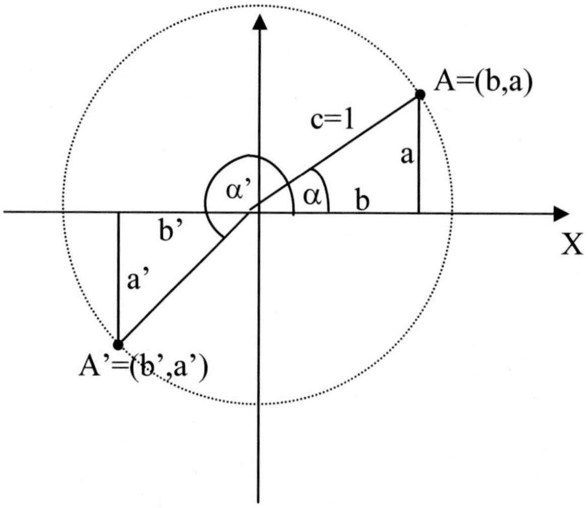

Figure 1.9: The unit circle and the definition of the trigonometric functions.

sin(α)=a; sin(α')=a';

cos (α)=b; cos (α')=b'

Figure 1.10 shows the functions sin(θ), cos(θ) for $0 \le \theta \le 360$ degrees, and for $0 \le \theta \le 2\pi$ radians. It can be seen that the values of the sin(θ) and cos(θ) oscillate between –1 and +1. Both functions have the same shape, but they are shifted with respect to each other.

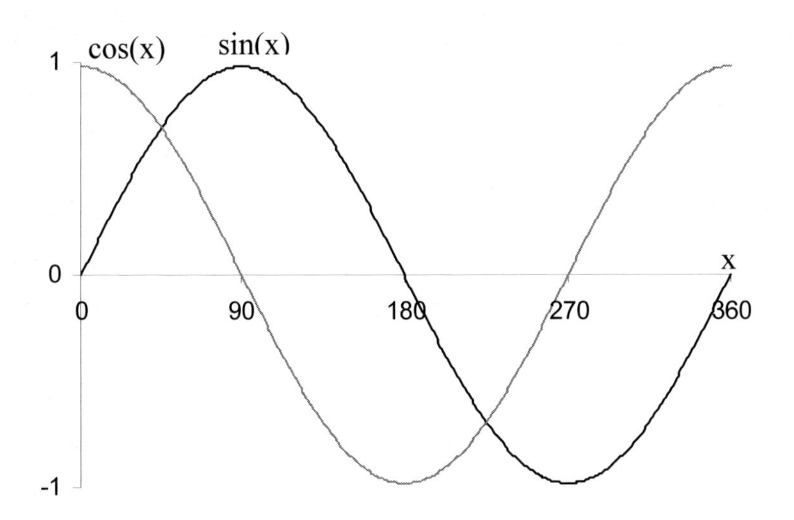

Figure 1.10: Sine and cosine.

1.3.3 The inverse trigonometric functions

Sometimes, you may know the value of the ratio between two sides of a right angle triangle, and you want to know what the angle is. To find this angle you can use the appropriate inverse trigonometric function, which are available on many calculators:

$$\alpha = \sin^{-1}\left(\frac{a}{c}\right); \qquad \alpha = \cos^{-1}\left(\frac{b}{c}\right); \qquad \alpha = \tan^{-1}\left(\frac{a}{b}\right) \qquad [1.4]$$

The notations in equations [1.4] correspond to figure 1.7.

If you know the coordinates of the point **A'**, as in figure 1.9, and you want to know the value of the angle α', you can still use equations [1.4] to find it. However, you should be aware that any ratio in the parentheses of the equations in [1.4] corresponds to two angles in the range $0 \le \theta \le 360$ degrees. For example, $\frac{a}{c} = 0.5$ correspond to θ=30 degrees, and to θ=150 degrees. The individual signs of the two coordinates of a point will uniquely determine the corresponding angle. For example, when $\frac{a}{c} = 0.5$ and a and b are positive yield θ=30 degrees. If a is positive and b is negative, θ=150 degrees.

1.3.4 Trigonometry in 3-D

So far, we have seen how trigonometry is applied to two dimensional situations. In order to treat three dimensional situations, we look for two-dimensional planes in those objects that would yield the solution by using trigonometry in two-dimensions.

Example

The dimensions of a hall are 16m x 12m x 15m, figure 1.11. What is the length of its long diagonal, and how big is the angle α that it makes with the floor?

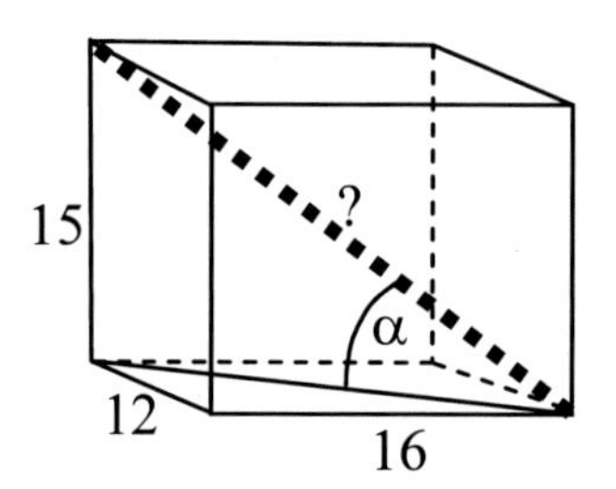

Figure 1.11

Solution

The long diagonal and the diagonal of the floor are found on one plane. The leftmost edge in the figure is also found on the same plane, and completes the two diagonals to a right angle triangle. The length of the floor's diagonal is $\sqrt{16^2+12^2}=20$ m. The length of the long diagonal is $\sqrt{20^2+15^2}=25$ m. The angle that the long diagonal makes with the floor is: $\alpha=\tan^{-1}\left(\frac{15}{20}\right)=36.9^0$

PROBLEMS

1. Draw an arbitrary right angle triangle, and measure the lengths of its sides and its angles. (a) From your measurements of the sides, calculate the sines, cosines and tangents of its angles. (b) Use your calculator to find the sine, cosine and tangent, directly from your measurements of the angles. (c) Compare the results of parts (a) and (b).

2. One of the angles in a right triangle is of 36^0. Using a calculator, find the ratio between (a) the side opposite this angle and the hypotenuse; (b) the side adjacent to the angle and the hypotenuse; (c) the side opposite the angle and the side adjacent to the angle.

3. The hypotenuse in a right triangle is 8 m long, and the side adjacent to one angle is 4.5 m. (a) Find the length of the other side. (b) Find the sinus of the angle between the hypotenuse and the 4.5m side.

4. Show that: (a) $\sin\alpha = \cos(90^0 - \alpha)$. (b) $\tan\alpha = \frac{\sin\alpha}{\cos\alpha}$

5. At noon, the sun is at the south and 75^0 above the horizon. Answer the following questions by drawing and by calculating:

(a) How high is the building if the length of its shadow is 5.5m?
(b) A south facing wall has an overhang of 60 cm. Find the length of the shade that the overhang casts on the wall.
(c) A south facing wall has a window 3' high. The basis of the window is 1' above the floor. Calculate the length of the sunbeam spot on the floor.

6. Following the notations of figure 1.9, find the angle for which:
a=3, b=5
a=3, b= -5
a= -3, b= -5
a= -3, b=5

7. In the following first draft (not to scale), the designer included various requirements of a building code. Now, more details of some elements have to be figured out. Calculate the lengths of a, b, and c, and the size of α.

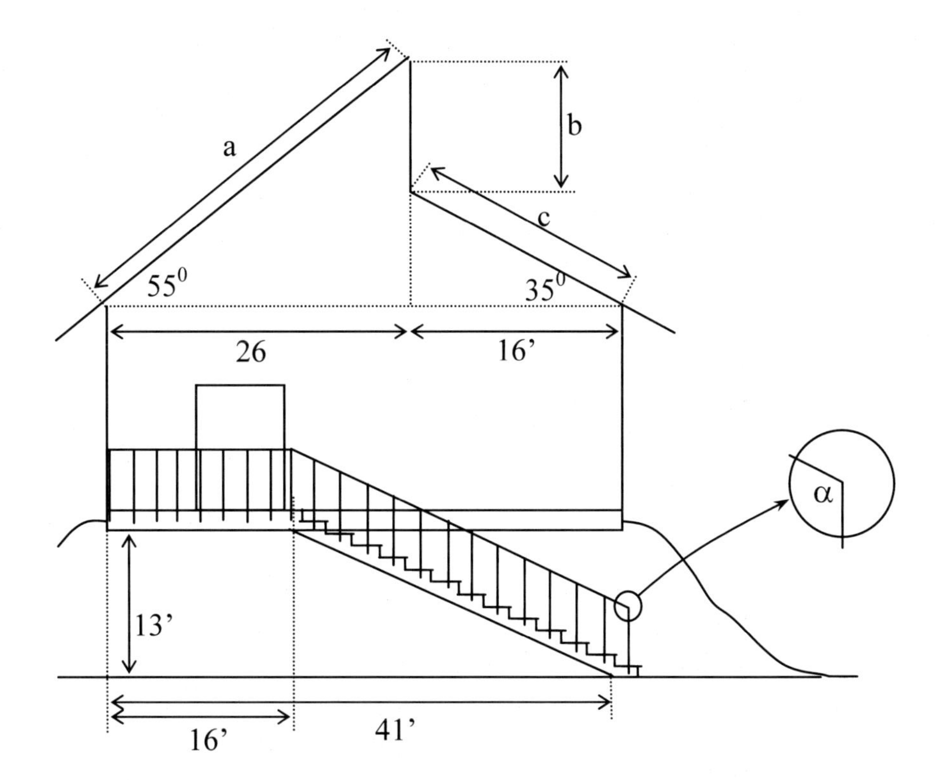

8. You have to design a staircase in a corridor. To minimize the "footprint" of the staircase, it has to be as steep as possible, but still satisfy the following constraints: (a) The entrance should be 6.5' high. (b) The horizontal width of each step should be at least 9". (c) Individuals 6' tall should not have to bend their head so that it does not hit the ceiling.

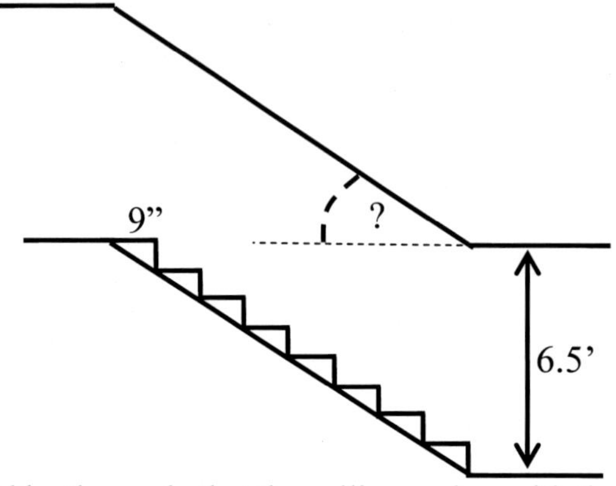

(a) What should be the angle that the ceiling makes with the horizontal?
(b) What would be the height of each step?

9. You have to design a roof in the shape of a pyramid. The side of the square base of the pyramid is 18 feet. The slope of the face of the roof has to be 60^0. (a) What is the height of the pyramid? (b) What is the length of its edge?

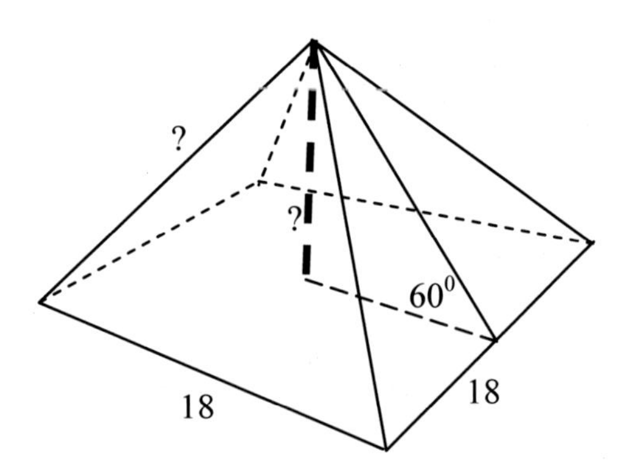

1.4 Units

A: I'll mail you the letter today. You'll get it in 5.
B: 5 what?
A: 5 days.
B: If, instead, you e-mail it, I'll get it in 5.
A: 5 what?
B: 5 seconds.

1.4.1 Conversion of units

So often in daily life, we must specify the units of numbers that we use. Otherwise, the information that we provide might be ambiguous. The same is true for physics. To make any sense, units must be attached to numbers that describe physical entities.

In physics, like in everyday life, the same entity may be expressed in a variety of units. For example, time may be expressed in seconds, minutes, hours, days, years, etc. We should know how to convert between two units that describe the same entity. Most often, conversion is accomplished through multiplication by a number, which is called a 'conversion factor'. For example, if we want to convert 3 hours into minutes, we use the fact that 1 hour is 60 minutes. We multiply 3X60=180, which means that 3 hours are 180 minutes. The conversion factor from hours to minutes is 60. If we want to convert 120 minutes into hours, we divide 120 by 60, which is the same as multiplying by 1/60. We get 120/60=2, meaning that 120 minutes are two hours.

Sometimes we encounter problems where we need to convert between combinations of units. Consider, as an example, the following problem.
In July 2008, the average price of gas in the US was 4.31 dollars (USD) per gallon, and in Great Britain it was 1.18 pound (GBP) per litre. The exchange rate at that time was 1 USD =0.501 GBP, 1GBP=1.996 USD. Where was the gas cheaper, and by what factor?

This is an example of a problem in which the unit of one entity (price) is a combination of units of other entities (money and volume). To answer the question, we'll have to convert from the price in the US to the price in Great Britain, ore vice versa.

"Per" means divide, so in general, $price = \dfrac{money}{volume}$. To convert the price, we convert each unit (money and volume) by itself, and maintain their relationship (divide). 1 gallon= 3.79 litre, 1 litre=0.264 gallon. So: The price of gasoline in Great Britain was: $\dfrac{1.18\,GBP}{1\,litre} = \dfrac{1.18 \times 1.996 USD}{0.264\,gallon} = \dfrac{8.92\,USD}{gallon}$, which is 2.07 times higher than the price in the US.

We use the same principles also in physics, when we have to convert a combination of units. We convert each unit of the combination by itself, while maintaining the relationships between the entities.

Another technique used in converting units is based on some basic ideas from fractions: First, if the numerator and the denominator of a fraction are equal to each other, the value of the fraction is 1. Based on this, we can say that $\dfrac{1\,\text{hour}}{60\,\text{min}} = \dfrac{60\,\text{min}}{1\,\text{hour}} = 1$. The second fact is that if we multiply any entity by 1, its value does not change. So, for example, $90\,\text{min} = 90\,\text{min}\cdot\frac{1\,\text{hr}}{60\,\text{min}}$. Third, based on the elimination rules for fractions, we can eliminate 'min' from the right hand side of the last expression to get

$$90\,\text{min} = 90\,\cancel{\text{min}}\cdot\frac{1\,\text{hr}}{60\,\cancel{\text{min}}} = \frac{90\cdot 1\text{hr}}{60} = 1.5\text{hr}.$$

Similarly, $1.5\text{hr} = 1.5\cancel{\text{hr}}\cdot\frac{60\,\text{min}}{1\cancel{\text{hr}}} = 1.5\cdot 60\,\text{min} = 90\,\text{min}$

1.4.2 Units and formulas

In everyday life, we use formulas naturally, without even noticing it. For example, when we want to buy 8 ounces of candy at 6 cents per ounce, we know that the cost would be 48 cents. We have used the formula

$$\text{cost} = (\text{price of one unit}) \times (\text{number of units})$$

We know that the numbers that we substitute in the formula must be in compatible units. Here we have used the two compatible units: cents-per-*ounce* for price-of-one-unit, and *ounces* for number-of-units. We know that we may use in that formula other sets of compatible units such as cents-per-*pound* and *pounds*. We also know that we should not use in the same formula cents-per-*ounce* together with *pounds*. Having pounds and ounces in one formula is incompatible, because these are two different units for the same entity. (If we were to use them, we would get that the cost of one pound of that candy is only six cents.)

In physics, we'll use dozens of entities that have various units. We will also use many formulas. In order to be sure that we always substitute in a formula only numbers that have compatible units, units are organized in **unit-systems**. All the units in a given unit-system are always compatible with each other. Because of that compatibility, general formulas in physics are the same for any system of units that we opt to use. (A general formula is a formula that does not include elements whose units are already fixed.)

The two major unit-systems that are used in physics are the SI (Standard International) system and the British system. The unit of time in both systems is the second (s in short). The unit of length in the SI system is the meter (m in short), and in the British system is the foot (ft in short). Each system has additional basic units, which will be introduced later, and units that are derived from the basic ones. For example, meter-square is the unit of area in the SI system, and foot-square is the unit of area in the British system. The formula: 'The area of a rectangle is equal to its length times its width' is valid in any system, as long as the units of the substituted values are compatible. In the SI system, length and width are expressed in meters, and the area in meter square. In the British system length and width are expressed in feet and the area in feet square.

It is ok to substitute units from different unit-systems in one formula, as long as they do not contain any incompatible common unit. For example, if a formula states that velocity equals distance divided by time, it is ok to substitute in it a distance in feet (British system) and a time in hours (a stand-alone unit). The velocity will be expressed in feet per hour, which is neither in the SI nor in the British system, but it is a legitimate unit of velocity.

PROBLEMS

1. Convert:
(a) 1 year into seconds
(b) 1 second into years
(c) 6 yards into feet
(d) 6 feet into yards
(e) 3 square feet into square inches
(f) 2 cubic meters into cubic cm
(g) 270 cubic feet into cubic meters
(h) $5.60 per hour into cents per minute
(i) 2 meter per second into cm per second
(j) 220 meter per hour into cm per second

2. Carpet from a roll of width 4 yards is sold at $11.98 per running foot. An apartment needs 65 square yards of this carpet. How much would it cost?

3. In January 2000, the price of gasoline in the US was 1.46 dollars (USD) per gallon, and in Great Britain it was 0.68 pound (GBP) per liter. The exchange rate at that time was 1 USD=0.618 GBP. Where was the gas cheaper, and by what factor?

4. A while ago, the price of a certain candy was 9 US cents per ounce in the U.S., and 1.20 Canadian dollars per pound in Canada. The exchange rate was 1 Canadian dollar equals 80 US cents. Which was cheaper?

2. MOTION IN ONE DIMENSION

2.1 Coordinate systems and the real world

-Mom, I'm late! Where is my other shoe?
-Did you look under your bed?
-Oh, thanks, you are the best!

Have you noticed that in order to describe the location of an object we always depend on a reference point (e.g. bed)? Once we have such a reference, we describe the spatial relationship between the object and that reference. In physics, we use coordinate systems for describing the location of objects. A coordinate system has an agreed upon reference point (the origin), agreed upon directions, (e.g. the directions of the axes in a Cartesian x-y-z coordinate system), and units for measuring distances. Once we decide on a coordinate system, we superimpose it on the part of space that we deal with. That will automatically assign numerical values, or coordinates, to the points of the space and to the points of the real objects that are found there. These coordinates can be used in equations and formulas that describe the physical processes under consideration.

It is completely up to us to decide how to superimpose a coordinate system on the real space. For example, consider a house whose highest point is 4m above street level. Let's call the highest point H and the street-level point below it L (figure 2.1a).

Figure 2.1: An object in the real world, and two arbitrary coordinate-systems that are used to describe it.

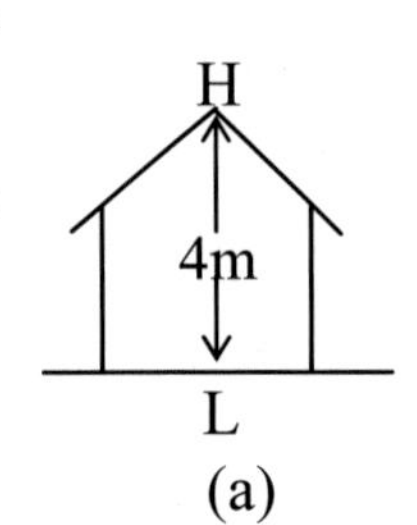

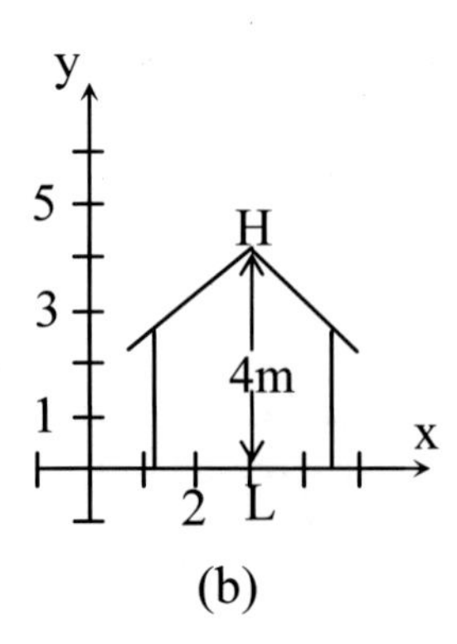

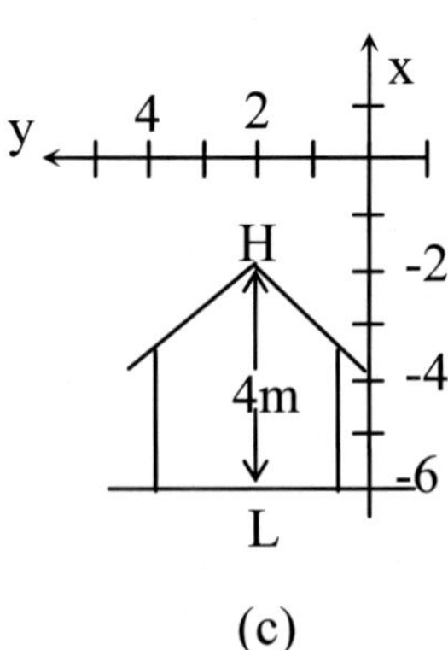

Figure 2.1b shows the same house with a superimposed coordinate system. The x axis points to the right and the y axis points upward. In this coordinate system, the coordinates of point H are (3,4) and the coordinates of point L are (3,0). In figure 2.c we see the same house with another superimposed coordinate system. Here, the x axis points upward, and the y axis points to the left. (Remember, it is up to us to choose the coordinate system that we want to use.) The coordinates of point H here are (-2,2) and the coordinates of point L are (-6,2). So, the same

physical object is described by different numbers in different coordinate systems. However, the height of the house remains 4m in all coordinate systems. The equations of physics always ensure that the real physical properties of an object, such as the height of a house, are the same, no matter what coordinate system is used to describe it.

We live in a three dimensional world. Therefore, we need a three dimensional coordinate system for describing processes that take place around us. However, many processes take place along a line, e.g., objects that fall straight down to the ground, or elevators that move up and down. These are called one-dimensional processes. Because it is up to us to choose the coordinate system, it is often more convenient to align one of the axes of a three-dimensional coordinate system along the line on which a one-dimensional process occurs. If we do so, we have to consider only one coordinate of the involved objects. For example, we can use only one axis to describe the motion of an elevator that goes up and down. This axis may be called x, or y, or any other name. It may point up or it may point down. Its origin may be fixed at any real point along the path of the elevator. After choosing the axis, we can describe the position of the elevator in space by giving its coordinate on this axis. If a one-dimensional motion of an object is in the horizontal direction, the axis that we choose would also be horizontal. If the object moves on a diagonal line, such as a car climbing up a hill on a straight road, the axis that we choose will also point in that diagonal direction.

Many processes take place in two dimensions. For example, collisions between the balls on a billiard table. Similarly to one-dimensional processes, the motion of objects on a plane can be described by a two dimensional coordinate system. The axes of the two dimensional coordinate system define a plane. This plane is fused with the real plane on which the objects move. For example, billiard balls will be described by a horizontal two-dimensional coordinate system. A fly walking on the wall will be described by a vertical two-dimensional coordinate system.

We start our discussion of motion of objects by focusing on one-dimensional motion.

2.2 Kinematics in one dimension.

2.2.1 Position, distance, and displacement in 1-D

The first thing that we need to do in order to describe one-dimensional motion of an object is to choose an axis. Choosing an axis means three things: First, we have to decide in what point of the real space we fix the origin of the axis. For example, should the origin of the axis coincide with the top of a building, the bottom of it, or any other point? Usually, this is a question of convenience, and it is completely up to us. Second, we have to decide in which direction of the real space we point our axis. The axis has to be aligned with the line on which the object is moving. We can only choose the polarity of the axis (e.g. pointing up or pointing down). Third, we have to decide on the unit that we use. For example, are we going to use meters, feet, or any other unit? After we have chosen the axis, the coordinate of the object with respect to this axis specifies its position in space at that given time. For convenience, we will call our axis the x-axis. That does not affect the generality of our discussions.

At any time, the **position** of the object in space is expressed by its **coordinate**. The position coordinate may be positive or negative. The sign indicates if the object is located on the positive side or on the negative side of the axis. For example, x=3m means that the object is located 3m from the origin, on the positive side of the axis. x=-3m indicates that the object is located 3m from the origin, on the negative side of the axis. Figure 2.2 shows the positions of three points A, B, and C marked on an x-axis. The coordinates of these points are $x_A=0$, $x_B=6m$, $x_C=-2m$. The coordinate x of an object tells us **where** is the object found on the axis that we use.

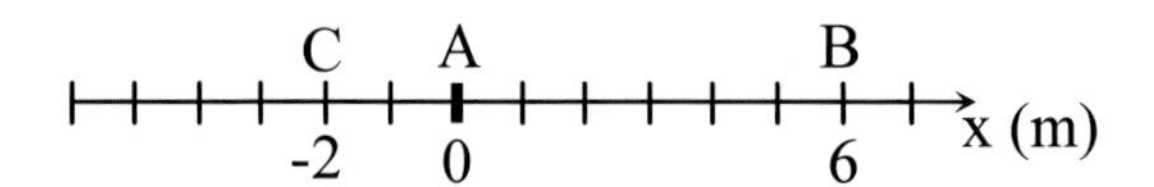

Figure 2.2: The coordinate of a point expresses its position.

The **distance** between two points on an axis can be found by subtracting their coordinates. As you can see in figure 2.2, the distance between A and B is $x_B-x_A=6-0=6m$; the distance between B and C is $x_B-x_C=6-(-2)=8m$. The distance between x_A and x_B is the same as the distance between x_B and x_A; the order of the points does not matter. Distance is always a positive number. Therefore, the distance between any two points A and B would be the absolute value of the difference of their coordinates: the distance (d) between A and B is given by:

$$d=|x_B-x_A| \quad [2.1]$$

In physics, we define, in addition to distance, another related entity– **displacement**, which will be denoted by **d**. The expression for the displacement is given by:

$$\mathbf{d}=x_B-x_A. \quad [2.1a]$$

Displacement describes the process of displacing an object from point x_A to point x_B. By comparing [2.1] with [2.1a] we see that unlike d, **d** may be positive or negative. However, both have the same absolute value: $|\mathbf{d}|=d$. The sign of a displacement from x_A to x_B is opposite to the sign of a displacement from x_B to x_A. In one dimension, the displacement **d** is positive when we move from x_A to x_B in the positive direction of our axis. The displacement **d** is negative when we move from x_A to x_B in the negative direction of the axis. For example, the displacement from A to B in figure 2.2 is 6m. The displacement from B to A is -6m. The displacement from C to A is 2m, and the displacement from A to C is -2m. The sign of the displacement is determined by the direction from x_A to x_B with respect to the axis that we use. To summarize: The absolute value of **d** tells us the magnitude of the displacement. The sign of **d** encodes the direction of the displacement.

In everyday life, we sometimes use the concept 'distance-covered'. If in figure 2.2 the object starts from point A, moves to point B, and from there moves to point C, the distance covered would be 6+8=14m. The distance-covered in this case is the sum of the distances of the individual segments. The distance covered is not equal to the distance between the initial and final points. Distance between initial and final points and distance-covered are equal to each other only if the object moves all the time in the same direction. The d in [2.1] stands for distance between two points. When the object moves in one direction, d indicates both the distance between initial and final points and the distance covered.

2.2.2 Vectors in 1-D

A scalar is a physical entity that has only magnitude. Time and temperature are two examples of scalars. A vector is an entity that has both direction and magnitude. For example, displacement of an object is a vector, because it specifies by how much (magnitude) the object is displaced from its initial point, and in what direction. As we have just seen in [2.1a], the magnitude of the displacement vector is the distance between the initial and final points. The direction of the displacement is encoded in the sign of the displacement.

When we deal with motion in one dimension, we can rely on an axis to specify that direction in space. We can also use this same axis to describe other vectors that happen to have the same spatial orientation as the axis. In such cases, a vector that points in the same direction as the axis is declared positive, and a vector that points in the direction opposite to that of the axis is declared negative. Thus, in one dimension, vectors can be described by positive and negative numbers. The absolute value of that number is the magnitude of the vector, and the sign of the number indicates the orientation of the vector with respect to the axis. In the following, a boldface letter will indicate a vector. The magnitude of a vector (in 1-D its absolute value) will be denoted by regular letters. For example, **A** denotes a vector and A denotes its magnitude |**A**|.

Two (or more) displacement vectors can be added to each other. Let **A** denote the first displacement and **B** the second. When we say **C**=**A**+**B** we mean that the **C** is one displacement, whose effect on the object is the same as first displacing it according to **A**, and then according to **B**. That means that first the object is moved a distance of |**A**| from its initial point, in the direction specified by the sign of **A**. The object's final position after the first displacement serves as its initial point of the second displacement **B**. From there, the object is moved a distance of |**B**|, in the direction specified by the sign of **B**. The displacement **C**, which is the sum of **A** and **B**, is from the initial point of **A** to the final point of **B**. In practice, **C** is simply the sum of the numbers that express **A** and **B**, including their signs. The sum of vectors is called the **resultant**. So, **C** is the resultant of **A** and **B**. Here are two examples (see figure 2.3). First, the first displacement is **A**=2m, and the second displacement is **B**=3m. In this case the resultant is **C**=2+3=5m. The

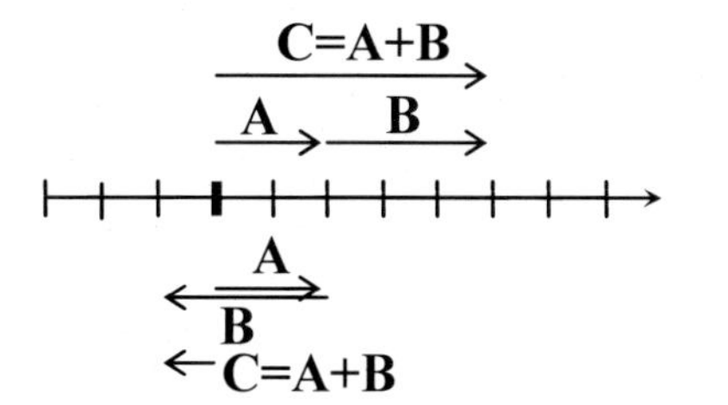

Figure 2.3: Two examples of adding displacement vectors: **C**=**A**+**B**.

magnitude of **C** is 5m, and it points in the positive direction of the axis. This is illustrated by the arrows above the x-axis in figure 2.3. A second example: The first displacement is **A**=2m, the second displacement is **B**=-3m. In this case the resultant **C**=2+(-3)=-1m. In this case, the magnitude of **C** is 1m, and it points in the negative direction of the axis. This is illustrated by the arrows bellow the x-axis in figure 2.3.

The same procedure is used for adding one-dimensional vectors of other entities, such as velocity or force. In one dimension, such vectors are added like regular numbers, which can be positive or negative. The magnitude of the number is the magnitude of the vector, and the sign of the number indicates the direction to which that vector is pointing relative to the axis.

Subtracting a vector is the same as adding a vector of the same magnitude and opposite direction.

Adding vectors in two and three dimensions are extensions of how vectors are added in one dimension. These topics will be discussed in an upcoming chapter.

2.2.3 Describing motion in 1-D

Motion involves two variables: position and time. The simplest way of describing the motion of an object is by an itinerary. An itinerary tells us where an object is found at specific times. In physics, an itinerary can be written as a table of two rows, one specifying the position of the object, in the form of its coordinate, and the other providing the corresponding time at which the object was at that point. Table 2.1 illustrates such a table.

Table 2.1: An itinerary

Position (where) x (m)	-2	0	2	4	8
Time (when) t (s)	0	10	20	30	40

It is very common in physics to express time as 'stop-watch' time. The 'stop-watch' is set when we start to observe the object, and keeps running as long as we keep observing.

An itinerary can be displayed graphically on an x-t coordinate system. Figure 2.4 shows graphically the itinerary of table 2.1.

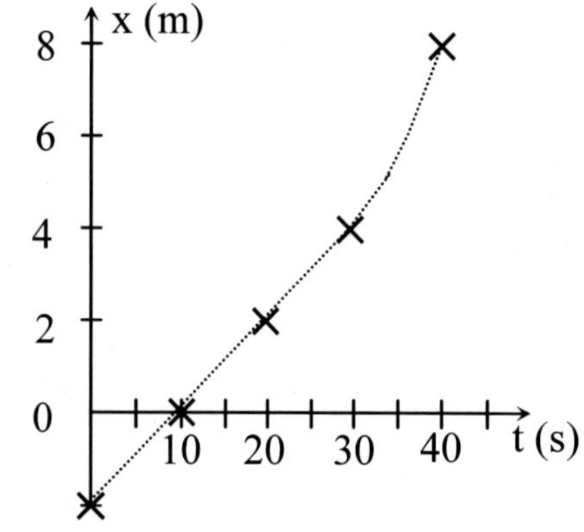

Figure 2.4: Graphical display of table 2.1

The X's in the figure correspond to the numbers in table 2.1. The table does not provide any information about what happens in between the specified times. If we assume that things 'go smoothly' between the specified times, we may interpolate between these points, as shown by the dotted line. In this sense, the graph provides more information than the table.

Another way of describing motion is by an **equation of motion**. An equation of motion gives the relationship between the position (coordinate) of an object and the time that it was at that position. For example, the first four entries in table 2.1 satisfy the equation of motion: $x=-2+0.2t$, where x denotes the position of the object at time t. x is in meters and t is in seconds. You can verify that this formula reproduces the first four entries in the table. The fifth entry does not satisfy this formula, and it would need another formula to describe it. Like the line in figure 2.4, this formula provides information also on points that are in between those given in the table. If we know the equation of motion of an object, we can construct from it an itinerary and a graphical description similar to those shown above.

In the following, we will use x_i to denote the initial position of the moving object. Time will be denoted by t. We will always use 'stop-watch' time, meaning that when we start measuring time ($t=0$), the position of the object is x_i. In the example of figure 2.4, $x_i=-0.2$m. The position of the object at time=t will be denoted by x.

2.2.4 Speed and velocity in 1-D

We know that if we covered a distance of 100 miles in 2 hours, our average speed was 50 miles per hour. To find average speed we divide the distance covered by the time that it took to cover that distance. We include the word 'average' because we were probably not moving at the same pace during the entire trip. We can express this relationship by a formula:

$$S_{avg} = \frac{D}{t} \qquad [2.2a]$$

where S_{avg} indicates average speed, D distance covered in time t.

Consider now the situation in which an object is **moving in one direction** on the x-axis. Let x_i be its initial position and x its final position. In this case, the distance covered would be given by $d=|x-x_i|$, based on [2.1]. If we substitute d in [2.2a] we get

$$S_{avg} = \frac{|x - x_i|}{t} \qquad [2.2b]$$

Since distance is by definition always positive, it follows from [2.2] that speed is also always positive.

Although in everyday use the words speed and velocity are interchangeable, in physics they are used to describe different concepts. Average velocity is defined as the displacement (from the initial point to the final point) divided by the time that it took to make that displacement:

$$\mathbf{v}_{avg} = \frac{x - x_i}{t} \qquad [2.3]$$

In this case, the numerator is the displacement, rather than the distance covered, as was the case in [2.2]. Displacement is a vector, and in one dimension, the sign of the displacement tells us in which direction, with respect to the axis, the object is displaced. Since the displacement can be positive or negative, $\mathbf{v}_{avg}$ can also be positive or negative. $\mathbf{v}_{avg}$ has the same sign as the displacement. Like displacement, velocity is also a vector. When the velocity is positive, the object moves in the direction to which the axis is pointing. When the velocity is negative, the object moves opposite to the direction to which the axis is pointing. In short, the sign of the velocity tells us in which direction the object is moving.

officer: Do you know why I stopped you?
driver: Have no idea.
officer: You were driving 48 miles per hour in a 25 miles per hour zone!!!
driver: I left home five minutes ago, so how can you say that I was driving for an hour?

Velocity and speed depend on the ratio of two variables (distance and time), but not on any one of those variables separately. One does not have to drive twenty-five miles in an hour to drive at twenty-five miles per hour. In fact, the speedometer in a car shows the ratio between the distance covered and the time interval that it took to cover it, but this time interval is very short–a fraction of a second. When velocity is found based on very short time intervals, it is called **instantaneous velocity**. The speedometer in a car shows the magnitude of the instantaneous velocity. When a car accelerates from rest to its cruising velocity, the speedometer shows the magnitude of the instantaneous velocity at any instant during that time. It starts by showing zero, and then its numbers grow as the instantaneous velocity grows. Then, as the car cruises, the reading does not change, indicating that the instantaneous velocity does not change. When the instantaneous velocity does not change with time, the motion is called **motion at constant velocity**.

Instantaneous velocity, which will be denoted by v, is expressed by the formula:

$$\mathbf{v} = \frac{x - x_i}{t} \qquad [2.4a]$$

The time interval t in [2.4a] must be very small, and the difference $x\text{-}x_i$, which is the displacement during that small time-interval, is also small. However, their ratio, which is the instantaneous velocity, may be of any magnitude.

When a car is moving on a straight line at a constant velocity, say 25 miles per hour, the average velocity for any time interval, short or long, will also be 25 miles per hour. Therefore, when an object is moving at constant velocity, equation [2.4a] can be used for time t of any length. Equation [2.4a] can be rearranged and written as:

$$x = x_i + v \cdot t \qquad [2.4b]$$

Equation [2.4b] can be used in various situations in which an object is moving in one direction on a straight line. In each situation, v indicates a different thing. In motion at constant velocity, v stands for the constant velocity. If the motion is not at constant velocity, v indicates the average velocity. In these two cases, there is no restriction on t: it may be large or small. When t is restricted to be very small, v stands for instantaneous velocity.

The unit of velocity is a unit of distance (or length) divided by a unit of time. The unit of velocity in the SI system is meter per second, and in the British system it is foot per second. Other common units of velocity are kilometer per hour and mile per hour, which do not belong to either of those two unit systems.

Solving numerical physics problems

One of the strengths of physics is that it employs mathematical formulas to solve problems. There are several steps that are usually taken when solving a physics problem.

(a) The general contents of the problem are analyzed and the physics area to which it belongs is identified. That step brings to the forefront the formulas that may be applicable to the problem. A sketch maybe drawn to help in visualizing the situation. A coordinate system is defined where needed.

(b) Numbers that are known about the situation of the problem are linked with the symbols that appear in those identified formulas. The variable/s that we have to find out (the unknown or unknowns of the problem) are linked to a symbol/s of the formulas.

(c) Units of all the variables have to be compatible. If they are all in the SI system, or they are all in the British System they are compatible. Incompatible numbers have to be converted to become compatible.

(d) The numbers and the identified unknown/s are substituted into the formulas and replace their linked symbols.

(e) The equations are solved for the unknowns. The relationships between the numerical solution and the real world situation are stated.

The following example illustrates those steps.

Example
An elevator starts its way up from rest at the ground floor. When it reaches the height of 2 m above the ground it starts to move at a constant speed of 1.5 m/s. How high above the ground will it be 10 seconds later?

Solution

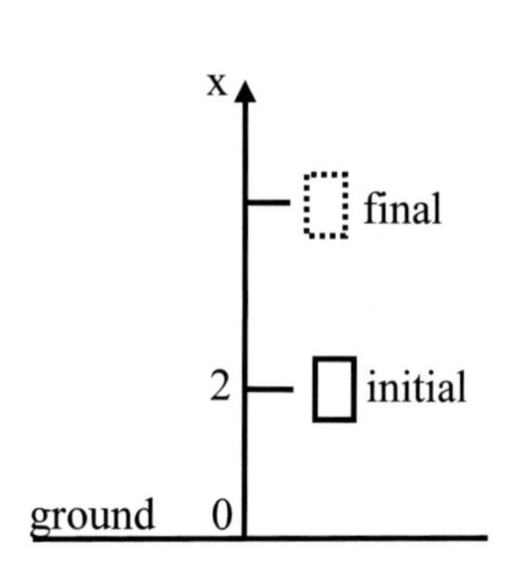

(a) This is a problem about an object that moves at a constant speed. Therefore formulas [2.4] apply. The object moves upwards, so we can use a vertical axis, which it is up to us to specify. We choose the positive direction upwards, and the origin at the ground (opposite figure).

(b) The elevator started to move at constant velocity 2 m above the ground. The variable that represents the initial position of the object in [2.4] is x_i. Linking the relevant number yields x_i=2m. The elevator moves up at a constant velocity of 1.5 m/s. That yields v=+1.5m/s. Note that the plus sign indicates moving upwards in the axis that we have chosen. The time of moving at constant velocity is 10 seconds, therefore t=10s. We have to find out how high above ground the elevator would be at t=10s. Our unknown is the final position x, which happens also to be the height above ground (see discussion of formulas [2.1]).

(c) All the units are in the SI system, so there is no need to convert.

(d) Substituting the numbers in [2.4b] gives : $x=2+1.5\cdot 10$.

(e) Solving the equation, which is in this case turns to be a simple evaluation of the RHS, yields x=17m. The unit of x is meter, because we are using the SI system. The final answer is: the height of the elevator 10 seconds after it started to move at constant velocity is 17 meters above the ground.

2.2.5 Acceleration in 1-D

When the velocity of an object changes with time, we say that the object accelerates, or that it moves in accelerated motion. An object can accelerate at different rates. When the rate at which that velocity changes is high, we say that the acceleration is large, and when the rate of the change is small, we say that the acceleration is small. Average acceleration is defined as

$$\mathbf{a}_{avg} = \frac{\mathbf{v} - \mathbf{v}_i}{t} \qquad [2.5]$$

$\mathbf{a}_{avg}$ stands for average acceleration, $\mathbf{v}_i$ is the initial velocity, $\mathbf{v}$ is the final velocity (i.e. at time=t), where t is the time interval during which the velocity changed from $\mathbf{v}_i$ to $\mathbf{v}$.

Instantaneous acceleration is the rate of velocity-change during a small time interval. It is expressed as:

$$\mathbf{a} = \frac{\mathbf{v} - \mathbf{v}_i}{t} \qquad [2.6a]$$

It is basically the same idea as in [2.5], but here the time interval t is very small. This is similar to the relationship between average velocity and instantaneous velocity ([2.3] and [2.4a]). The similarity goes even further. Motion in constant acceleration is motion in which the instantaneous acceleration does not change with time. Equation [2.6a] expresses also motion in constant acceleration, but the meanings of the letters have to change accordingly. If '**a**' stands for the constant acceleration, t is the time interval during which the initial velocity v_i became v. The time interval t may be of any length. This formula can be rearranged to:

$$\mathbf{v} = \mathbf{v}_i + \mathbf{a} \cdot t \qquad [2.6b]$$

Equation [2.6b] does not contain the variable x, which is the position of the object. In order to be able to fully describe accelerated motion we have two more formulas that contain x and x_i. These formulas are:

$$x = x_i + \mathbf{v}_i t + \frac{1}{2}\mathbf{a}t^2 \qquad [2.7]$$

and

$$v^2 = v_i^2 + 2 \cdot \mathbf{a} \cdot (x - x_i) \qquad [2.8]$$

where x is the position at time t, x_i is the initial position, v_i is the initial velocity, and the constant acceleration is **a**.

Equations [2.6], [2.7], and [2.8] give us the relationships that hold between the variables that describe **motion in constant acceleration**, and we use them to solve problems of motion in constant acceleration. We use equations [2.4] when the motion is at constant velocity (no acceleration).

The unit of acceleration is the unit of velocity change divided by the unit of time. The unit of velocity change is the same as the unit of velocity, which is the unit of distance divided by the unit of time. Combining all these elements we get (Square brackets [..] indicate the unit of the entity that they enclose). [acceleration]=([distance]/[time])/[time]. This can be written as [acceleration]=[distance]/[time]^2. In the SI system, the unit of acceleration is m/s^2. In the British system, the unit of acceleration is ft/s^2.

Like velocity, acceleration is a vector. In motion in one dimension, acceleration can be positive or negative. We say that positive acceleration points in the direction of the axis and negative acceleration points against the axis. There are three ways to determine the sign of the acceleration. All three ways yield, of course, the same result.

1) The sign of the acceleration is determined based on the signs and on the magnitudes of the initial and final velocities, as provided by equation [2.6a]. If the final minus initial velocity is positive, the acceleration is positive. If the final minus initial velocity is negative, the acceleration is negative.

2) If the magnitude of the velocity (the speed) is **increasing**, the sign of the acceleration is the **same** as the sign of the velocity. If the magnitude of the velocity is *decreasing*, the sign of the acceleration is *opposite* to the sign of the velocity.

3) The third way for determining and interpreting the sign of the acceleration will be given in 2.3.2.

2.2.6 Acceleration of free fall (g)

The term 'free falling objects' refers to objects that are falling straight down or that have been thrown straight up. Other than the gravitational attraction of the Earth, the effects that other forces, such as air resistance, might have on the motion of free falling objects are assumed negligible. The instantaneous velocity of free falling objects varies continuously with time. Therefore, free fall is one kind of accelerated motion. Galileo Galilei (1564-1642) studied systematically free fall. He used indirect methods that involved objects sliding on inclined planes, and probably also direct observations on objects dropped from the leaning tower of Pisa. His conclusions, which were reconfirmed many times since, have been that if the effects of air on falling objects were taken away, free fall would be motion at constant acceleration, and that that acceleration is the same for all objects. The common symbol that now denotes the acceleration of free fall is g.

The magnitude of g is approximately $9.82 m/s^2$, and it varies slightly from location to location. g is greater at the Earth's poles than at the equator. As we climb to higher elevations, g decreases slightly. The presence of nearby mountains also affects the value of g. However, an important observation was made: g is the same for all objects that are free falling at the same location. g does not depend on the size of the object, its shape, or the material of which it is made. At the same location, g does not depend on the instantaneous velocity of the free falling object, meaning that free fall is motion at constant acceleration.

The following example illustrates how g can be found experimentally.
The tower of Pisa is 55m high. A piece of lead is dropped from its top and reaches the ground in 3.35 seconds. We need to find the acceleration during this fall.
First, we have to choose an axis along the trajectory of the falling lead. The axis may point up or down. For illustrative purpose, we'll solve the problem first with an axis that points up, and then with an axis that points down (figure 2.5). The origin of both axes is chosen at the top of the tower.
We first match the numbers and information provided in the problem with the notations of the equations for accelerated motion: For the left axis we have: Initial position of object: $x_i=0$. Final position: $x=-55m$. Time to reach final position: $t=3.35s$. Initial velocity: $v_i=0$ (starting from rest). The unknown: $a=?$
The equation that contains all these variables is [2.7]:

$$x = x_i + v_i t + \frac{1}{2}at^2$$

When we substitute the numbers in this equation we get:

$$-55 = 0 + 0 \cdot 3.35 + \frac{1}{2} a \cdot 3.35^2$$

Solving for a gives $a=-9.8m/s^2$.

Let's solve it now by using the right axis in figure 2.5, which points down. Initial position of object: $x_i=0$. Final position: $x=+55m$. Time to reach final position: $t=3.35s$. Initial velocity: $v_i=0$. The unknown: a=? When we substitute it in [2.7], which is applicable to any coordinate system, we get:

$$+55 = 0 + 0 \cdot 3.35 + \frac{1}{2} a \cdot 3.35^2$$

Solving for a gives $a=+9.8m/s^2$.

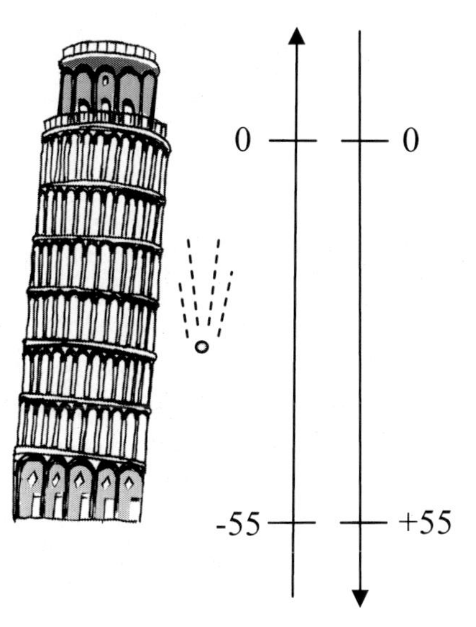

Figure 2.5: The Tower of Pisa and two axes

The letter 'a' in these problems stands for the acceleration of free fall, as determined by this experiment. So, a=g. The question now is this: First we got $g=(-)9.8m^2$ and then $g=+9.8m/s^2$. Which is the correct solution? The answer is that both are correct. Acceleration is a vector. The sign of a vector indicates which way it is pointing to, **with respect to the chosen axis**. In the first case, the axis points up, therefore, $-9.8m/s^2$ indicates that the vector g points down. In the second case, the axis points down, therefore $+9.8m/s^2$ indicates that g points down. So, the two apparently different answers have the same meaning in the real world: g points down.

Example

An elevator started its way up from rest at the ground floor, moving in constant acceleration. When it reached the height of 2 m, its velocity was 1.5 m/s. Find: (a) Its acceleration. (b) The time it took to climb the 2 m.

Solution

The elevator moves at constant acceleration on a straight line, therefore the applicable formulas are [2.5]-[2.8]. We choose the axis such that the origin is at the ground, and the positive direction is up. Linking the information to the symbols of the formulas we get: $x_i=0$, $x=2m$, $v=+1.5m/s$. The English phrase "started from rest" is expressed by $v_i=0$. In part (a) we have to find the acceleration, so a=? In part (b) we have to find the time, so t=? All the information is in SI units, so there is no need to convert. The solutions will also be in SI.

In order to solve (a) we need to substitute the known information in an equation, such that the only unknown in it would be a. This will be accomplished if we use formula [2.8]. We get: $1.5^2=0^2+2\cdot\mathbf{a}\cdot(2-0)$. Solving yields: $\mathbf{a}=0.56m/s^2$. The acceleration is $0.56m/s^2$ pointing upwards.

In order to solve (b) we need to substitute the known information in an equation such that the only unknown in it is t. We can do it by using formula [2.6b]. We get: $2=0+0.56\cdot t$. Solving yield t=3.57 seconds. The time to climb 2 meters is 3.57 seconds.

2.3 Newton's Laws in 1-D

2.3.1 The Law of Inertia (Newton's First Law)

Three of the most common safety devices in cars are the seat belt, the air bag, and the headrest. Why are they needed?

When a moving car comes to a sudden abrupt stop, e.g. by hitting head-on a tree or an oncoming car, the driver's body, because of its inertia, keeps moving forward. It hits the steering wheel and suffers trauma. The driver is not 'thrown' onto the steering wheel. The steering wheel, which is solidly attached to the car's frame, comes to a stop together with the rest of the frame. The driver's body, which is not solidly attached to the car's frame, continues its motion and hits the steering wheel. The seat belt corrects this problem by securing the driver to the seat. The seat belt prevents the driver's body from continuing its motion due to its inertia, and hitting the stopped steering wheel.

So, why is an air bag needed? The seat belt holds the torso of the driver in place, but it is not wrapped around the driver's head. Because of its inertia, the driver's head would continue its forward motion while the torso is being held in place by the seat belt. That may cause neck injuries, and the head may hit the windshield. The airbag prevents the head from moving forward due to its inertia.

The headrest is there to protect the driver when the car is hit from behind. In a car that stops normally, such as for a red light, the driver's body is also at rest. If the stopped car is hit from behind by another car, the frame of the stopped car together with its attached seats are suddenly pushed forward. The driver's torso is pushed by the back of the seat. The driver's head, because of its inertia, wants to stay at rest as the torso is being pushed forward. That may traumatize the neck. The headrest is there to push the head forward together with the torso, so that the neck is not strained.

The seat belt and the air bag are there to counter the inertia of the body and the head when they are in motion. The headrest is there to counter the inertia of the head when it is at rest. These situations are examples of the more general Newton's First Law (the Law of Inertia):

Any object at rest will stay at rest, and any moving object will continue to move at the same speed and on the same straight line, unless a net force acts on it.

Inertia is that property that all objects have, which makes them 'want' to stay at rest, or to continue moving at a constant velocity. The car safety devices illustrate special cases of the Law of Inertia. Another example is when you are in a car that makes a turn, and you feel that you are 'pushed' sidewise towards the door. Because of its inertia, your body wants to keep moving straight ahead, while the car changes direction. The car is the one that 'cuts' the intended motion of your body, and that created the perception that your body is pushed against the door.

The Law of Inertia can be used to reach some conclusions about the existence of forces that may not be so conspicuous otherwise. Here are some examples.

First, that famous apple that fell from the tree and hit Newton's head. At a certain moment, the stem that connects an apple to the tree weakens. Then, the apple falls down. But why does the apple not stay up there, with the weaken stem, or even without any stem at all? According the Law of Inertia, the fact that the apple does not stay at rest indicates that a force is acting on it. This is the force of gravity. The force of gravity acts on all objects on the Earth. No object can hang in mid-air without something to support it. When such a support is taken away, the object starts moving down due to the force of gravity that the Earth exerts on it.

The fact that the Earth orbits around the Sun, which means that the Earth is moving on a line which is not straight, implies that a net force acts on the Earth. And indeed, the Earth moves in its orbit because of the force of gravity that the Sun exerts on it.

When you try to push a heavy piece of furniture, it may not move. We have here an object at rest and a force that is acting on it, but the object stays at rest. According to the Law of Inertia, the object can stay at rest only if the net force that acts on it is zero. The pushing force is definitely different from zero. So, the only conclusion is that there is another force acting in the system and the net of that force and the pushing force is zero. This other force is the friction force between the furniture and the floor. If the friction is eliminated, the pushing force will cause the furniture to move.

When you roll a ball on the floor it will slow down and eventually come to a stop. Slowing down means that the ball's motion is not at a constant speed. According to the Law of Inertia, that can happen only if a net force is acting on it. In this case, no active force is pushing the ball, but friction forces are acting on it, and they are responsible for changing its constant speed.

2.3.2 Newton's Second Law

When a driver pushes the accelerator-pedal, the engine delivers a greater force and the car accelerates. The final velocity of the car will depend on the force provided by the engine and on the time during which that force was acting on the car. A greater force shortens the time needed to reach a certain final velocity. In other words, a *greater force* causes the car to move at a *greater acceleration.*

Let's compare now two trucks at rest that need to reach the same final velocity. The trucks are identical except that one is loaded and the other is not.

Under those circumstances, it would take the loaded truck longer to reach the desired final velocity. The inertia of the loaded truck is greater than that of the empty one. The amount of inertia that an object has is called the mass of the object. The greater the mass, the greater is the inertia of the object. When two identical forces act on two objects of different masses, the object with the ***greater mass*** would have the ***smaller acceleration.***

It would be very useful to have a quantitative relationship (a formula) between the ***force*** that acts on an object, the ***mass*** of that object, and its ***acceleration***. In order to have such a formula, we first need to know how to express magnitudes of forces, masses, and acceleration.

We already know how to express the magnitude of acceleration. A unit of acceleration is a unit of length divided by a unit of time squared (e.g. m/s^2), and the acceleration of any object can be expressed quantitatively by such units. In order to be able to quantify forces and masses we have to define units of force and of mass, and prescribe how those units are used to express any arbitrary force and mass, respectively. We will do it in steps. First, we will see how units of force and mass are defined. Then, we will see how the magnitudes of any arbitrary force and mass could be determined, based on those units.

The kilogram (kg) is a unit of mass, and the Newton (N) is a unit of force. By definition, one kg is the mass of one liter ($10^{-3}m^3$) of water. By definition, one Newton is the magnitude of any force that would accelerate one liter of water by $1m/s^2$. So far we have defined the kg and the Newton as applied to one liter of water.

We now generalize to concept of 1 kg to other objects: Any object that when pushed by a force of one Newton is accelerated at one m/s^2 has a mass of one kg.

Next, we define how the magnitude of any arbitrary force could be determined: To determine the magnitude of any unknown force F, we apply it to an object of mass m=1kg, and measure the resulting acceleration a. F, m, and a are related through $F = m \cdot a$. Since m and a are known, we can determine the value of F: $F=1 \cdot a$.

Next, we define how the mass of an arbitrary object could be determined: To determine the unknown mass m of any object, we apply on that object a force F=1N, and measure the resulting acceleration a. The mass m is determined from $F = m \cdot a$: m=1/F.

Those procedures allow us, in principle, to quantify the mass of any object and the magnitude of any force. All we have to do is to apply the unknown force on a mass of 1 kg and measure the acceleration, and to apply a force of 1 N on the unknown mass and measure the acceleration. Now that, in principle, we can quantify all forces and all masses, what would be the relationship between any arbitrary force that would be acting on any arbitrary mass, and the resulting acceleration? Newton's Second Law provides the answer.

Newton's Second Law states that when a force of magnitude F acts on an object of mass m, the object would be accelerated by the acceleration a, where F, m, and a are related by: $F = m \cdot a$

The direction of the force is the same as the direction of the acceleration. To indicate that force and acceleration are vectors that point in the same direction, they are expressed by bold letters:

$$\mathbf{F} = m \cdot \mathbf{a} \qquad [2.9]$$

Equation [2.9] is known as Newton's Second Law.

Newton's Second Law ([2.9]) has been validated, directly and indirectly, in numerous experiments. Because of that, we can use [2.9] to determine the magnitude of any unknown force, by applying it to a known mass and measuring the acceleration. Similarly, we can determine the magnitude of any unknown mass based on the acceleration imparted to it by a known force. We are not restricted to use a unit mass to determine a force, and a unit force to determine a mass.

Newton's Second Law holds in all practical, macroscopic situations. However, when an object moves at a very high speed that approaches the speed of light ($3x10^8$m/s), or when the object is very small (atomic size or less), equation [2.9] has to be replaced by other formulas.

We understand intuitively what is meant by the 'direction of a force'. For example, when we push a chair, we understand what is the direction of our pushing force. In general, the direction of a force that causes an object at rest to move is the direction in which that object moves. In other words, in such cases, the direction of the force is the same as the direction of the displacement that it causes. If an applied force is acting on an object where other factors are also present (other factors means: the object is already moving, and/or other forces are acting on the object), the direction of the applied force is the same as the direction of the displacement that it would have caused, if the other factors were not present (if the object were not moving and/or if the other forces were taken away).

The direction of the force is the same as the direction of the net displacement that it causes. The force and the displacement that it causes, or would have caused, could be represented on the same axis. When the force and the displacement are represented on the same axis, the sign of the force is the same as the sign of its associated displacement.

In section 2.2.5 we discussed the acceleration of free fall (g). We said that the direction of g is downwards. This is because the force of gravity, which causes objects to move at acceleration g, pulls all objects down, toward the ground.

Example
What is the force that causes an object of mass 5kg to move at an acceleration of $3m/s^2$ in the direction 30^0 north of west?

Solution
Following the notations of [2.9] F=?, m=5kg, $a=3m/s^2$. Substitution yields F=15N. The direction of **F** is the same as that of **a**, which is the direction of the net displacement that **F** causes, namely, 30^0 north of west.

The standard kilogram (NIST).

The original definition of the kilogram was as the inertia of one liter of water. From practical reasons it was agreed later that a certain cylinder of platinum-iridium kept at the International Bureau of Weights and Measures at Sèvres, France, will serve as the standard kilogram. Identical copies of that block are found also in other institutions around the world. The kilogram (kg) is part of the SI system. With this new standard, the volume of water that has inertia of one kg is 1.000027 liter.

In the British system, acceleration is expressed in ft/s^2. The unit of force is the pound (lb), and the unit of mass is the slug. Equation [2.9] is, of course, the same in the SI and in the British systems. Accordingly, a force of one pound will cause a mass of one slug to accelerate at $1ft/s^2$. 1lb≈4.45N; 1N≈0.225lb; 1slug≈14.6kg; 1kg≈0.0685slug

Newton's first law enables us to determine if two forces have the same magnitude. All we have to do is to have each of them pull an object in opposite directions–like in a 'tug-of-war'. If the two forces have the same magnitude, the object will not move.

Newton's First Law is a special case of Newton's Second Law. When the net force that acts on an object is zero, according to Newton's Second Law [2.9], the acceleration would also be zero. Zero acceleration means no change in velocity. No change in velocity means that if the object was at rest, it will stay at rest, and if it was moving at a certain velocity it will keep moving at that velocity. This is what Newton's First Law is stating.

2.3.3 Adding forces in 1-D

When two or more forces act on the same object, the object will be accelerated as if one force was acting on it. That one force is the sum, or the resultant, of all those forces. As we have already discussed in the case of the displacement vector, in 1-D a vector is represented by a number. The absolute value of the number is the magnitude of the vector, and the sign of the number indicates the direction of the vector with respect to the selected axis. Forces are added in the same way that displacement vectors are added. In 1-D, we just add the numbers that represent the force vectors. The absolute value of the sum is the magnitude of the resultant force. The sign of the sum indicates the direction of the resultant force.

Example

Three forces act on an object of mass 3kg. Two of the forces, whose magnitudes are 4N and 6N, point to the south. The third force, whose magnitude is 16N, points to the north. (a) what is the resultant of these three forces? (b) What would be the acceleration of the object?

Solution

(a) Let's chose the axis in the north south direction, where northward is positive. With respect to this axis, the forces are: $\mathbf{F}_1$=-4N, $\mathbf{F}_2$=-6N, and $\mathbf{F}_3$=+16N. The resultant force $\mathbf{F}$ is: $\mathbf{F}=\mathbf{F}_1+\mathbf{F}_2+\mathbf{F}_3$= -4-6+16=6N. The magnitude of the resultant is 6N, and it points to the north, because it is positive.

(b) According to Newton's Second Law, $\mathbf{F}=m\cdot\mathbf{a}$, we get: $6=3\cdot a$, yielding $a=2m/s^2$. The acceleration **a** is also positive, hence it points to the north.

2.3.4 Newton's Third Law

According to Newton's Second Law, if an object is accelerating, a force must be acting on it. Imagine that you sit at a desk, and your chair has wheels. Now, you push the desk with your hands. The chair will move back from the desk. According to Newton's Second Law, a force must be acting on your chair, because its velocity has changed. However, you did not push the chair, you pushed the desk. Where did the force on the chair come from?

Newton observed similar situations and concluded that whenever an object A exerts a force on object B, B will exert a reaction force on A. The two forces, the action that acts on B and the reaction that acts on A, have the same magnitude and opposite directions. This is Newton's Third Law. Succinctly, it can be formulated as 'every action has a reaction of equal magnitude and opposite direction'.

In the example above, you and the chair played the role of object A, and the desk played the role of object B. The forward-pointing force that you applied on the desk affected the desk. The desk did not move after all because its friction with the floor prevented it from doing so. The backward-pointing reaction force, which the desk exerted on you, caused your motion. The friction between the chair and the floor was small, and the reaction force that came from the desk could overcome it and cause your backward motion.

Newton's Third Law applies to all objects and to all forces. Here are few examples.

When a gun is fired and the bullet is pushed out in one direction, the bullet exerts a reaction force on the gun. This reaction force pushes the gun in the opposite direction, causing its recoil.

If a piece of rocket's fuel is ignited in the open, flames, which are hot jets, burst in all direction. Consider now a rocket on the launch pad. Because of the shell of the rocket, jets, which would have burst in all directions, are now redirected and stream through the tail of the rocket. These jets have been pushed downwards by the shell of the rocket. The shell has exerted downwards forces on those jets. According to Newton's Third Law, the jets will exert reaction upwards forces on the shell. Those reaction forces will push the rocket upwards, as it really happens.

There are situations in which one part of object A exerts a force on another part of the same object A. If we apply Newton's Third Law to such a situation it will say: If A exerts a force on A (replacing B in the original formulation), then A (again, replacing B) exerts a reaction force of equal magnitude and opposite direction on A. The net force on the two parts of A is the sum of the active force and the reaction to it, which adds up to zero in this case. According to Newton's Second Law, if the net force on A is zero, the acceleration of A would be zero. The meaning of it is that an object at rest cannot start moving by pushing a part of itself. For example, you cannot lift yourself by pulling your hair up, no matter how hard you try. Don't try it...

2.4 Forces

There are many forces in nature. When any of these forces act on an object, the response of the object would be determined by Newton's Laws. In order to be able to figure out that response, we need to know more details about the properties of the forces. In the following we discuss the properties of some of the most common forces.

2.4.1 The force of gravity on Earth

The force of Earth's gravity pulls all objects on earth down, towards the center of the earth. 'The force of gravity that acts on an object' is synonymous with 'the weight of the object'.

Galileo has established that all objects in free fall have the same acceleration g, which is approximately $9.8m/s^2$. If we combine this fact with Newton's Second Law (equation [2.9]), the force of gravity that acts on the object, or the weight of the object, is:

$$w = m \cdot g \qquad [2.10]$$

where w is the weight of the object, m is its mass, and g is its gravitational acceleration.

Because the force of gravity is everywhere around us, it can serve as a convenient way for determining the mass of an object: m=w/g. We weigh the object, express its weight in Newton (in SI) or pounds (in British units), and divide it by g ($9.8m/s^2$ in SI or 32.1 ft/s^2 in British units). We get the mass in kg in SI or in slugs in the British system.

2.4.1.1 Weight, mass, and amount of material

A practical question at the beginning of civilization was how to measure amounts of various materials. In order to trade in commodities like wheat or rice, our ancestors needed ways of measuring the amounts of what was being traded. An earlier practice, which is being used to our days, has been to measure amounts by volume. The volume of a material is proportional to the amount of that material. Two gallons of milk have twice the amount of milk that there is in one gallon. Units of volume were agreed upon and were used to measure amounts of material.

Quite early, it was discovered that the amount of material in an object is also proportional to its weight. Units of weight were established and were used side by side with units of volume to measure amount of material. Balance-scales determine the weight of the object on one hand of the scale by equating it to the weight of standard pieces on the other hand. Springs are calibrated by standard weights and are widely used in spring-scales to measure weight, thus quantifying amounts of material.

It was found that mass, which measures the amount of inertia of an object, is also proportional to the amount of material in that object. The inertia of two liters of ice is twice the inertia of one liter of ice. That difference will show itself, according to Newton's Second Law (equation [2.9]), through the accelerations of these two objects due to any force that acts on them. Therefore, mass can also be used to quantify amount of material.

So, the three different properties of objects: volume, weight, and mass can be used to quantify the amount of material that the object has. These properties are related to each other, but the relationships depend on the kind of material that makes up the object. The volume of two pounds of lead is twice the volume of one pound of lead. However, the volume of one pound of lead is different from the volume of one pound of feathers.

Weight density and mass density of a material are properties that depend only on the kind of that material; they do not depend on the amount of the material. Weight density (D) is defined as the weight per unit volume of that material:

$$\text{Weight density} = \frac{\text{weight}}{\text{volume}}$$

$$D = \frac{W}{V} \qquad [2.11]$$

Similarly, mass density of a material (ρ) is defined as the mass per unit volume of that material:

$$\text{mass density} = \frac{\text{mass}}{\text{volume}}$$

$$\rho = \frac{m}{V} \quad \text{or} \quad m = \rho \cdot V \qquad [2.12]$$

The weight density and mass density of various materials are listed in tables. These tables are useful when one has to find out what is the weight of an object, if its volume and the material of which it is made are known. Similarly, if the weight and kind of material are given, the volume of that object could be found.

Example

What is (a) the volume, (b) the weight, (c) the mass of a 1.5"x3.5"x8' pine lumber beam? (Incidentally, in the US, the "two by four" stud is a staple in wood frame buildings. Its actual cross section is 1.5"x3.5". The actual cross sections of standard lumber beams in the US are smaller than their nominal "dimensions".)

Solution

Using the notations of [2.11] we get:

(a) The volume V of the stud is (1.5/12)'x(3.5/12)'x8'=0.29ft^3.

(b) The weight density of pine wood is D=34.4lb/ft^3, which yield w=34.4x0.29=10.0 lbs.

(c) Based on [2.10], the mass is
m=w/g=10.0lb÷32ft/s^2=0.31slugs=0.31x14.6=4.5 kg.

Substance	*Mass Density (kg/m^3)*	*Weight Density (lb/ft^3)*
Aluminum	2,700	169
Concrete	2,300	145
Iron	7,800	490
Ice	917	57
Water	1,000	62.4
Wood (pine)	550	34.4
Mercury	13,600	846
air (at 0^0C and 1 atm.)	1.29	0.081

Table 2.2: Densities of some substances

The pound and the kilogram

In every day's applications, some countries are using the kilogram and others are using the pound in order to quantify amounts of materials. For example, in some countries food is sold by the kilogram and in others by the pound. Although the pound measures the weight of the material and the kilogram measures its inertia, both can be used to quantify amounts of material. This is because the amount of material is proportional both to its weight and to its mass. When considering amounts of material, and only in those cases, it is appropriate to say that 1 pound is approximately equivalent to 0.45 kilogram, and that 1 kilogram is approximately equivalent to 2.2 pounds.

2.4.2 The Universal Force of Gravity

Any two objects attract each other by a gravitational force, the same way that the Earth attracts any object on its surface. The strength of the gravitational force (F_G) depends on the masses of the two objects (m_1 and m_2) and on the distance (r) between their centers. This is expressed by the formula:

$$F_G = G\frac{m_1 \cdot m_2}{r^2} \qquad [2.13]$$

where $G=6.6726x10^{-11}Nm^2/kg^2$ is the universal gravitational constant.

The gravitational force between common objects around us is very small, because their masses are relatively small. On the other hand, the gravitational attraction between the Earth and common objects on it is consequential, because of the large mass of the Earth.

The weight of an object is by definition equal to the gravitational force that acts on that object. Based on [2.13], that force, and hence the weight, would depend on r, the distance between the center of the Earth and the center of the object. As an object is moved up, away from the Earth, its weight will decrease. For example, the weight of an object at a distance of two Earth's radii from the center of the Earth is only one quarter of its weight on the surface of the Earth.

The gravitational attraction of the Earth keeps the moon and man-made satellites in their orbits. The gravitational attraction between the sun and the planets keeps the latter in their orbits around the sun.

2.4.3 Elastic forces

If we stretch a rubber band a little and then release, it will eventually return to its original shape. If we stretch it more, it will get to a point that when released it will contract, but it will not return to its original shape. If we stretch it more, it will reach its breaking point and break. The first region, in which the rubber band could return to its original shape, is called the elastic region. Many materials have elastic regions. In some, like rubber, we can see the stretching with the naked eye. In others, the stretch is small and can be seen only under a magnifying glass or a microscope. A spring is a simple device that illustrates many of the processes that occur in elastic systems.

A spring can be stretched or compressed from its undisturbed state. The more we stretch or compress a spring, the more force we need to apply. If we want to keep a spring in a stretched or a compress state, we need to keep applying an external force that will balance the elastic force of the spring. It was found that the elastic force of a spring is proportional to the change in its length.

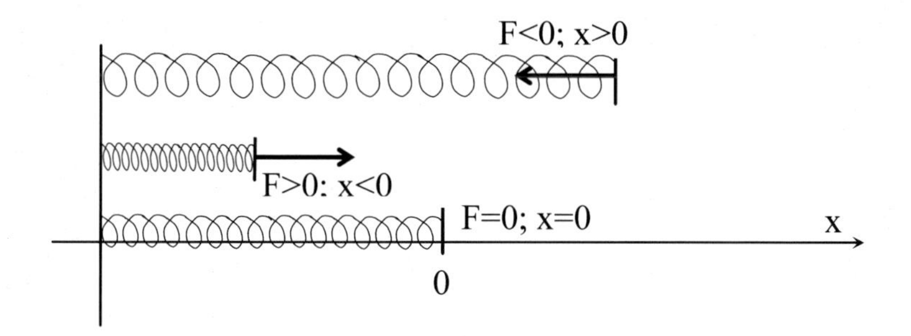

Figure 2.6: Three states of a spring: undisturbed (bottom), compressed (middle), and stretched (top). The arrows indicate the direction of the elastic force exerted by the spring.

In order to describe processes that involve a spring, we first assign to it an x-axis. The origin of the axis coincides with the free end of the undisturbed spring. The other end of the spring is fixed at some point on the negative half of the axis. The spring stretches and compresses along the axis, and the position of its free end is expressed by its x coordinate (figure 2.6).

In the undisturbed state, the end of the spring is at x=0 (figure 2.6). In stretched position, the coordinate of the end of the spring is positive. When the spring is compressed, its position coordinate is negative. The direction of the elastic force provided by the spring is against the direction of the x-axis when the spring is stretched, and with the direction of the axis when the spring is compressed. The magnitude of the elastic force that the spring generates (F) is proportional to the displacement of its end from the origin (x). Formula [2.14], which is called Hooke's Law, encompasses all these features:

$$\mathbf{F} = -\mathrm{k} \cdot \mathbf{x} \qquad [2.14]$$

k is the spring constant. It describes how 'soft' or 'hard' the spring is. The 'harder' the spring, the larger is the spring's constant k. The minus in the formula ensures that the relationship between the signs of F and x is as it should be, as illustrated in figure 2.6.

2.4.4 Normal forces

The force of gravity that acts on a vase wants to pull it to the ground. When we put the vase on a table, the table won't let it pass through; if it were not for the table, the vase would be accelerating to the floor. According to Newton's second Law, the fact that the vase is at rest on the table, in spite of the gravitational force that acts on it, is an indication that another force balances the effects of the force of gravity. The source of the second force is the table. The table exerts a force on the vase. That force acts on the vase and balances the gravitational force that acts on the vase. The magnitude of this force is equal to the weight of the vase, and it points upward.

Now, let's consider Newton's Third Law. According to it, if the table exerts a force on the vase, the vase exerts a force of the same magnitude and opposite direction on the table. In other words, a force that is equal to the weight of the vase is now acting on the table. Figure 2.7 shows the three forces that operate in the vase-table system. The force that the table exerts of the vase and the force that the vase exerts on the table are called normal forces. (Normal here means perpendicular to the surface of contact between the two objects).

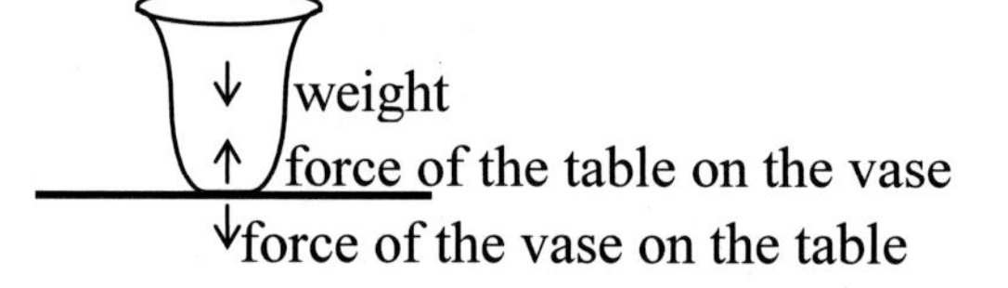

Figure 2.7: The three forces involved in a vase on a table.

In more general terms, let two objects, A (the vase) and B (the table), be in contact with each other, and let there be a force F (weight of vase) that tries to push A into B. The direction of F is perpendicular to the surface of contact between A and B. Under these circumstances, B will exert a reaction force on A equal to F in magnitude and opposite in direction. A will also 'transfer' F to B. The two forces that act between A and B perpendicularly to their common contact surface are called normal forces.

Solids objects are 'territorial', so to say. A solid object does not 'like' other objects to enter its space. The reason for the 'territorial behavior' of solids is that their molecules are held together by spring-like forces. When the vase tries to penetrate the table, the molecules of the table have to be pushed away to make room. Like in springs, it takes a force to push the molecules away from each other. As the vase tries to push the molecules away, they exert a force on the vase, the same as a stretched spring exerts a force on the object that causes it to stretch. As long as the spring does not break, it will keep exerting its elastic force on that object. Similarly, the table will keep exerting its normal force on the vase as long as it does not break from the weight of the vase. This force prevents the vase from passing through the table. The vase will stretch the table to the point that the elastic reaction force of the table (the normal force) is equal to the vase's weight. The normal force is a reaction force that adjusts itself to the active force (the weight).

2.4.5 Friction forces

When we apply a small horizontal force on a heavy box lying on the floor, it won't move. As we increase the force, there comes a point that the box will start to move. As we keep increasing the force, the box will increase its velocity. If we then stop pushing, the box will slow down, and eventually come to a complete stop.

According to Newton's Second Law, the fact that the box did not move when we pushed it at the first stages indicates that another force was acting on it. That force, which is called 'static friction force' (SFF) exactly balanced the pushing force that we applied. The word static indicates that this friction force was acting when the box was not moving. The SFF was directed opposite the pushing force that we applied. The SFF was able to adjust itself to the active pushing force up to a certain level. When this critical level was reached, the SFF could not increase any more. When the pushing force kept growing, it overcame the critical friction force, and the object started to accelerate. A friction force was present also at this stage, when the box was accelerating due to the pushing force. This friction force is called dynamic friction force, to indicate that it acts when the object is in motion. When we stopped pushing, the dynamic friction was still present, and it caused the box to slow down and to come to a complete stop. Figure 2.8 illustrates the various types of friction forces that were present in the different stages of pushing the box.

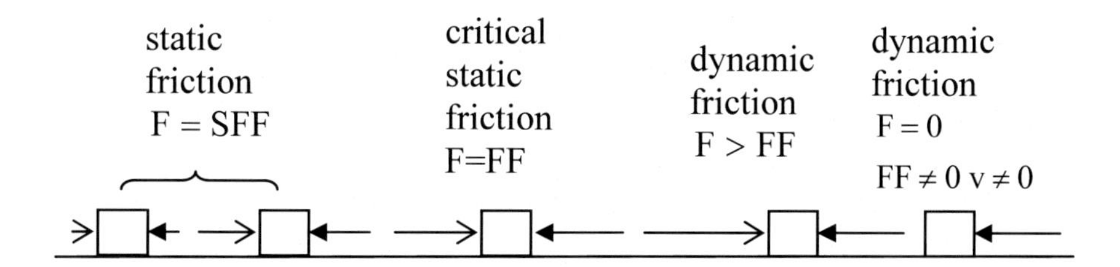

Figure 2.8: Various types of friction force (FF). The pushing forces (F) are denoted by ⟶ and the friction forces by ⟵

The direction of the friction force in cases of static friction is against the applied (pushing) force. The direction of the friction force when the object is moving is against the velocity.

The magnitude of the critical friction force depends on two factors: First, it depends on the properties of the contact-surfaces between the two objects. This property, which is expressed by the friction coefficient and is usually denoted by the Greek letter μ, is given in tables. The second factor is how tight the two objects are pressed against each other. The force that presses the two objects towards each other is called the normal force (NF). The critical friction force, which is called 'friction force' (FF) for short, is given by:

$$FF = \mu \cdot NF \qquad [2.15]$$

The normal force is not shown in figure 2.8. If the object is sliding on a horizontal surface and the pushing force is also horizontal (as shown in the figure), the normal force is the weight of the object. In other situations, the normal force, which pushes the object to the surface on which it would slide, has to be determined on a case by case basis. The critical friction force does not depend on the size of the contact area.

Equation [2.15] applies also to cases of dynamic friction. In these cases, FF stands for the dynamic friction force; μ stands for the dynamic friction coefficient, which is usually a little less than the corresponding static friction coefficient; and NF is the normal force, which is the same as in the static and the critical cases. The dynamic friction force does not depend on the speed at which the object moves and on the size of the contact area.

Friction forces exist when two solids slide on each other. When an object moves in a liquid or in a gas, e.g. a boat in the water, or a parachutist in the air, or when one part of liquid slides on another part, forces similar to friction operate. These are viscosity forces. The main similarity is that both forces are not action forces. They are reaction forces that respond to other forces or to motion. The main difference between friction forces and viscosity is that viscosity forces depend on the shape of the moving object and on its velocity–two properties on which friction forces do not depend.

Example
A box of mass 200 kg rests on the floor. The static friction coefficient between the box and the floor is 0.2, and the dynamic friction coefficient is 0.18. (a) What is the smallest horizontal force needed to move the box? (b) What would be the acceleration of the box if that minimal force continues to push the box after it has started to move? (c) What would be the acceleration of the box if that force is removed while the box is still moving?

Solution
(a) The minimal horizontal force has to overcome the critical friction force, given by [2.15]. The μ that we have to use is the coefficient of static friction μ=0.2. The normal force that presses the box to the floor is the weight of the box: NF=w=m·g, yielding NF=1960N. Substituting in [2.15] yields FF=392N. The active force is F_A=392N.

(b) As the box moves, two forces act on it in the horizontal direction. First, the active minimal force F_A, which is according to (a) F_A=392N. The second force is the dynamic friction force, which acts against F_A. The dynamic friction force is also given by [2.15], but now μ=0.18, the dynamic friction coefficient. Substituting into [2.15] yields FF=352.8N. The net horizontal force is F=F_A-FF =392-352.8=39.2N. This net force acts on the box, whose mass is m=200kg. Based on Newton's Second Law 39.2=200·a, yielding a=0.196 m/s^2. This acceleration is in the direction of the net force, which is the direction of the applied force, and also the direction of the motion of the box.

(c) As the applied force is removed, the box continues to move due to its inertia. It moves against the dynamic friction force, which is the only horizontal force now acting on it. From Newton's Second Law we get: $352.8=200\cdot a$, yielding $a=1.76\ m/s^2$. This acceleration is in the direction of the dynamic friction force and against the velocity of the box.

2.5 Statics in one dimension

NO FORCE - NO ACCELERATION

Newton's laws deal with the response of an object to the net force that acts on it. These same laws apply also to parts of an object. Any part of an object will respond to the net force that acts on it according to Newton's Laws.

Although Newton's Laws deal with acceleration of objects due to forces that act on them, building designers have particular interest in those aspects of the Laws that deal with zero acceleration. Zero acceleration is required of buildings and other structures that are not meant to move. It is required from the structure as a whole and from each of its elements. Zero acceleration means zero net force. In buildings, the weight of any element is usually known. The reaction forces that act between elements, and may affect the stability of the structure, are usually not as explicit as the weights. For a structure not to move, the vector sum of the active and reactive forces that act on any of its elements must be zero. Let's analyze now the active and reactive forces that operate in some common structural-elements.

NO ACCELERATION - NO FORCE

Consider a statue on a column (figure 2.9). This structure may be divided into two elements: the statue (the egg) and the column. Each of these two elements obeys Newton's Laws. Let's focus first on the statue. Let the weight of the statue be w=500N and the weight of the column be negligible. We know that a gravity force of w=500N that points down acts on the statue, as shown in figure 2.9a. Based on the discussion in 2.4.3 (normal forces), a pair of normal forces F_1 and F_2 are formed due to the contact between the statue and the column. The magnitude of each of them is $F_1=F_2=w=500N$. The upward pointing normal force F_1 acts on the statue, and the downward pointing normal force F_2 acts on the column.

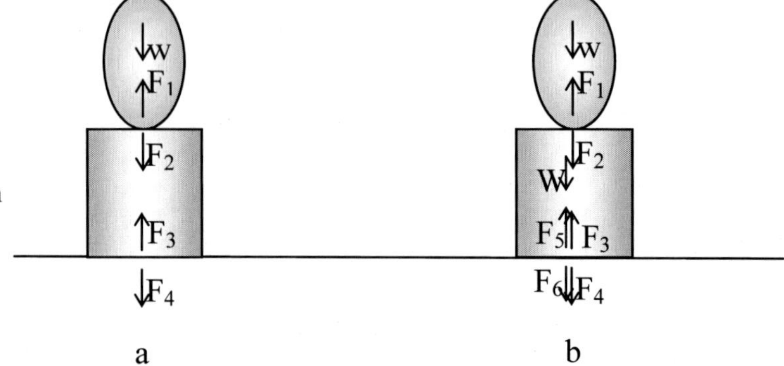

Figure 2.9: Reaction forces involving columns. The forces are shown inside the structure-elements on which they act.

Now, let's focus on the column. Since the column is not moving in spite of the normal force F_2 that acts on it, there must be another reaction force F_3, coming from the ground, that balances F_2. F_3 points upward and its size is 500N. And last, according to Newton's Third Law, if the ground acts on the column by F_3, the column acts on the ground. The downward force on the ground is F_4=500N. Therefore, we got $F_1=F_2=F_3=F_4=w=500N$.

The column serves as a 'conduit' that transfers the weight of the statue to the ground. In general, a column would transfer from its one end to the other an external force that acts in the direction of its long axis. The column would be compressed by that force. As you can see in figure 2.9.a, F_2 and F_3 are compressing the column. In addition to being a conduit, a column also separates the elements between which it is located. (The column keeps the statue and the ground apart). This is accomplished by the reaction forces that act from the column on the separated elements (F_1 and F_4 in figure 2.9a).

Figure 2.9b shows the same statue on a column whose weight is W=800N. All the forces that were present in 2.9a, where the weight of the column was negligible, are still present in 2.9b. In addition, the weight of the column is transferred to the ground, the same as the weight of the vase was transferred to the table in section 2.4.3. This is accompanied by the creation of a normal pair of forces $F_5=F_6=W=800N$. The total force that acts on the ground is now 500+800=1300N.

A very similar situation to a column that carries a weight occurs when a cable attached to the ceiling supports a weight. Figure 2.10 shows the forces that participate in a chandelier of mass 100kg that hangs from the ceiling on a cable of negligible weight. The elements of this structure are the chandelier and the cable. The chandelier is at rest in spite of the force of gravity $w = m \cdot g = 100 \cdot 9.8 = 980N$ that acts on it. This is an indication that the cable exerts on the chandelier a force F_1 that balances the weight: $F_1=w$. F_1 acts on a chandelier and points upward. According to Newton's Third Law, the chandelier exerts on the cable a reaction force F_2 that points down: $F_2=F_1=w$. The cable is not moving in spite of the F_2 that acts on it. This indicates that the ceiling exerts on the cable a force F_3 that points upward: $F_3=F_2$. According to Newton's Third Law, the cable exerts on the ceiling a force F_4 that points down such that $F_3=F_4$. Overall, we get $F_1=F_2=F_3=F_4=w$, similar to the case of the statue and the column. The difference is that the column was compressed by a pair of forces $F_2=F_3$ while the cable is stretched by a pair of forces $F_2=F_3$. The column was pushing apart the objects at its ends, while the cable pulls the objects toward each other.

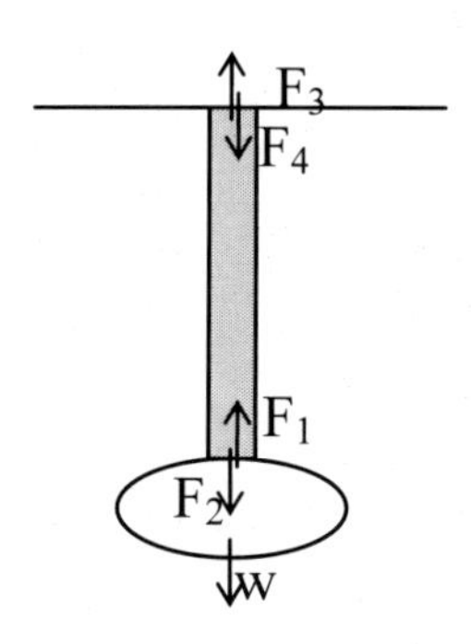

Figure 2.10: Reaction forces involving a cable. The tails of the arrows indicate the structure-element on which a force is acting.

Compressing forces and stretching forces are sometimes called **tension**. When tension acts on the two ends of an element (column or cable), it is relayed to other segments in the element. Figure 2.11 shows a cable divided into three arbitrary segments A, B, and C. Tension forces T act on the ends of the cable. The fact that A is at rest in spite of T indicates that B exerts on it a force of size $T_1=T$ to the right. According to Newton's Third Law, A exerts on B a force $T_2=T_1$ to the left. Since B is not moving in spite of the T_2 that acts on it to the left, C must exert on it a force $T_3=T_2$ to the right. According to Newton's Third Law, B exerts on C a force $T_4=T_3$ to the left. Hence, $T=T_1=T_2=T_3=T_4$. In conclusion, when tension forces T act at the ends of a cable, the same pairs of tension forces act on any segment inside the cable. The same thing happens in columns.

Cables under stretch are similar to springs under stretch. Columns under compression are similar to springs under compression. Identical pairs of forces act at the ends of the element and on every section inside it.

2.6 Dynamics in one dimension

Dynamics deals with the relationships between forces that act on objects and the

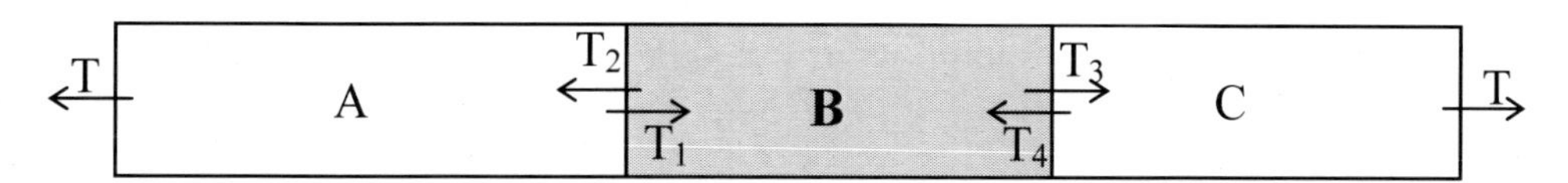

Figure 2.11: Distribution of tensions in a cable. $T=T_1=T_2=T_3=T_4$. The tails of the arrows are located in the segment on which the force is acting.

motion of those objects. According to Newton's Second Law, $\mathbf{F} = m \cdot \mathbf{a}$, or $\mathbf{a} = \frac{\mathbf{F}}{m}$. Based on this, if we know the net force F that acts at any time on an object of mass m, we know what is its acceleration a at that time. From that acceleration, we can find by how much the velocity of the object would change in the next short time interval. If we know the position and velocity of the object at the beginning of that time interval, we can find out what would be its position and velocity at the end of that short time interval e.g. we can use formulas [2.6-2.8]. So, in principle, from knowing the mass and the initial position and velocity of an object, and from knowing the net force that acts on it at any time, we can figure out its position and velocity at any time thereafter. In other words, we can find the equations of motion of that object. In the following, we will discuss the equations of motion of objects that are subjected to different forces. The appropriate equations of motion will be provided without proof, and the general characteristics of the motion will be described.

2.6.1 Net force=0

When the net force that acts on an object is zero, its acceleration would also be zero. That means that if the object is at rest, it will remain at rest. If it is found at position x_i at t=0, and if its velocity at that time is v, v will not change. This is motion at constant velocity v that can be described by: $\mathbf{x} = \mathbf{x_i} + \mathbf{v} \cdot t$. The sign of v indicates the direction of the motion. We have already discussed this kind of motion in 2.2.4.

2.6.2 Net force=constant (non-zero)

When a net force F that does not change with time acts on the object, its acceleration a would also not change with time. The direction of a is the same as the direction of F. That would be expressed by the sign of a. The magnitude of a would be: $a = \frac{F}{m}$. This is motion in constant acceleration, whose equation of motion are [2.6], [2.7], and [2.8].

Motion in constant acceleration occurs in a number of common situations.

- First, when the only force that acts on the object is a dynamic friction force. In such cases, the direction of the force is against the direction of the velocity. The friction will slow down the object until it comes to a complete stop. At that point, the dynamic friction force, which is a reaction force, will vanish, and the object will remain at that point.
- A second common example is in free fall. The only force that acts on the object is the force of gravity: $F = m \cdot g$. The acceleration is therefore $a = \frac{m \cdot g}{m} = g$. The direction of the constant acceleration is down. If the object starts to move up, with ever decreasing speed, it

will reach a highest point, which depends on its initial position and initial velocity. Then, it will move down with ever increasing speed, till it hits the ground. At this point free fall is over. If an object starts to move down, it will keep moving down at ever increasing speed, till it hits the ground.

- A third common example is motion on an inclined plane. This motion is also affected by the force of gravity, but because of the inclined plane, the acceleration, which depends on the angle of inclination, is less than g. This acceleration points down the inclined plane. In all other respects, the motion is similar to free-fall.
- A fourth common example is a combination of free fall or motion on an inclined plane with the presence of friction forces. Since the force of gravity that acts on an object and the dynamic friction force of that object are constants (if the object keep sliding on the same surface), the sum of the forces on the object is also constant. The motion would be in constant acceleration. The effect of the friction would be to reduce the speed compared to the same motion without friction. It should be remembered that the dynamic friction force points against the instantaneous velocity of the object, while the free-fall acceleration always points down.

Example

The brakes of cars that go downhill on steep and long mountain roads may fail. To help drivers that find themselves in such a predicament, many of those roads have emergency braking exits. The driver can drive the malfunctioning car to a side dirt road that goes uphill. The surface of the road provides a friction force that together with the force due to the uphill incline should stop the car. The overall stopping force provided by such a road is proportional to the weight of the car.

How long should such a road be, if the stopping force that it provides is twice the weight of the car, and the speed of the car entering it is 80 mph?

Solution

Since it is up to us to assign a coordinate system to a situation, we'll choose the x axis along the inclined plane. The motion of the car could be fully described with respect to this axis. The origin of the axis is at the foot of the emergency road, and L is the point where the car should come to a complete stop. v denotes the car's velocity, and F the total slowing force of the emergency road.

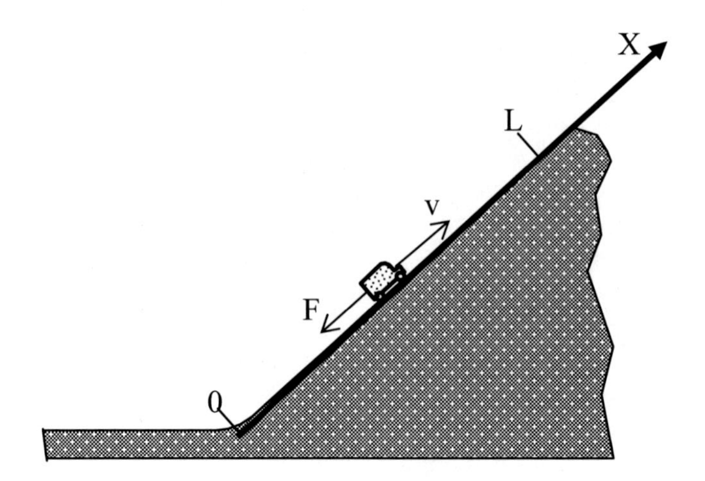

Let the weight of the car be w. The total stopping force is F=2·w=2·m·g, where m is the mass of the car. The acceleration a of the car, based on Newton's Second Law, would be $a = \frac{F}{m} = \frac{2 \cdot m \cdot g}{m} = 2g$. Since F points in the negative direction of our axis, it is negative, and so is a: a=-2g=-19.6 m/s^2. Using the notations of section 2.2.5, at the foot of the slope we have x_i=0, and v_i=+80 mph. The plus sign of v_i is because the car is going in the positive direction of our axis. At the point where the car has to come to a complete stop we have x=L and v=0. The units of v_i and a are in-compatible. We convert v_i to get 80 mph=80·1,600/3,600 m/s=35.5 m/s. When we insert that information into [2.8] we get:
$0^2=35.5^2+2\cdot(-19,6)\cdot(L-0)$, which yields L=32 m.

2.6.3 *Oscillations*

Harmonic motion

There are many phenomena that repeat themselves in cycles. The time that it takes to complete one cycle is called the period of that phenomenon, and it is denoted by T. For example, the period of the Earth's cycle around the sun is T=365 days (approximately). The period of a news session on CNN's Headline News is T=0.5 hours.

Another entity that is used in describing periodic phenomena is frequency. Frequency measures the number of cycles per unit time. The relationship between the period (T) of an event and its frequency (f) is:

$$f = \frac{\text{number of cycles}}{\text{time to complete them}} = \frac{1}{T} \quad [2.16]$$

For example, in the case of the CNN's news broadcast, $\text{frequency} = \frac{2\,\text{cycles}}{1\,\text{hour}} = \frac{1\,\text{cycle}}{0.5\,\text{hours}} = \frac{1}{T}$.

Frequency in the SI and in the British system is expressed by the same unit, Hertz (Hz), which is per-second.

Many objects move in periodic ways. Based on Newton's Second Law, any periodic motion has to be caused by a force, because the velocity vector of the objects changes with time. Different periodic motions are caused by different forces. The force provided by a stretched spring ($\mathbf{F} = -k \cdot \mathbf{x}$, [2.14]) is an example of a family of forces, called harmonic forces, that can cause periodic motions. Figure 2.12 shows a spring that a mass is attached to its free end. Three states of the spring are shown. At the bottom, the spring is in its undisturbed length. Its free end is at the origin of the axis, and it does not exert any force on m. Above it, the compressed spring is shown. Its free end is in the left half of the x-axis, and the force that it exerts on the mass points to the right. At the top, the spring is stretched. Its free end is in the right half of the x-axis, and the force that it exerts on the mass points to the left. (The spring is shown shifted upward for clarity. In reality, the spring stays all the time on the x-axis.)

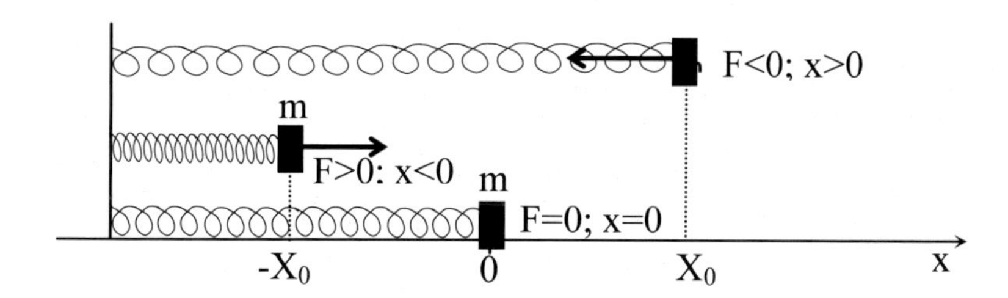

Figure 2.12: Oscillations of a mass attached to a spring

Imagine that we first stretch the spring in the positive direction to the position $x=X_0$, and then let go. Due to the force that acts on the mass, it will be accelerated to the left for as long as it is to the right of the origin. The magnitude of the force will decrease as the mass approaches the origin, according to the formula $\mathbf{F} = -k \cdot \mathbf{x}$. At the origin, x=0, and therefore F=0. However, the mass won't stop there, because it has accumulated velocity. It will continue moving to the left. As it enters the left half of the x-axis, the force that the spring exerts on it will point to the right. That will slow the mass down, and eventually it will stop briefly at $x=-X_0$. Then, it will start returning to the right, cross the origin, and reach $x=X_0$. From there, the entire process will repeat itself in cycles.

So, when a harmonic force, expressed by the formula $\mathbf{F} = -k \cdot \mathbf{x}$, acts on a mass m, the mass will oscillate back and forth around the origin of the x-axis. This kind of motion is called harmonic motion. The period T of the oscillations is given by:

$$T = 2\pi\sqrt{\frac{m}{k}} \qquad [2.17a]$$

and the frequency is:

$$f = \frac{1}{T} = \frac{1}{2\pi}\sqrt{\frac{k}{m}} \qquad [2.17b]$$

T is called the natural (or normal) period of the system, and f is the natural (or normal) frequency. According to [2.17a], if the mass increases, so does the period. If the mass decreases the period decreases too. Because of the square root, the relationship between T and m is not linear. T is proportional to the square root of m. For example, in order to double T, m has to be quadrupled. The formula also indicates that increasing the spring constant k decreases the period, and decreasing k increases the period. Again, the relationship is not linear. (Increasing k means using a 'stiffer' spring).

It is interesting to note that the period T (and the frequency f) do not depend on the initial stretch X_0 of the spring. It does not matter if we stretch the spring a little or a lot–it will take the mass the same time T to complete one cycle.

The initial displacement of the mass depends on the external force that we have to apply in order to bring the mass from the unperturbed position (x=0) to its initial position $x=X_0$. Based on [2.14] this force is given by $\mathbf{F} = -k \cdot \mathbf{X_0}$.

The equation of motion of that mass, after its release from $x=X_0$ is given by:

$$x = X_0 \cdot \cos(\frac{2\pi}{T} t) \qquad [2.18a]$$

The velocity of the mass as a function of time is given by

$$v = -X_0 \cdot \frac{2\pi}{T} \cdot \sin(\frac{2\pi}{T} t) \qquad [2.18b]$$

and the acceleration as function of time is given by

$$a = -X_0 \cdot \left(\frac{2\pi}{T}\right)^2 \cos(\frac{2\pi}{T} t) \qquad [2.18c]$$

Where x is the position of the mass at any time t, and T is the period as given by [2.17]. (The trigonometric functions $\cos(\alpha)$ and $\sin(\alpha)$ were defined originally for angles in triangles. Here we don't have any triangle. We treat cos and sin as functions, that when we substitute a number for their argument, they give us back another number–the cosine or sine of the number that we have substituted. The arguments of the sin and cos in [2.18] are in radians. Equations [2.18] enable us to calculate the position x of the oscillating mass at any time after its release. It tells us also that the oscillations take place between $x=X_0$ and $x=-X_0$. The maximum displacement X_0 is called the amplitude of the oscillations.

Equation [2.18a] describes a motion in which at the beginning (t=0) the object is found at $x=X_0$. If at t=0 the object is found at x=0, the equation of motion would be:

$$x = X_0 \cdot \sin(\frac{2\pi}{T} t) \qquad [2.18d]$$

The velocity as function of time would be

$$v = X_0 \cdot \frac{2\pi}{T} \cdot \cos(\frac{2\pi}{T} t) \qquad [2.18e]$$

and the acceleration as function of time:

$$a = -X_0 \cdot \left(\frac{2\pi}{T}\right)^2 \cdot \sin(\frac{2\pi}{T}t) \qquad [2.18f]$$

Example

The vibrations of the ground during an earthquake can be described as a superposition of harmonic functions like those of equations [2.18], having a variety of T's and X_0's. Sometimes, those vibrations can be approximated by just one harmonic function having a period T and amplitude X_0. Assume that that is the case in the following situation: the maximum displacement of the ground from its normal place was 0.3m, and 4 cycles of equal strength were noticed in 20 seconds.

What was (a) the maximum velocity of the ground? (b) What was the maximum acceleration of the ground?

Solution

Since the initial state of the ground was at rest, we'll use equations [2.18d]-[2.18f]. (At time t=0 the first equation [2.18d] gives x=0, and so it correctly describes the initial state of the ground).

The amplitude of the motion is X_0=0.3m, because this is the greatest value of the displacement x in the current situation.

The period of the motion is T=20second/4cycles=5 sec.

By substituting these values in [2.18e] and [2.18f] we get:

The velocity v of the ground as a function of time is : $v = 0.3 \cdot \frac{2\pi}{5} \cdot \cos(\frac{2\pi}{5}t)$,

and the acceleration a of the ground as a function of time is:

$$a = -0.3 \cdot \left(\frac{2\pi}{5}\right)^2 \cdot \sin(\frac{2\pi}{5}t)$$

The highest value of the velocity is when cos=1, which is

$v_{max} = 0.3 \cdot \frac{2\pi}{5} = 0.37 \mathrm{m/sec}$.

The highest value of the acceleration is when sin=1, which is:

$a_{max} = 0.3 \cdot \left(\frac{2\pi}{5}\right)^2 = 2.37 \mathrm{m/sec^2}$. Expressed in units of g:

a_{max}=2.37/9.8=0.24g.

Equations [2.18] suggest that once started, the oscillations will continue forever. This is, of course, an idealization, because in reality the amplitude of the oscillation will decrease with time, and eventually the mass will come to a stop. Friction and viscosity forces that are present in the system cause that slowing down and eventual stop. This is called **damping**. Equations [2.17] and [2.18] are applicable for situations where damping is small.

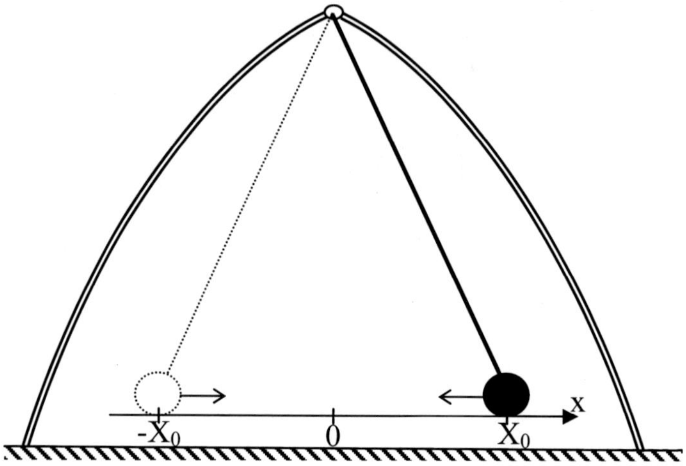

Figure 2.13: A swing-set (or a pendulum) and the forces that act on it.

Other examples of objects that move in harmonic motion are the pendulum and the swing-set. To describe the motion of a swing-set, let's assign to it an x-axis.

The unperturbed state of the swing-set coincides with the origin, and it swings back and forth along the x-axis between $x=X_0$ and $x=-X_0$ (figure 2.13). Actually, the swing-set swings along an arc of a circle, but we will assume that the circle under consideration is very shallow, and it can be approximated by the straight x-axis.

First, the swing-set is moved all the way to the right. There, a force in the negative direction of the x-axis acts on it. This force is the resultant of the force of gravity and the reaction force that comes from the cable of the swing-set. If the swing-set is let go, this resultant force will drive the swing-set in the negative direction of the x-axis. At x=0, the driving force will diminish to zero, but due to its accumulated velocity, the swing-set will move to the other side. As it moves along the negative side of the x-axis, a force in the positive direction will oppose that motion. Eventually, the swing-set will stop briefly, and due to that force, it will start its swing back. The entire situation is very similar to the oscillating spring, which we have just discussed.

Resonance

Imagine now Big Bob sitting in a swing-set, and Tiny Tina trying to push him. As Tiny Tina pushes a little push, Big Bob swings a little. Tiny Tina is not happy. She wants Big Bob to swing real far, but he is too heavy for her. Then she gets a brilliant idea. She pulls him as much as she can and let go. She waits till he swings back, and at the farthest points of his return, when he stops in mid air, she pushes him back. This push is not much, but it is enough to make him swing a little farther back and forth. She applies her little push several times, as he reaches the farthest point of his return. Every time he swings a little farther back and forth. After few more cycles, Big Bob is swinging vigorously, to their delight. This is an example in which a small force, when acting repeatedly and in synchrony with a natural period of a system, can cause large oscillations.

In physics, we use general terms to describe such processes. First, there is a system that can oscillate at its own natural frequency (e.g. the oscillations of the swing-set with Big Bob). (Some systems, like a swing set, can oscillate at one normal frequency. Other systems, like a guitar string, can oscillate at many normal frequencies.) Then, there is a periodic external force (e.g. the periodic pushing of Tiny Tina). **Resonance** occurs when the frequency of the external force equals a natural frequency of the system. In resonance, the oscillations of the system grow due to the applied external force. If the damping of the system is negligible, a small persistent external force at resonance can cause big oscillations in the system. A situation like this occurred at the Tacoma Narrows Bridge. As wind blew across the bridge, eddies were formed. The frequency of those eddies was in resonance with a natural frequency of the bridge. The steel bridge started to oscillate, the oscillations grew steadily, and they reached a level where the bridge collapsed, November 7, 1940).

Collapse of the Tacoma Narrows Bridge, November 1940. Top: Twisting of the structure. Bottom: the collapse. (© MSCUA)

In most real situation, there is a damping force in the oscillating system. The damping force would slow the motion of the system, thus prolonging its period and slowing it down to a stop. After the external force has stopped pushing, the swing set will come to a stop due to the damping caused by friction and air resistance. If the damping force is strong, it may prevent the system from oscillating even once. The system will 'crawl' from its initial position to its undisturbed state.

In resonance, the frequency of a periodic external force matches exactly a natural frequency of the system. That is when the external force would have the maximum effect on the system. If the external frequency is slightly off the natural frequency, the effect of the external force will diminish slightly. If the two frequencies are far off, the external force will have small effect on the system.

When an external periodic force is in resonance with a damped system, a steady state evolves after a while. The system will keep oscillating at a certain fixed amplitude for as long as the external force continues. In the case of Big Bob and Tiny Tina, if the axis of the swing-set is very rusty, the pushes of Tiny Tina will have to overcome the effects of the friction force. At a certain point, all her push will go towards overcoming the friction force, and the amplitude of his swing will not grow any more.

Structure engineers have to consider not only static forces that act on a structure but also possible vibrations that may develop in it. Each object has several natural frequencies at which it can oscillate. If the frequency of an external perturbation matches a natural frequency of the structure, the structure may enter into resonance. The extent of the ensuing vibrations will depend on the strength and duration of the perturbation, on the stiffness of the structure, and on damping forces that exist within the structure. If a resonance is anticipated, it may be possible to change the natural frequency of the structure by adding masses at critical points, or by re-distributing its weights. That may prevent the resonance by eliminating the match between the frequencies of the structure and of the external perturbation. In other cases, stiffening the structure, reducing the coupling between the external force and the structure, adding damping elements, or designing the structure so that it can tolerate strong vibrations are options that are considered.

PROBLEMS

Sections 2.1-2.2

1. Find the distance between two points A and B on the x-axis whose x-coordinates are:

a) A= +3.4m, B= +7.9m
b) A=+3.4m, B=-7.9m
c) A=-3.4m, B=+7.9m
d) A=-3.4m, B=-7.9m

2. Draw a horizontal x-axis. An object is placed at the origin. It is then pushed 2 cm to the right, then 4 cm to the left, then 5 cm to the right. Mark the position of the object after each of these three displacements (choose a convenient scale on the axis).
What is the displacement vector (magnitude and sign) of each of those three pushes?
What is the resultant displacement vector (magnitude and direction)?

What is the total distance covered in the three pushes?

3. A car covers the distance of 180 miles in three hours. Find the average speed of the car in: a) miles per hour. b) km per hour. c) meter per second.

4. An airplane flies from New York to Los Angeles in 5 hours and 35 minutes. What is its average velocity if the distance is 3,500 miles?

5. A sprinter can run 100 m in 9.69 seconds. What is his average speed in m/s and mph.

6. What is the average speed of a runner who completes the marathon is three hours. Express your result in mph and m/s. (A marathon is 42 km, 195 m or 26 mi, 385 yd). (You can find conversion data in section 2.1).

7. The motions of three objects are given by three equations of motion: The equation of the first object is: $x=3t$; that of the second object is $x=-3t$, and of the third object the equation is $x=4+3t-0.5t^2$, where t is in seconds and x is in meters. For each object do the following: (a) Write the "itinerary" of the object for the first four seconds of the motion (for integer seconds only). (b) Draw a x-t plot of the first four seconds. (3) Draw an x-axis, and mark on it the positions of the object in the first four seconds.

8. City B is 35 km to the east of city A. City C is 30 km to the east of city B. An x-axis was drawn to show this situation (below). Using this x-axis, the x-coordinate and the velocity vector of four cars were recorded as follows:
+20 km, +30 km/hr
+20 km, -30 km/hr
-20 km, +30 km/hr
-20 km, -30 km/hr
Mark the position of each car by a dot and its velocity by an arrow on the axis.

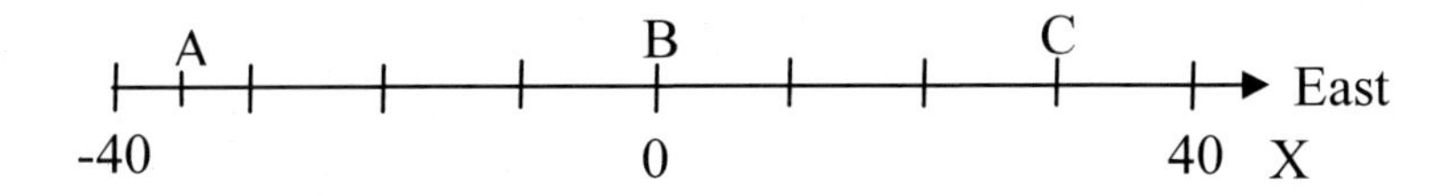

9. In problem 8, state the position of each car and its direction of travel with respect to city A in plain English (for example: located that many km to the east of A, and moving away from A at that many km/hr).

10. An airplane reaches cruising altitude 75 km to the east of the airport. It then keeps cruising eastwards at 800 km/hr. How far from the airport will it be 2.5 hours later?

11. A car travels 0.2 m in 0.01 sec. What is its instantaneous velocity?

12. Two trains leave the train station at the same time, moving in opposite directions. One moves at 30 m/s, and the other at 40 m/s. Write the equation of motion for each train, using the same x-axis.

13. A car maker claims that his car can accelerate from rest to 60 miles per hour in 6.2 seconds.
What is the acceleration of the car? Express your answer in SI units and in British units.
What is the distance that the car travels in those 6.2 seconds?

14. A car travels at 45 m/s. The driver then accelerates the car at 2.3 m/s^2 along 150 m. What is the final velocity of the car?

15. An airplane must reach a velocity of 75 m/s in order to take off. What is the shortest runway that it can use if it accelerates at a constant acceleration of 12 m/s^2?

16. On its way from the ground up, an elevator accelerates at 1.2 m/s^2 for 2 seconds. It then continues to climb at a constant speed for 33 second. It then slows at an acceleration of 1.2 m/s^2 for 2 seconds.
What distance did it cover in the first 2 seconds?
What was its final velocity at the end of the 2 seconds?
What distance did it cover in the 33 seconds, as it was moving at a constant speed?
What distance did it cover in the last two seconds?
Did it come to a stop after those two seconds?
What was its final height above the ground?
What was its average speed on the way up?

17. The way that an elevator accelerates and slows down in a certain multistory building is programmed into the system. It does not depend on the floor, or on the number of floors that the elevator has to travel between stops. In one non-stop trip from the ground, the elevator stopped at the tenth floor after 31 seconds. In another non-stop trip from the ground it stopped at the fifteenth floor after 44 seconds. What is the speed of the elevator when it climbs at constant speed, if the height of each floor is 2.7m?

Ignore air resistance in the following three problems.

18. A rock is dropped from a bridge and reaches the water below in 3.2 seconds. How high is the bridge above the water?

19. A rock is thrown from a cliff and reaches the water, 25 m below, in 1.9 seconds. What was the initial velocity of that rock? Was it thrown up or down?

20. A bullet is shot straight up. It leaves the gun at a speed of 290 m/s. As it falls back, it hits the gun.
What was the speed with which it hit the gun?

What was the highest point that it reached?

Section 2.3

21. The mass of a car is 2,500 kg. What force has to be provided by the engine so that the car accelerates at 5 m/s^2?

22. What is the magnitude of the gravitational force that acts on a 1 kg object?

23. A force of 25 lbs. accelerates an object at 75 ft/s^2. What is the mass of that object?

24. Four forces act on an object of mass of 25 kg. Two of the forces, of magnitudes 17N and 29N point to the east, and the other two, of magnitudes 12N and 32N point to the west.
a) What is the magnitude of the resultant force?
b) What is the direction of the resultant force?
c) What would be the acceleration of the object?
d) A fifth force is now added. What should be its magnitude and direction so that the object stays at rest?

Section 2.4

25. What is the weight of a 60 kg person?

26. What is the mass of a 180 lbs. person?

27. Using equations [2.10], [2.11], and [2.12], show that $D = \rho \cdot g$.

28. In order to build a rectangular swimming pool on the roof of a high-rise building, the designer needs to know the weight of the water that it will hold. The size of the pool is 12 ft x 30 ft, and the depth of the water is 3 ft. What is the weight of the water?

29. What is the weight of a cylindrical column of concrete whose height is 3m and the diameter of its base is 36 cm?

30. The radius of the earth is 6.36×10^6m, and its mass is 5.98×10^{24} kg (approximately). Find the weight of an object whose mass is 10 kg by using a) [2.10], and b) [2.12].

31. The radius of the moon is 1.74×10^6m, and its mass is 7.4×10^{22}kg (approximately). Find the weight on the moon of an object whose mass is 10kg.

32. When a force of 12N is applied to a spring, it is stretched by 4cm. What is the spring's constant?

33. A mass of 8kg is hung from the free end of a spring whose constant is 250N/m. The other end of the spring it attached to the ceiling. By how much will the spring stretch?

34. When a mass of 14 kg is hung from a spring whose length is 48 cm, it stretches to the length of 49.7 cm. A second mass of 20 kg is then added. What would be the final length of that spring?

35. A sofa of weight 150 lb. rests on the floor. A horizontal force of 50 lb. is the smallest force that can make the sofa slide. (Assume $\mu_{static}=\mu_{dynamic}$)
a) What is the static friction coefficient between the sofa and the floor?
b) What would be the acceleration of that sofa if pushed by a horizontal force of 60 lb?
c) What would be the smallest force needed to push that sofa if a person whose weight is 180 lb. lies on it?

36. A refrigerator of mass 75 kg rests on the floor. The static friction coefficient between the refrigerator and the floor is 0.15. (Assume $\mu_{static}=\mu_{dynamic}$)
a) What is the smallest horizontal force needed to move the refrigerator?
b) What would be the acceleration of that refrigerator when pushed by a horizontal force of 185N?

Section 2.5

37. You heard the saying "the strength of a chain is determined by its weakest link". Can you show that when a chain is stretched, all its links are stretched by the same force? (If some links were stretched stronger than others, a weak link that were stretched by a lesser force might hold, while a stronger link that was stretched by an even stronger force might break, contradicting the saying)

38. An elevator of mass 600 kg hangs on its cable.
(a) What is the tension in the cable?
(b) The elevator starts to accelerate up at 0.1 m/sec^2. What is the tension in the cable now?
(c) The elevator then moves at constant velocity of 3 m/sec. What is the tension in the cable?
(d) The elevator then slows down at an acceleration of 0.1 m/sec^2. What is the tension in the cable now?

39. A bartender slides a glass of beer on the bar towards a patron. The velocity of the glass as it leaves the hand of the bartender is 4 m/s, its mass is 0.6 kg, and the dynamic friction force that acts on it is 3N. Use an axis to describe the process.
(a) What is the acceleration of the glass? Is it positive or negative? Why?
(b) The glass stopped by itself on the bar. How far did it travel?
(c) How much time did it travel?
(d) What is the dynamic friction coefficient between the glass and the bar?

40. A truck of mass 3,000-kg accelerates from rest to 24 m/s in 6 seconds, when the engine provides its maximum force.
(a) What is the acceleration of the truck?
(b) What is the force provided by the engine? (Assume that friction is negligible).
(c) A rock is tied to the back of the truck. The mass of the rock is 1200 kg, and the friction coefficient between the rock and the ground is 0.1. What is the static friction force between the rock and the ground?
(d) The truck now moves with the engine at its maximum force. What is the net force that acts on the truck+rock system?
(e) What is the acceleration of the truck+rock system?

Section 2.6

41. An employee gets her paycheck every other Friday. What is the frequency of that event?

42. A car's engine completes 2000 revolutions per minute.
(a) What is the frequency of the revolutions in Hz?
(b) What is the period of a revolution?

43. When a mass of 3 kg was hung from a spring, the spring was stretched by 8 cm. Let's call this the equilibrium state of the system.
(a) Find the spring's elastic constant.
The mass was then pulled down for additional 2 cm, and then let go. Assume that the ensuing oscillations are not damped. Find:
(b) The amplitude of the oscillations.
(c) The period of the oscillations.
(d) The position of the mass with respect to its equilibrium position after 3 seconds.
(e) The maximum velocity of the mass.
(f) The velocity of the mass after 3 seconds. Was the mass moving up or down?
(note: to solve parts (d) and (f) you'll need a calculator that has sine and cosine functions. You'll have to set the mode of the calculator to radians, or to convert radians to degrees. 1 radian=57.2958 degrees.)

44. A beam in a structure vibrates 30 cycles in 10 seconds.
(a) What is the frequency of these vibrations?
(b) What is the period of these vibrations?

45. A chandelier of mass 15 kg hangs from the ceiling on a wire that behaves like a spring of constant k=2,366 N/m.
(a) What is the normal oscillation frequency of the chandelier?
(b) A boxer jumps with a jump rope on the floor above, right over the chandelier. He jumps two times every second for several minutes. Might the chandelier be at risk due to resonance? (Justify by calculation). If there were such a risk, what could you do as a designer to eliminate it?

46. The Citicorp Tower in New York City is a 915-feet-high skyscraper, opened in 1977. Part of the land on which it stands belonged to St. Peters Lutheran Church. One of the stipulations in the land sale agreement called for a hospitality plaza, and for open sky above a new church to be built on a portion of the lot. To accommodate those requirements, the building was built on four 17.5 foot columns located at the center of each face, and a central core. The tower itself is ten floors above street level. The natural lateral oscillation period of the building is 6.5 seconds. According to the building code, it has to withstand wind pressure of 40 pounds per square foot, and hurricane wind gusts of 100 miles per hour. The building may enter into resonance with strong winds. To counter that danger and to reduce the sway of the building under normal wind conditions, a tuned-mass-damper was installed in a room at the top floor. A mass of 400 ton is supported by hydraulic oil bearings, and by horizontal springs attached to the main structure of the building. The oscillation period of the mass-springs system can be tuned to 6.25±1.25 seconds, and the damping rate is adjustable from 8% to 14%. As the building starts to oscillate back and forth due to a wind, the mass would tend to stay at rest because of its inertia. That compresses the springs on one side of the mass, and tenses the springs on the other side. Consequently, the springs cause the mass to oscillate too. The combined effects of those oscillations and the viscosity forces of the dampers, similar to those in car shock absorbers, damp the oscillations of the building.

Citigroup Center (aka CitiCorp Tower)

Courtesy: Boston Properties

(a) The damping mechanism is activated whenever sensors sense that a maximum acceleration of 0.003g has been reached in two consecutive oscillation cycles. How much is the sway of the top of the building in that time period?

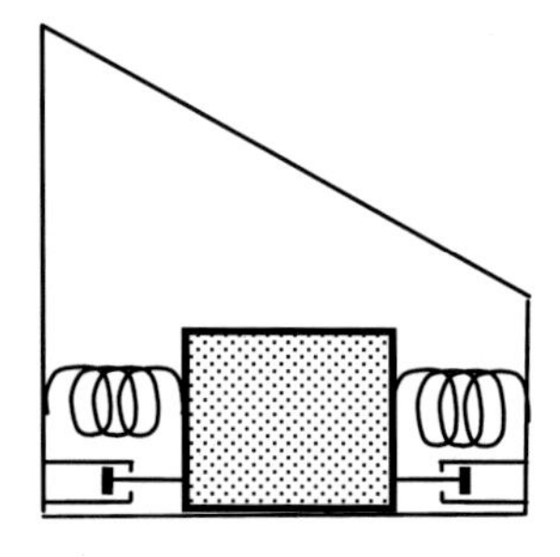

(b) The damping mechanism is most efficient when the natural oscillation frequency of the building, which is 6.5 seconds, is the same as the oscillation frequency of the mass-springs system. Assume that the oscillations in one horizontal direction (e.g. north-south) are independent of those in the perpendicular direction (east-west). Assume that two identical springs, one on each side of the mass, control the oscillations in that horizontal direction (e.g. one spring to the north of the mass and one spring to the south of the mass). What is the spring constant (k) of such a spring?

3. FROM 1 TO 2 AND 3 DIMENSIONS

3.1 Representing vectors

3.1.1 Representing displacement vectors

In a previous chapter we introduced the concept of displacement, and said that this is a vector that describes by how much an object is displaced from its initial position to its final position (the distance), and in what direction. An x-axis was aligned in space in the direction of that displacement, and the magnitude of the displacement was described on this chosen axis. We were able to describe a process that takes place in three dimensions by a one-dimensional axis. There are situations that this simple approach cannot be used. For example, we cannot describe on the same axis two objects that move along different lines. We have to use a two dimensional (2-D) coordinate system to describe general 2-D situations, and a 3-D coordinate system to describe general 3-D situations.

A 2-D coordinate system, say an x-y system, describes the position of a point in 2-D by its x and y coordinates. Figure 3.1 shows a coordinate system and a point **A**, whose coordinates are x=4, y=3 (all units are in meters).

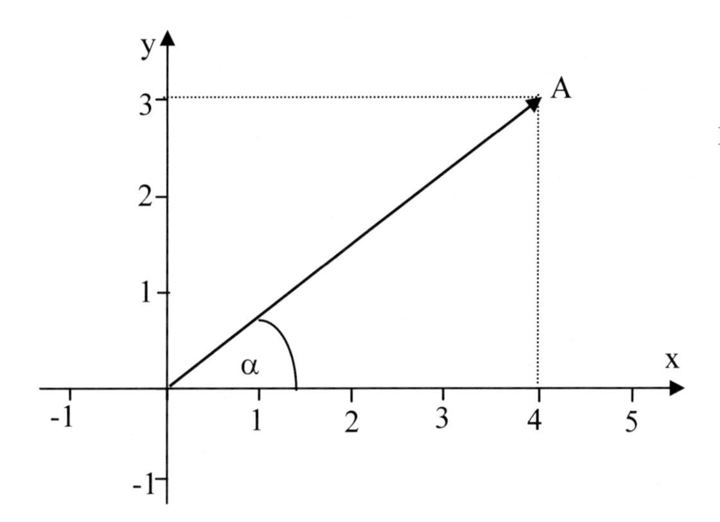

Figure 3.1: Relationships between position and displacement-vector

The act of displacing an object from the origin to point A is described by a corresponding displacement vector. This displacement vector, marked by the arrow, has a magnitude and a direction. Based on Pythagorean Theorem [1.3b], the magnitude of the vector is $\sqrt{4^2+3^2}=\sqrt{25}=5$ m. The direction of the vector can be expressed by the angle α that it makes with the x-axis. According to

[1.4], we get $\alpha = \sin^{-1}\left(\frac{3}{5}\right) = 53.13$ degrees. So, the displacement vector of figure 3.1 can be described in two ways. First, by magnitude and direction: We can say that its magnitude is 5m and its direction is 53.13 degrees with respect to the x-axis. Second, by components: We can say that in order to get to the final point we first moved 4 m in the direction of the x-axis, and then 3 m in the direction of the y-axis. We say that the x-component of the displacement is 4 m and the y-component is 3 m.

In this example, the coordinates of the tip of the displacement vector are the same as its components. This is true only when the initial point of the displaced object was at the origin. However, we can displace an object from any point in the x-y plane. For example, figure 3.2 shows the same displacement vector as in figure 3.1 operating on an object that is located initially at (1,1).

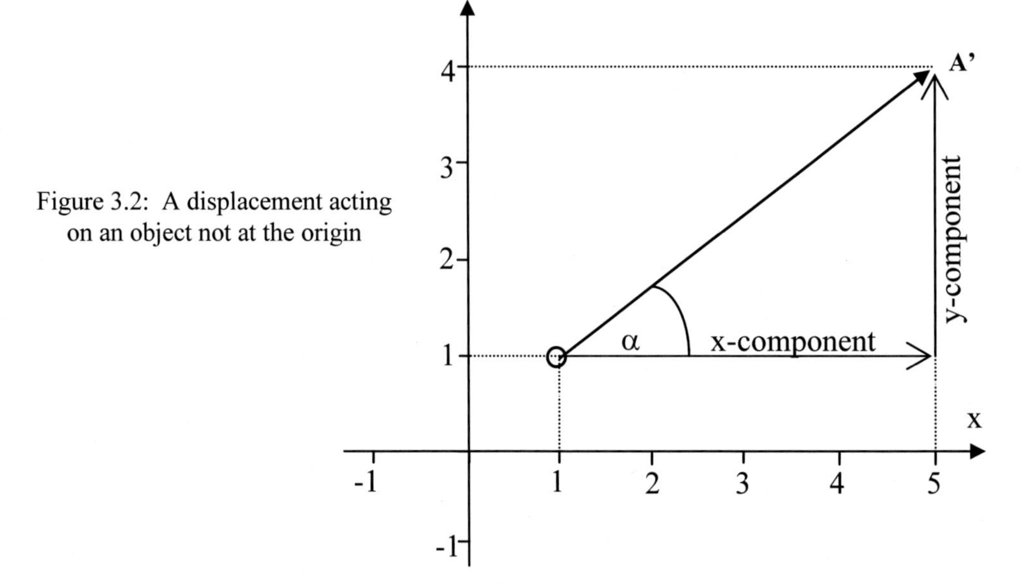

Figure 3.2: A displacement acting on an object not at the origin

The final position of the object would be at A'=(5,4), which is different from A=(4,3) of figure 3.1. However, the magnitude of the displacement vector is 5m, and its direction is 53.13 degrees to the x-axis, as in figure 3.1. The x-component of the vector is 4 m, and the y-component is 3 m, as in figure 3.1.

As can be seen from figures 3.1 and 3.2, the x-component of the vector is equal to its projection on the x-axis, and the y-component is equal to its projection on the y-axis.

3.1.2 Representing any vector

The two ways that we have just described for representing displacement vectors– by magnitude and direction or by components–can be applied to any vector such as velocity, acceleration, or force. By definition, a vector is a physical entity that has magnitude and direction. The direction of any vector can be expressed with respect to the coordinate system that we have chosen, the same as the direction of the displacement vector. We assign units to the axes according to the type of vector that we represent. For example, if we deal with forces, the units on the axes would be newtons or pounds. If we deal with velocities the units on the axes would be m/s or f/s, etc.

It is possible to go from representing any vector by magnitude and direction to representing it by components, and vice versa. This can be done graphically, as we did for displacement vectors, and by formulas. These formulas are the familiar trigonometric formulas with different letters (the same sire with a different attire). Figure 3.3 shows a vector **V** whose magnitude is V and whose direction angle with respect to the x-axis is α. The components of this vector are V_x and V_y.

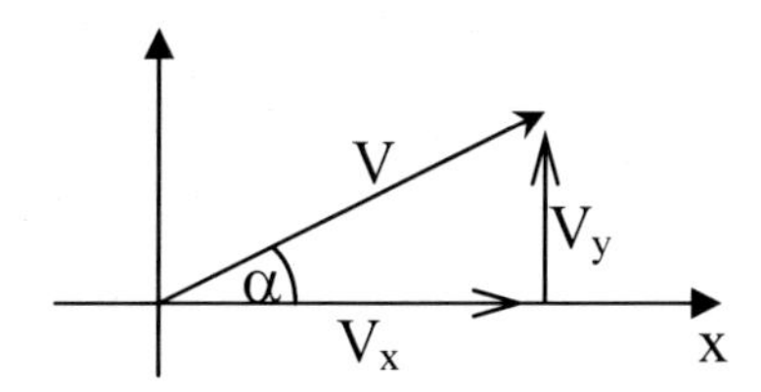

Figure 3.3: Magnitude, direction, and components of the vector **V**.

Formulas [3.1] are used when we know the magnitude and direction of a vector **V** and want to find its components.

$$V_x = V \cdot \cos\alpha \qquad [3.1a] \quad \text{(same as [1.2b])}$$

$$V_y = V \cdot \sin\alpha \qquad [3.1b] \quad \text{(same as [1.2a])}$$

Formulas [3.2] are used when we know the components of **V** and want to find its magnitude and direction.

$$V = \sqrt{V_x^2 + V_y^2} \qquad [3.2a] \quad \text{(same as [1.3b])}$$

$$\alpha = \tan^{-1}\frac{V_y}{V_x} \qquad [3.2b] \quad \text{(same as [1.4])}$$

3.1.3 Adding vectors graphically

A displacement vector may be viewed as an instruction to move an object from its initial position to a related final position. The sum of two displacements is that one displacement whose effect would be the same as having the object moved first according to the first displacement, and then according to the second displacement.

Imagine that you are the captain of one of the teams competing in a TV game 'The Navigator'. The starting point is at the bank of a lake, and each team gets a boat that can sail at 5 miles per hour. You get also a compass, paper, and basic drafting tools. The assignment that each team gets reads: "*If you sail 8 miles 30 degrees North of East you'll reach a patch of water lilies. If from there you sail 6 more miles 60 degrees North of West, you'll find yourself over a sunken boat. The first team to reach that point wins*".

You are about to tell your team to start sailing in the direction 30 degrees North of East, which is straight to the water lilies, when the math freshman in your team stops you: "the shortest distance between two points is a straight line, so let's sail straight to the final target". You are not sure that the rules allow it, but the law student in your team assures you that after reading the assignment very carefully, there is nothing there against it. Now it is up to you to tell your crew in which direction to go, and how far. Fortunately, you know your physics, and you realize that this is a simple problem of adding displacement vectors.

As illustrated in figure 3.4, you draw a coordinate system whose x-axis points to the East and y-axis points to the North. You decide that the origin represents the starting point. First, you draw the first displacement vector. You draw a line, 30 degrees North of East. You decide on a scale, and mark on the line an arrow whose length represents 8 miles. The tip of that arrow represents where the water lilies should be. Then, you draw the second displacement vector. From the tip of the first vector you draw a line 60 degrees North of West. You mark on it an arrow whose length represents 6 miles. That is where the sunken boat should be. Now, you draw the sum of these two displacements: You connect the starting point with the tip of the second vector. You measure the angle that that line makes with the East-West axis. It turns out to be 67 degrees North of East. You tell your crew to sail in that direction. Then you measure the length of that displacement. Based on the scale that you have used for the other two displacement vectors, you figure that the length of the sum is 10 miles. Since you know that your boat is sailing at 5 miles per hour, you tell your crew that the estimated sailing time is 2 hours. (Based on $t = \frac{x - x_i}{v} = \frac{10\,\text{miles}}{5\text{mph}} = 2\text{hr}$).

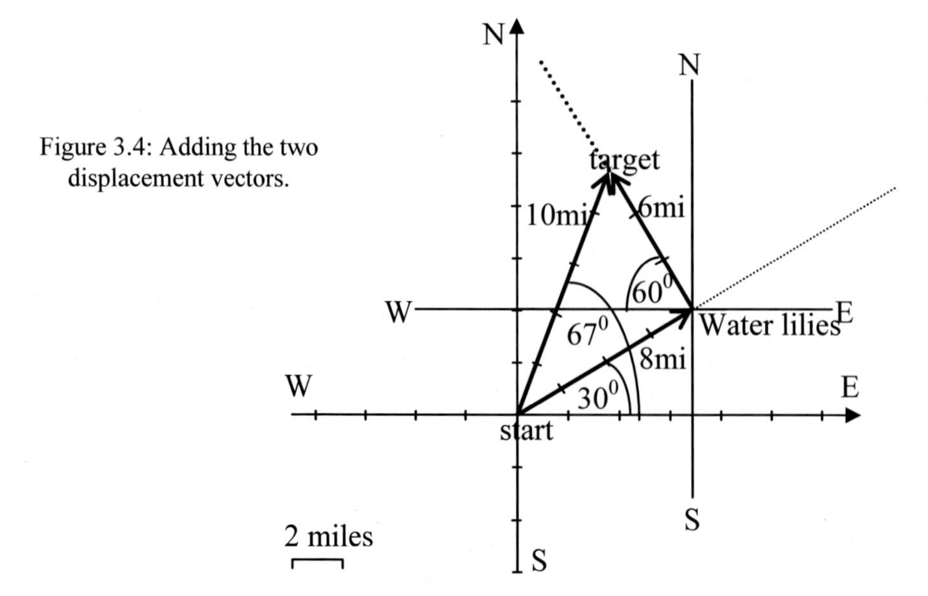

Figure 3.4: Adding the two displacement vectors.

The general procedure for adding two displacement vectors **A** and **B** to get their sum (resultant) **C**, (**C**=**A**+**B**) is as follows: Draw **A** to scale in its direction. From the tip of **A** draw **B** to scale in its direction. Connect the tail of **A** with the tip of **B**. This is the resultant **C**. The same procedure applies to adding any two vectors of the same type, such as two forces. The only difference is that the scale used represents the units of the vectors involved. For example, m/s may be used as a unit of velocity on the axes and on the vectors themselves when velocity vectors are added.

3.1.4 Adding vectors by formulas

Formulas for adding vectors are based on the procedure described above (3.1.3), for adding vectors graphically. In these formulas, vectors are represented by their components. Figure 3.5 shows two force vectors, $A=(A_x,A_y)=(4,1)$ and $B=(B_x,B_y)=(2,4)$, whose sum (resultant) is the vector $C=(C_x,C_y)$. The components of the vectors are in newtons. The process of adding the vectors has been identical to that of figure 3.4. First A is drawn in the direction, as provided by its components, and to scale. Then, B is drawn from the tip of A in direction and to scale. Then, C is obtained by connecting the tail of A to the tip of B. If you look carefully, you'll notice that the x-component of C is the sum of the x-components of A and B, (C_x=6=4+2). The y-component of C is the sum of the y-components of A and B, (C_y= 5=1+4). This can be expressed in general as:

$$\begin{aligned} &\mathbf{C} = \mathbf{A} + \mathbf{B} \\ &\text{means} \\ &C_x = A_x + B_x \\ &C_y = A_y + B_y \end{aligned} \qquad [3.3]$$

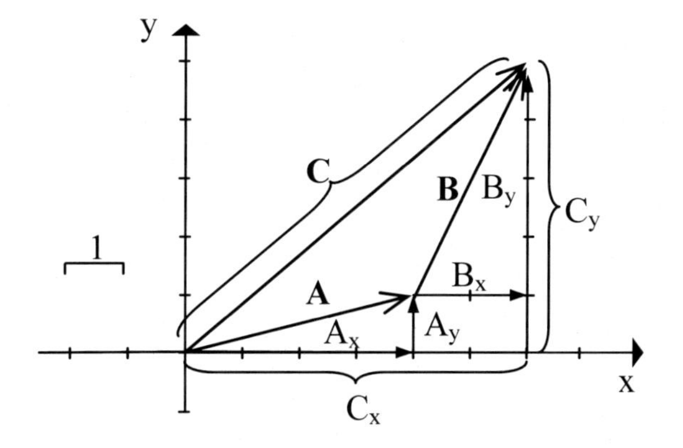

Figure 3.5: Adding the vectors **A** and **B** to get the vector **C**. The relationships between the components of the vectors are shown.

Thus, we have found that the components of C are (6,5) newtons. If we want to find the magnitude and direction of C we use equations [3.2]. The magnitude of **C** is $C = \sqrt{6^2 + 5^2} = \sqrt{61} = 7.81\text{N}$.

The angle that **C** makes with the x-axis is $\alpha = \tan^{-1}\frac{5}{6} = 39.8^0$.

So far, we talked about adding vectors. How do we subtract vectors? In order to answer this question, let's reconsider the meaning of subtracting numbers. Saying that 3-3=0 is the same as saying that 3+(-3)=0. We added 3 and -3. Minus three is the number that when added to 3 give zero. So, in numbers, subtraction of a number is the same as adding its negative, and the sum of a number and its negative is zero. The same principles apply in vectors. Subtracting a vector means adding its negative. Based on [3.3] we know what should be the negative of a vector. If a vector is expressed by its components $\mathbf{A}=(A_x,A_y)$, then the negative of that vector would be: $-\mathbf{A}=(-A_x,-A_y)$.

As you can see in figure 3.6, the negative of a vector has the same magnitude as the vector itself and it points in the opposite direction.

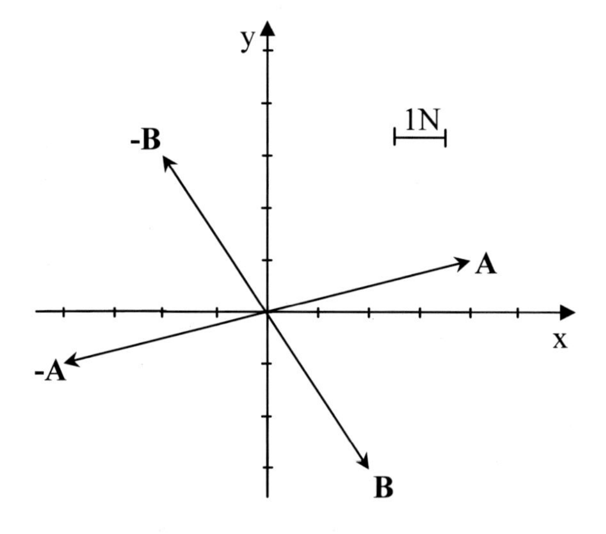

Figure 3.6: The relationship between vectors and their negatives. The negative of a vector has the same magnitude as the vector itself, but they point in opposite directions. The components of a vector and its negative have opposite signs (correspondingly)

Example

Two forces, F_1=40N pointing 30^0 north of east, and F_2=50N pointing 40^0 north of west, act on the same object. Find the magnitude and direction of the sum (resultant) of the two forces.

Solution

We choose a coordinate system whose origin is at the object, its x-axis points east, and its y-axis points north. In order to be able to add vectors we have to use [3.3]. However, F_1 and F_2 are given by their magnitudes and directions, while [3.3] requires their components. The components of the two forces are shown by dotted arrows in the upper coordinate system of the figure. To calculate the components we use formulas [3.1]

For F_1 we get: $F_{1X}=40\cdot\cos30^0=34.6N$, and $F_{1Y}=40\cdot\sin30^0=20N$.

For F_2 we get: $F_{2X}=50\cdot\cos140^0=-38.3N$, and $F_{2Y}=50\cdot\sin140^0=32.1N$. Please note that F_{2X} is negative, while F_{2Y} is positive, as should be according to their directions with respect to their corresponding axes in the figure. The correct signs of the components are obtained automatically if we insert in equations [3.1] the angle that the vector makes with the positive direction of the x-axis (in this case 140^0 rather than the 40^0 that F_2 makes with the negative direction of the axis.)

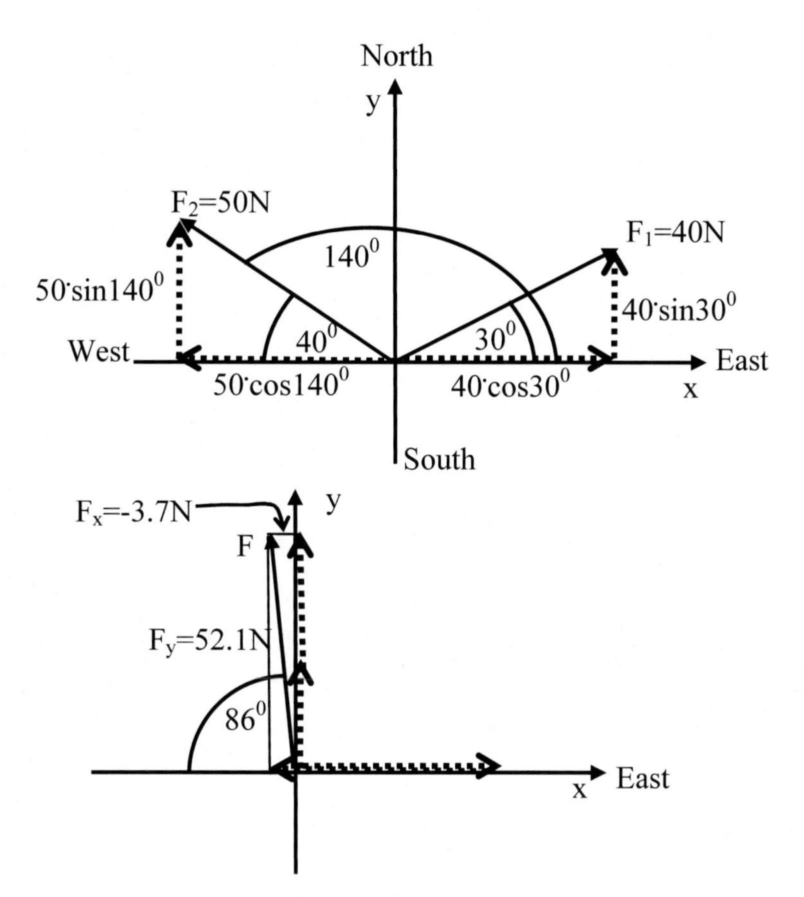

Now we can use the values of those components in order to calculate the components F_X and F_Y of the resultant F:
$F_X=F_{1X}+F_{2X}=34.6-38.3=-3.7N$; and $F_Y=F_{1Y}+F_{2Y}=20+32.1=52.1N$.
The resultant F is drawn based on its components in the lower coordinate system. It is found in the second quadrant.
Its magnitude is found based on [3.2a]:
$F=\sqrt{F_X^2+F_Y^2}=\sqrt{(-3.7)^2+52.1^2}=52.2N$.
Its direction is found based on [3.2b]:
$\alpha=\tan^{-1}\left(\frac{F_Y}{F_X}\right)=\tan^{-1}\left(\frac{52.1}{3.7}\right)=86^0$. This angle is with respect to the negative direction of the x-axis, therefore it is 86^0 north of west. When using [3.2b], it is better to use absolute values of the components, and to decide on the quadrant based on the signs of the components.

3.2 Force and motion in 2-D and 3-D

3.2.1 Two dimensions

When we talked about motion in 1-D, we said that we can describe the motion of an object by an itinerary, a time table, or a formula. In all these cases, the x coordinate, or the displacement from the origin, is given as a function of the time t. In 2-D, displacement is given by two coordinates, or by components of the displacement vector. Therefore, the motion can be described by specifying the x-component of the displacement as a function of t, and the y-component of the displacement as a function of t. For example, the two equations $x=5t$ and $y=35-2t^2$, where x and y are in meters and t is in seconds, describe the motion of a certain object in an x-y plane. For any given time t these equations give the corresponding x and y coordinates of the object, or its position in two dimensions.

The velocity and acceleration of an object, as well as the force that acts on it in 2-D can also be described by components. If we need to know the magnitude or direction of any of these vectors, we can always find them by using equations [3.2].

Newton's Second Law $\mathbf{F} = m \cdot \mathbf{a}$ ([2.9]) can also be expressed by components:

$$\begin{aligned} F_x &= m \cdot a_x \\ F_y &= m \cdot a_y \end{aligned} \qquad [3.4]$$

Equations [3.4], which express the same ideas as [2.9], tell us that the x component of the force is equal to the mass of the object times the x-component of the acceleration, and similarly for the y-components.

If more than one force is acting on an object, F_x is the net of the x-components of the forces in the x-direction, the same as we saw in chapter 2, where all the forces were acting along one axis. Similarly, F_y is the net of all the y-components of the forces. This is in agreement with equations [3.3], which tell us that in order to add vectors we have to add all their x-components together and all their y-components together.

Newton's Third Law relates the action and reaction forces that exist between two interacting bodies A and B. It states that the force $\mathbf{F}^A$ that A exerts on B is of the same magnitude and in opposite direction of the force $\mathbf{F}^B$ that B exerts on A. In components notation, Newton's Third Law is expressed as:

$$\begin{aligned} F_x^A &= -F_x^B \\ F_y^A &= -F_y^B \end{aligned} \qquad [3.5]$$

3.2.2 Three dimensions

The same vectors that describe the motion of an object in 2-D, namely, displacement, velocity, acceleration, and force are used also in three dimensional motion. Here again, Newton's laws apply separately to each component. So, equations [3.4] will have now an equation for the z-component: $F_z = m \cdot a_z$, and equations [3.5] will have $F_Z^A = -F_Z^B$.

The magnitude of a 3-D vector is related to its components by the Pythagorean Theorem, which in three dimensions will be:

$V = \sqrt{V_x^2 + V_y^2 + V_z^2}$, where V is the magnitude of the vector and V_x,V_y, and V_z are its components. The direction of a vector in 3-D is specified by two angles. Since in the following we will concentrate only on motions and situations that can be fully described in 2-D, we won't pursue here the 3-D issue any further.

3.2.3 Translation and rotation

When we dealt with motion in one dimension, the object's motion was restricted to a straight line, and all the forces were acting along that line. Under these conditions, it does not matter where are the actual points at which the forces are attached to the object. For example, in a 'tug-of-war', the actual point at which each team-member holds the rope does not matter. All that matters is the sum of the forces that pull one way compared to the sum of the forces that pull the other way. In two dimensions, the points at which the forces are attached to the object are important. Consider the two situations shown in figure 3.7. In both, two forces of equal magnitudes and opposite directions are acting on a box. In 3.7a, the two forces act on the same line and in 3.7b they are acting off-line. In 3.7a, the forces will not cause the object to move. In 3.7b, the forces will cause the object to rotate, as indicated by the dotted lines. However, the rotation in 3.7b is such that the object as a whole does not move to the right or to the left. Its center stays at the same point all the time.

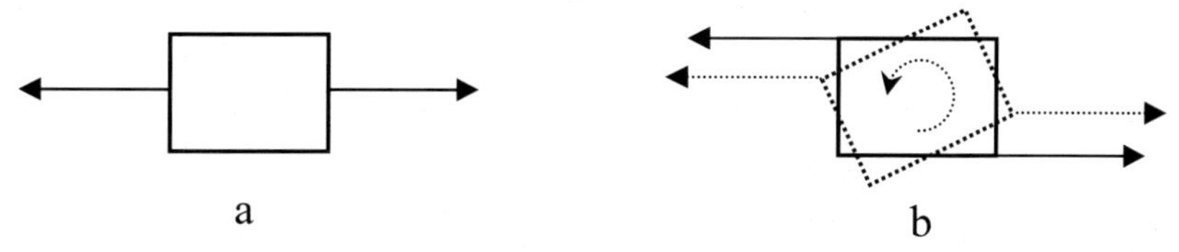

Figure 3.7: (a) Two forces of equal magnitudes and opposite directions that act on the same line do not move the object. (b) Two forces of equal magnitudes and opposite directions that act off-line cause the object to turn around, while leaving its center at the same point (Solid lines initial state, dotted lines final state).

Figure 3.8 shows a situation in which two different off-line forces $\mathbf{F_1}$ and $\mathbf{F_2}$ act on an object. The object will move in a combination of translation and rotation.

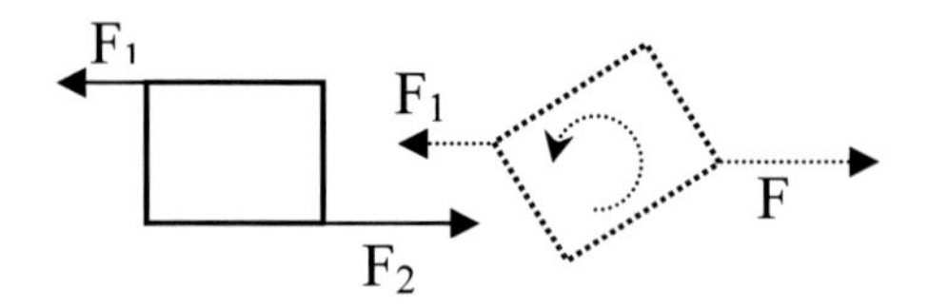

Figure 3.8: Two off-line forces that are different in their magnitudes will cause an object to be both translated and rotated. Solid lines: initial position; dotted lines: position after some time.

When applied to a solid 2-D object at rest, Newton's Second Law (equations [3.4]) deals only with the translation of the center of the object. So, when the sum of the external forces is zero, the translational acceleration of the center of the object is zero, and there is no translational motion. Rotational motion, as in figures 3.7b and 3.8 can still occur if the forces act off line. When the sum of the external forces is not zero, there will be translational acceleration of the object's center, given by [3.4]. Depending on the details, the object may move in a pure translational motion, or in a combination of translational and rotational motions.

3.2.4 Torques

When we push something harder, its acceleration increases. In general, in translational motion, the greater the force the greater is the response (acceleration) of the translated object. What is the analog of force when it comes to rotational motion e.g., when we open a door?

Force is only one of the factors that affect rotational motion. When we push a door, the harder we push the faster the door opens. However, the point where we push the door and the angle between our force and the door also affect the rate at which the door turns around its axis (its hinges). Figure 3.9 analyzes the effects that the magnitude of the pushing force, its point of attachment to the door, and the angle between the force and the door have on the rotation of the door.

Figure 3.9 illustrates what you already know from everyday life. The

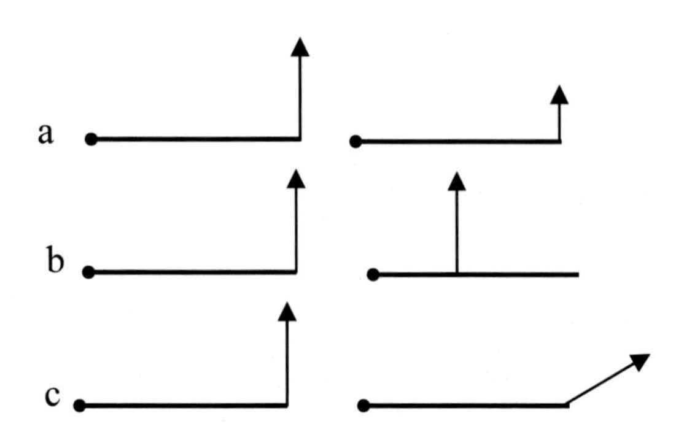

Figure 3.9: Top views of a door, its axis (hinge), and forces that act on it in various situations: (a) The force on the left door is greater than that on the right. Both are at the same distance from the axis, and make the same angle with the door. (b) The force on the left door is farther from its axis than the force on the right. The forces have the same magnitude and the same angle with the door. (c) The forces make different angles with the doors, but their magnitudes and distances form the axis are the same.

The forces in the left parts of a, b, and c are more effective in making the door rotate than their counterparts on the right.

effectiveness of a force on the rotation of an object (door) increases with the magnitude of the force. The effectiveness also increases with the arm of the force. **Arm** is defined here as the distance between the axis of rotation of the object and the point of the attachment of the force. In addition, the force is most effective when it makes 90^0 with the arm. The force has no effect at all when it is in the arm's direction (0^0 or 180^0). All these properties are combined into one entity, called **torque**. Figure 3.10 shows the elements involved in the definition of torque. An object can rotate around an axis. A force **F** is attached to the object. The arm r is the distance between the point at which the force is attached to the object and the axis. The angle θ is between the direction of the force and the direction of the arm. The magnitude of the torque $|\tau|$ is defined as:

$$|\tau| = |F \cdot r \cdot \sin\theta| \quad [3.6]$$

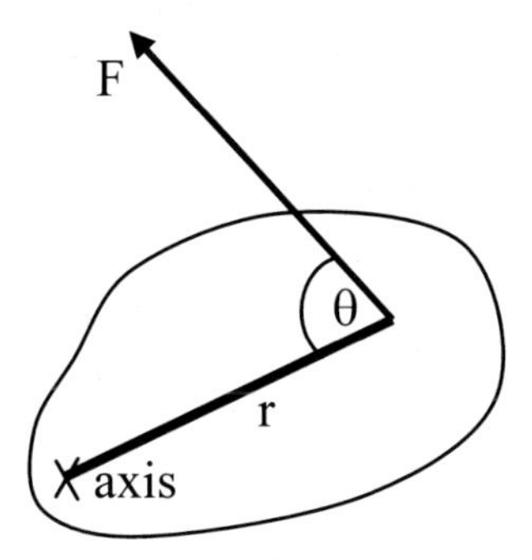

Figure 3.10: The elements of torque.

According to [3.6], the torque increases if the force F increases. The torque increases if the arm r increases. The torque is at its maximum when the force is 90^0 to the arm. The torque decreases gradually to zero as the angle approaches 0^0 or 180^0. All these features are what you would expect from the discussion about the door (figure 3.9). Please note that the magnitude of the torque [3.6] is always positive.

Example

A child pushes the edge of a 60 cm wide door with a force of 82 N that makes an angle of 30^0 with the door. What is the magnitude of the torque applied to the door?

Solution

Following the notations of [3.6], F=82 N, r=0.6 m, θ= 300^0. Substitution yields τ, the torque is 24.6 N·m.

The magnitude of a torque due to a non-zero-force may be zero in two cases. First, when the sine of the angle between the arm and the force is zero. Second, when the arm of the force (r) is zero. That happens when the force is attached to the axis.

An object can rotate around an axis clockwise (CW) or counter-clockwise (CCW). The sign of a torque is determined by the sense of rotation that it will cause the object to rotate. It is customary to define torques that cause CW rotation as negative, and torques that cause CCW rotation as positive. .

In order to find out in which way an object will rotate, we add all the torques that act on it, considering their signs. If the sum of all the torques is positive, the object will rotate CCW. If the sum of all the torques is negative, the object will rotate CW. **If an object is not rotating, the sum of all the torques that act on it is zero**. This is the rotational analog of Newton's First Law (if the object is not translating, the sum of all the forces that act on it is zero).

We can use equation [3.6] to calculate the torque if we know the length of the arm r, the magnitude of the force F, and the angle θ that the force makes with the arm. Sometimes, both the force and the arm are given in an X-Y coordinate system. Figure 3.11 illustrates how torques are calculated in such cases.

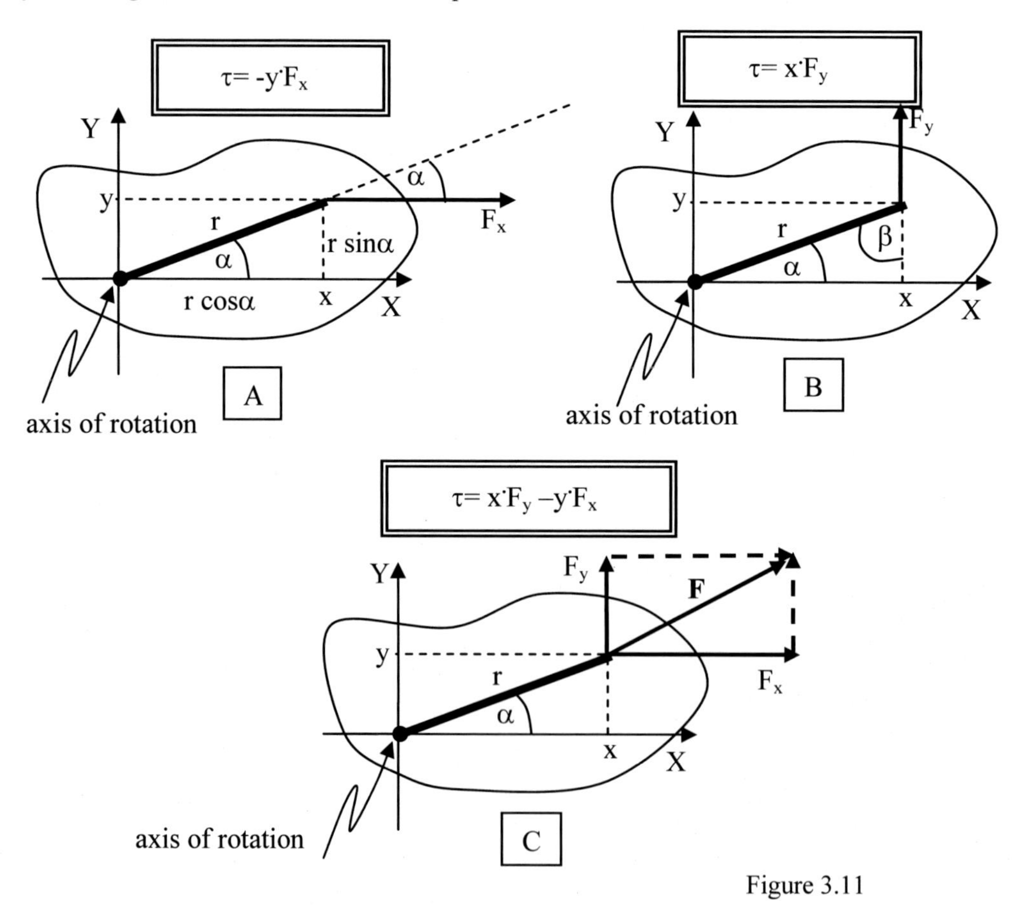

Figure 3.11

Figure 3.11A shows a force F_x that points in the X direction, the axis of rotation of the arm r is at the origin, and the arm makes an angle α with the X axis. In this case, the angle between the force and the arm θ=α. Substituting in

[3.6] gives $|\tau| = |F_x \cdot r \cdot \sin\alpha|$. But $r \cdot \sin\alpha = y$, where y is the projection of the arm on the Y axis (or the Y component of the arm). In addition, this is a clockwise torque, therefore it is negative. For this force we get: $\tau = -y \cdot F_x$. In words: The magnitude of the torque of a force that points in the X direction equals to the force times the projection of the arm on the Y axis.

Figure 3.11B shows a force F_y that points in the Y direction. In this case, the force makes an angle $\beta = 90^0 - \alpha$ with the arm. But $\sin(90^0 - \alpha) = \cos\alpha$, therefore [3.6] gives for this torque: $\tau = F_y \cdot r \cdot \cos\alpha$. Since $r \cdot \cos\alpha = x$, we get $\tau = x \cdot F_y$. Here the torque is counter clockwise, so it is positive. In words: The magnitude of the torque of a froce that points in the Y direction equals to the force times the projection of the arm on the X axis (or the X component of the arm).

Figure 3.11C shows a general force, whose components are F_x and F_y, that acts on an arm that makes an angle α with the X axis. The projections of the arm on the X and Y axes are x and y, respectively. These are also the coordinates of the point of attachment of the force. By combining the results for the previous two forces, the torque in the general case is:

$$\tau = x \cdot F_y - y \cdot F_x \qquad [3.6a]$$

The sign of the torque τ in [3.6a] automatically satisfies our convention: positive for CCW and negative for CW.

Example

A weight of 40 N hangs from a pole that sticks out diagonally from a wall (see drawing). The string is 0.5 m from the wall. What is the torque with respect to the lower end of the pole that the weight exerts on the pole?

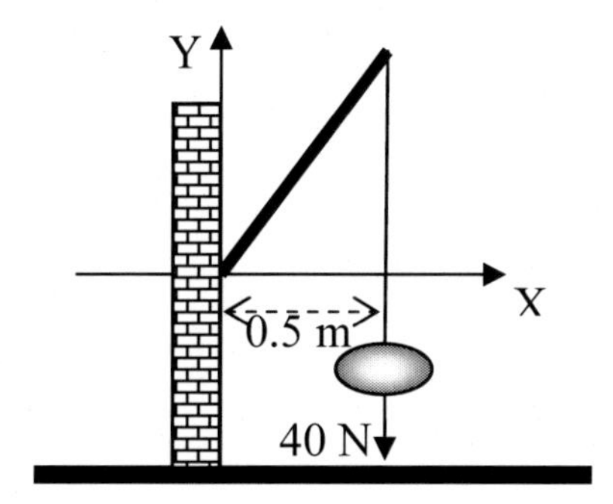

Solution

Try first to use [3.6]: F=40 N, r is not given, θ is not given, τ=? So, the straight-forward approach does not work. However, if we introduce an X-Y coordinate system whose origin is at the axis of rotation, we get $F_x=0$, $F_y=-40$ N, x=0.5 m, y is not given, and τ=? Substituting in [3.6a] we get: $\tau = -0.5 \cdot 40 + 0 = -20$ N·m.

Example
A door, 0.8 m wide, is 30^0 open with respect to the wall. A 22 N force that makes and angle of 40^0 with respect to the wall acts at the edge of the door. What is the torque of that force?

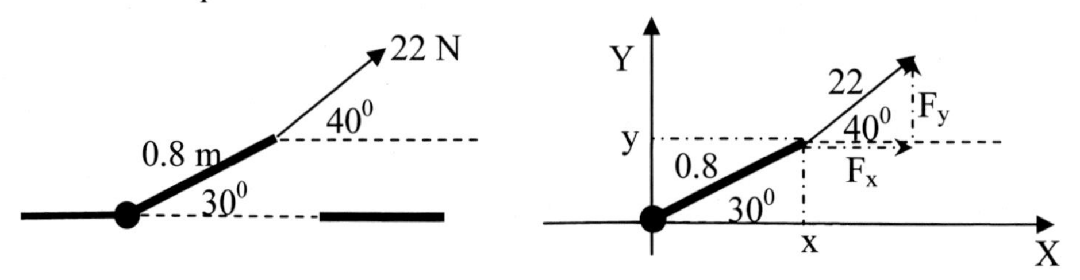

Solution
The left side of the figure shows the door and the force. In the right side, an X-Y coordinate system is added. The components of the force and the arm are shown: x=0.8cos30=0.6928m; y=0.8sin30=0.4; F_x=22cos40=16.8529; F_y=22sin40=14.1413. Substituting in [3.6a] yields τ=3.06 N·m.

A second way: The angle between the force and the arm is 10^0, or 170^0. In either case, based on [3.6], $|\tau| = 22 \cdot \sin 10 = 3.06$. The torque is CCW, therefore positive. τ=3.06 N·m, as in the first solution.

PROBLEMS

Section 3.1

1. Choose a coordinate system such that the x axis points to the east and the y axis points to the north. Solve the following problems graphically.

(a) The magnitude of a displacement vector is 120 m, and it points 60 degrees north of east. Find its components.
(b) The magnitude of a displacement vector is 120 m and it points 60 degrees north of west. Find its components.
(c) The components of a displacement vector are (50, 60) meters. Find its magnitude and direction.
(d) The components of a displacement vector are (-50,-60) meters. Find its magnitude and direction.

2. Solve the previous problems, which you solved graphically, by formulas [3.1] and [3.2].

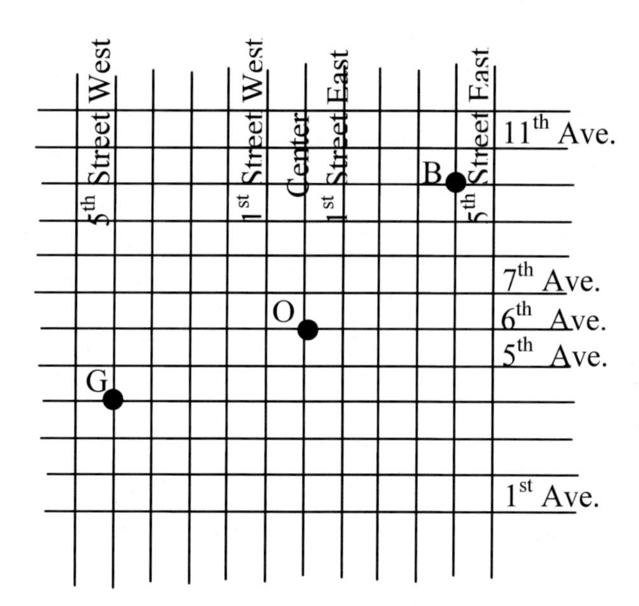

3. The streets of Generic City form a square grid. Each city block is 100m x 100m. Avenues are oriented east-west and streets are north-south (see map). City center is at the intersection of 6th avenue and Center Street. The numbering of the streets starts from City Center. An observation deck (O) is located above City Center. First Bank Tower is at the intersections of 4th street East and 10th Avenue (B in the map). Global Tower (G in the map) is at the intersections of 5th Street west and 4th Avenue.

Add a coordinate system whose origin is at City Center, its x-axis points to the east, and its y-axis points to the north. Wherever applicable, give your answer with respect to the axes of the coordinate system, and with respect to the real world. Derive your answers to the questions by both graphical and numerical methods. Approximate the location of the objects in the questions as the intersection of their corresponding street and avenue. The city lies on flat land.

(a) What is the azimuth at which the bank (B) is seen from the observation deck (O)?
(b) What is the azimuth at which Global Tower (G) is seen from the observation deck (O)?
(c) What is the distance between: O and B? O and G? G and B?

4. One cab goes east from G, turns north to O, turns east, and then turns north to B. A second cab goes east from G, and turns north to B. Mark their routes on the map.
(a) What is the total x component of the displacement of each cab?
(b) What is the total y component of the displacement of each cab?

5. In an improvised two-dimensional tug-of-war, three ropes are tied at one knot. Each of the three contestants applies a force of 900N on the end of his rope. The rope of the first contestant is directed straight to the east. The rope of the second contestant is directed 70^0 north of west. The rope of the third contestant is

directed 70^0 south of west. Make a drawing of this situation and answer the questions by both graphical and numerical methods.

(a) What is the resultant of the forces that act on the knot of the three ropes?
(b) Who wins the contest?

6. Here is some information about a motor-boat: (i) The speed of the motor-boat in standing water is 3 m/s. (ii) When that boat is in a river and its motor is not working, it is swept away at the velocity of the water. (iii) When the boat is in a river and its motor is working, the two causes of motion (the motor and the flow of the water) affect the boat independently of each other.

That boat is in a river that flows to the east at 2.5 m/s, and the motor is working.
Calculate the displacement of the boat in 10 second of sailing due to its motor alone.
Calculate the displacement of the boat in 10 seconds due to the water flow alone.
Draw those two displacement vectors and the total displacement of the boat (the resultant) for the following four cases.
1) The boat sails downstream.
2) The boat sails upstream.
3) The boat sails so that its bow points to the north (across the river).
4) The boat sails so that its bow points 30^0 north of east (downstream but diagonally to the bank).

For each of the drawings (1 through 4) answer the following questions: An observer at the starting point on the bank watches the boat.
How far would the boat be from the observer after 10 seconds?
What would be the azimuth from the observer to the boat after 10 seconds (degrees with respect to the east)?
What would be the magnitude of the velocity of the boat as measured by the observer?
What would be the direction of the velocity of the boat as measured by the observer (degrees with respect to the east)?

7. Repeat the previous problem numerically.

Section 3.2

8. A force of 100 N makes an angle of 20^0 with the x-axis.
(a) What are the components of that force?
(b) What would be the components of a force of the same magnitude and opposite direction?

9. A force of 80 N points 30^0 with respect to the x-axis. The force acts on an object of mass 4 kg.
(a) What is the magnitude and direction of the acceleration of that object? (Treat this problem as a 1-D situation.)

(b) What are the magnitudes of the x-components of the force and the acceleration?
(c) What are the magnitude and direction of the y-components of the force and the acceleration?

10. The mass of a train-car is 3,000 kg. The car can roll only on the rail tracks. A force of 2,000 N, which is directed 20^0 to the rails, pushes the car. What is the acceleration of that car? Neglect friction forces.

11. Two forces act on a box of mass 4.5 kg that rests on a horizontal plane. The first force, of 33 N points to the east, and the second of 44N points to the north. Find the acceleration of the box. Neglect friction.

12. Calculate the torque (magnitude and CCW (+) or CW (-)) in each of the following situations. An ellipse indicates an object, a bold arrow indicates a force of 10N, and the dark circle indicates the axis of rotation of the object. The length of the arm (objects (a) through (h)) is 5m.

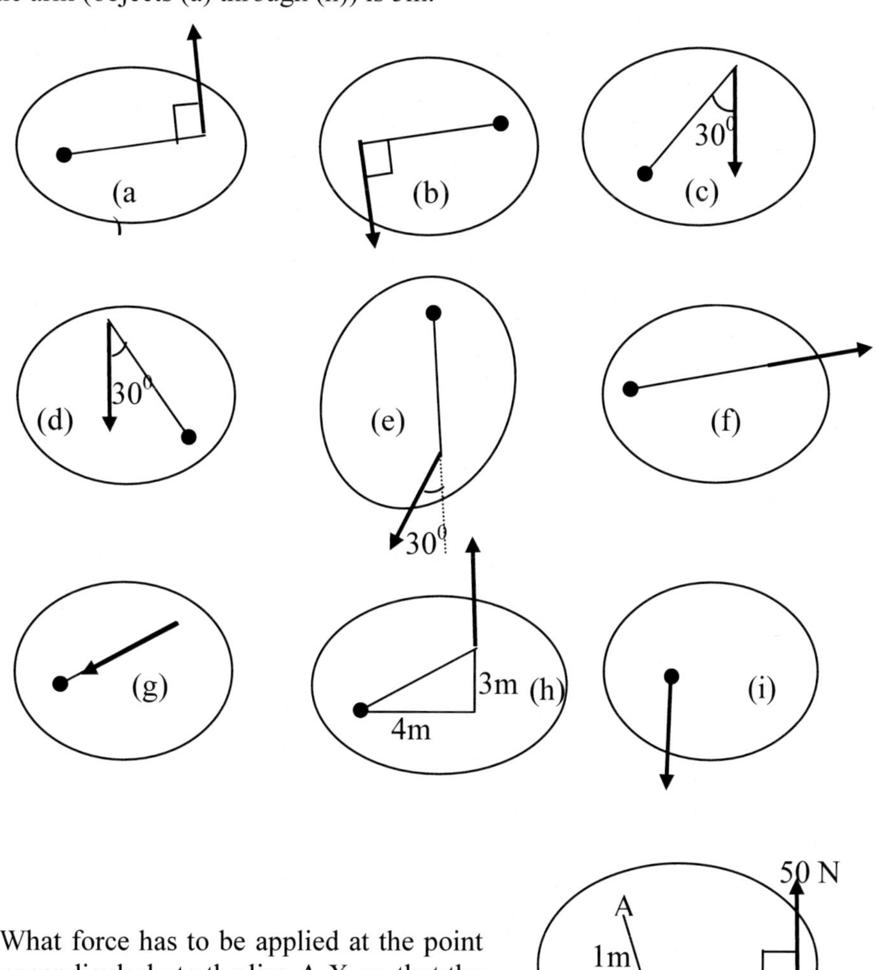

13. What force has to be applied at the point A, perpendicularly to the line A-X, so that the object does not rotate?

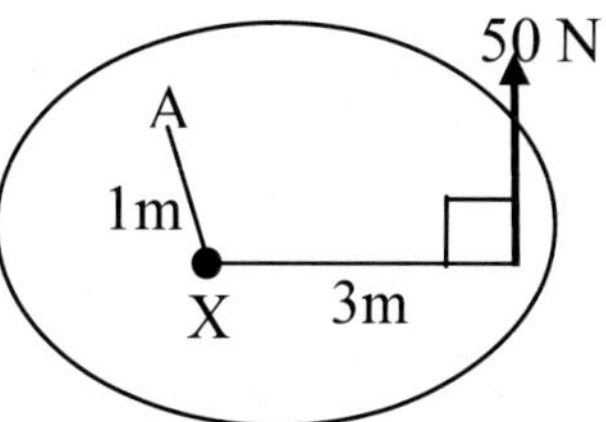

14. Three forces are acting on an object that can rotate around an axis, as shown in the figure. Will the object rotate CW or CCW? Justify your answer.

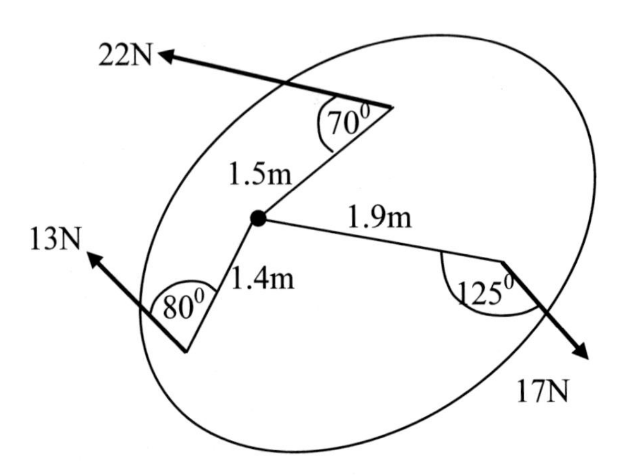

15. A beam is connected diagonally to the top of a 1.7m post. The lower end of the beam is 1.1 m above the ground and 1.3 m from the post. A bird whose weight is 3 N lands at the top of the beam and walks down to its lower end. What is the torque with respect to the top of the beam that the bird exerts on the beam when it stands at its (a) top, (b) lower end?

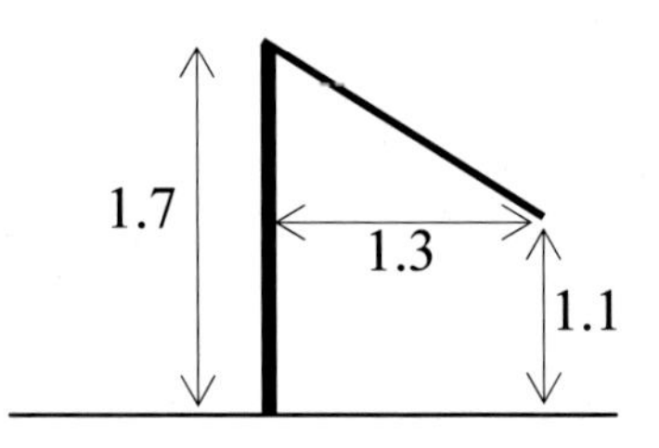

16. The same structure as in problem 15. Now the bird flew away, and a horizontal force of 3 N is exerted on the lower end of the beam. What is now the torque on the beam?

4. STATICS

4.1 Conditions for static equilibrium

Every element in a structure is subject to many forces all the time. First, the force of gravity acts on the element, trying to pull it down. Then, there are reaction forces from neighboring elements. Other active forces, such as those caused by winds or due to heating and cooling of the structure, may affect the structure element as well. A structure designer has to ascertain that every element of the structure will stay in place, in spite of all the forces that may act on it. This requirement translates into two sets of conditions:

First, each element should not be translated from its original resting position. For this to be guaranteed, the sum of all the forces, active and reactive, that act on the element should be zero. This is a direct consequence of Newton's Second Law (If the force is zero, the acceleration is zero. If the acceleration is zero, a resting object will stay at rest). When expressed by components, this requirement can be formulated as: the sum of the x-components of the forces on each element of the structure should be zero, and similarly for the y and z-components. This **first condition for equilibrium** is expressed by equations 4.1:

$$\Sigma F_x=0 \qquad \Sigma F_y=0 \qquad \Sigma F_z=0 \qquad [4.1]$$

The Greek letter Σ (sigma) means "the sum of". Equations [4.1] deal with a structure element that three-dimensional forces act on it. In the following, we will discuss only situations that involve two-dimensional forces, and therefore we will not have to consider the third equation of [4.1], for the z-component.

The second requirement is that any element of the object should not rotate around any axis. This requirement translates into the second condition of equilibrium, stating that the sum of the torques on any element, stemming from active and reactive forces, should be zero. The **second condition for equilibrium** is formulated by equation 4.2:

$$\sum \tau = 0 \qquad [4.2]$$

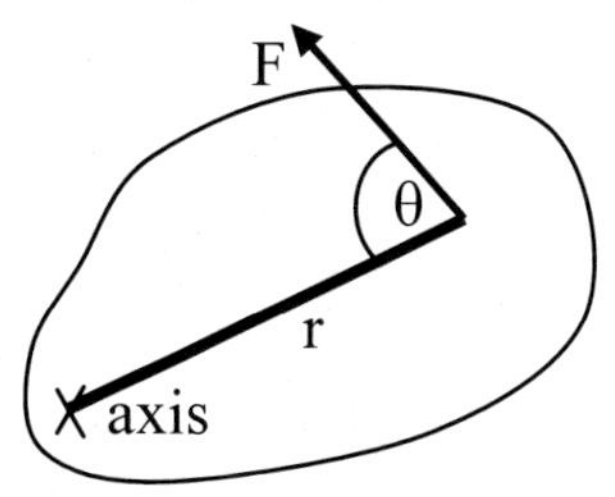

Torque was defined previously as:

$$|\tau| = |F \cdot r \cdot \sin\theta| \qquad [3.6]$$

Or

$\tau = x \cdot F_y - y \cdot F_x$

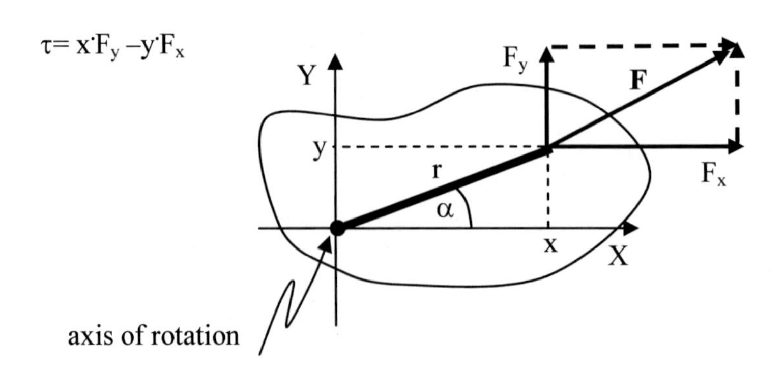

Here, again, we limit ourselves to torques that are caused by two-dimensional forces. The torques in [4.2] may be positive or negative. By convention, positive torques cause CCW rotation, and negative torques cause CW rotation.

Comment: Equation [4.2] states that at equilibrium the sum of the torques that act on the object is zero, regardless of the axis of rotation that we have chosen. So, for differently chosen axes of rotation, we may get different expressions for equations [4.2]. In certain situations, it is possible to replace one or two of equations [4.1] with additional one or two equations [4.2]. For example, if all the forces point in the x-direction, it is possible to express the equilibrium conditions as two equations [4.2], each of which with a different axis of rotation. The two chosen axes of rotation must have different x coordinates. The solution of those two [4.2] equations would automatically satisfy $\sum F_x = 0$. If forces point in both x and y directions, the solution of three [4.2] equations for three different axes of rotation that are not on the same straight line would automatically satisfy $\sum F_x = 0$ and $\sum F_y = 0$.

4.2 The Center of Gravity (c.o.g)

4.2.1 Properties of the Center of Gravity.

We can divide a book, or any object, to many imaginary elements. A gravitational force acts on each of those elements. So, a multitude of gravitational forces are involved in pulling the book, or any other object, to the ground.

It is possible to balance the book on the tip of one finger. The book can be in different positions and still be supported by the finger. For example, the book can be facing up, facing down, on its side, etc. A single reaction force, which comes from the finger, balances the multitude of those gravitational forces. All this is possible because any solid object has an imaginary point, called **the center of gravity** (c.o.g.). Although the force of gravity acts on each and every element of

x

the object, its effect is as if the entire weight of the object is concentrated at the c.o.g. **When considering the static equilibrium of an object of any shape, it is justified to treat its entire weight as if it were concentrated at its center of gravity.**

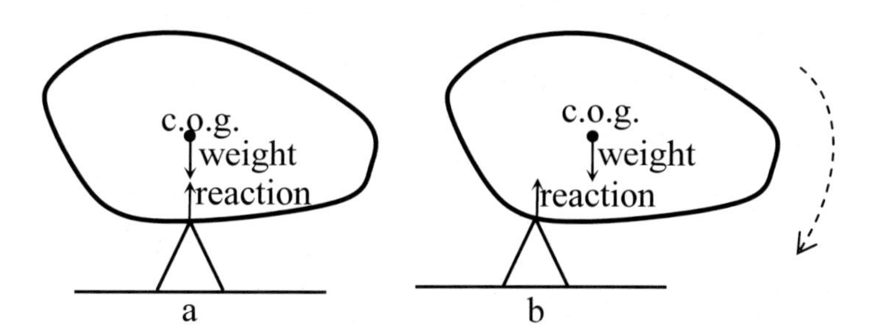

Figure 4.1: (a) The reaction to the weight is directed through the center of gravity of the object. The object is at rest. (b) The reaction to the weight is off the c.o.g. The reaction and the weight form a couple of forces whose torque topples the object.

When an object is balanced on one point of support, the reaction force from that point is directed to its c.o.g. (figure 4.1a). The force of gravity appears to act on its c.o.g. and to pull the object down. These two forces cancel each other, thus satisfying the first condition for equilibrium ([4.1]). The second condition of equilibrium ([4.2]) is satisfied because the torque of each of the two forces around an axis at the point of support is zero. (The forces act along the arm, thus the sine of their angles is zero).

If the reaction force from the supporting point is not directed to the center of gravity (figure 4.1b), a non-zero torque is formed, and topples the object.

In general, when the center of gravity is over the basis of the object, that object will stay in place. If the center of gravity is not over the basis, the object will topple. In figure 4.2, can you mark the part of the rock in which the center of gravity is not located?

Figure 4.2: The Balanced Rock in the Garden of the Gods, Colorado. (Courtesy of The Trading Gods)

The center of gravity of simple uniform objects such as squares, rectangles or circles is at their geometrical center. There are formulas for calculating the location of the c.o.g. of more complex objects. These formulas are based on the idea that any object can be divided into small elements of mass Δm. Each of these small elements will have its own torque with respect to any arbitrary axis around which the object will rotate. The c.o.g. is the point with respect to which the sum of the torques of all the mass elements adds up to zero (Figure 4.3a).

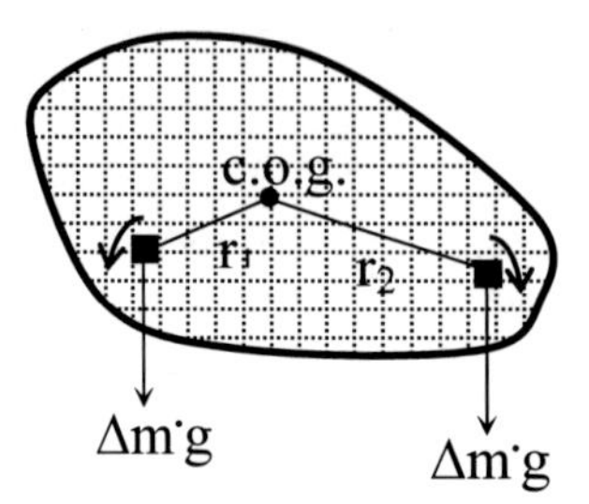

Figure 4.3a: An object is divided into small mass elements Δm. The sum of the torques around the center of gravity of all these masses is zero. Two such small masses are shown.

It is possible to find the c.o.g. of objects experimentally. The object is first hung from any arbitrary point, and the line from that point straight down is marked. Then, the object is hung from another arbitrary point, and the line straight down from that point is marked too. These two lines intersect at the c.o.g. (figure 4.3b).

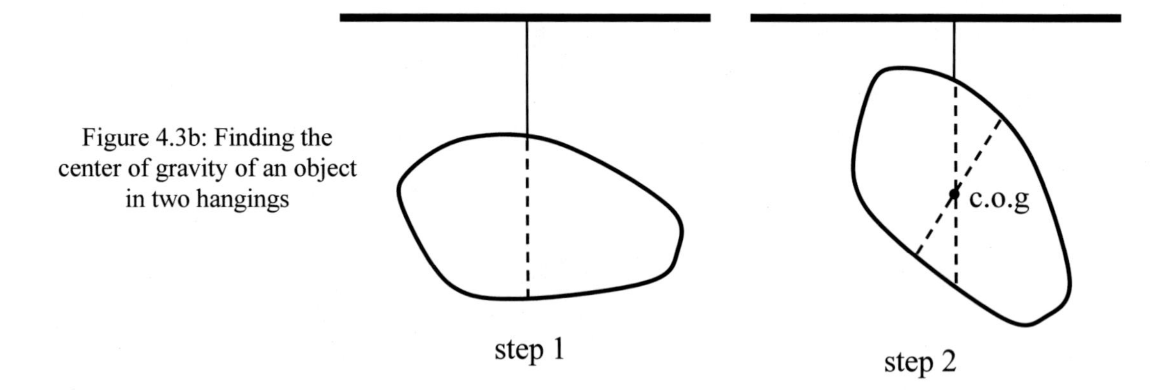

Figure 4.3b: Finding the center of gravity of an object in two hangings

4.2.2 Calculating the center of gravity.

It is possible to support an object by applying a single upward force at its center of gravity (c.o.g.), e.g. figure 4.1a. Based on that concept, we can calculate the location of the center of gravity of an object. Let's see how it is done for a few simple objects. The first object consists of three small masses on a mass-less bar (figure 4.3c). The object is assigned to an arbitrary x-axis. The masses are m_1, m_2, and m_3, and their coordinates are x_1, x_2, and x_3, respectively. The weights of those masses point down, and the supporting force F is attached at $X_{c.o.g.}$, which is the coordinate of the c.o.g. on the x-axis that we have chosen. Since the object is at equilibrium, it satisfies the equilibrium conditions [4.1] and [4.2], which give:

$F-(m_1g+m_2g+m_3g)=0$ (for the sum of the vertical forces)

and

$F \cdot X_{c.o.g} -(m_1g \cdot x_1+m_2g \cdot x_2+ m_3g \cdot x_3)=0$ (for the sum of the torques around an axis-of-rotation at the origin.)

The first equation yields $F=(m_1+m_2+m_3)g$, meaning that the magnitude of the supporting force is equal to the weight of the object.
Substituting this F in the second equation yields:

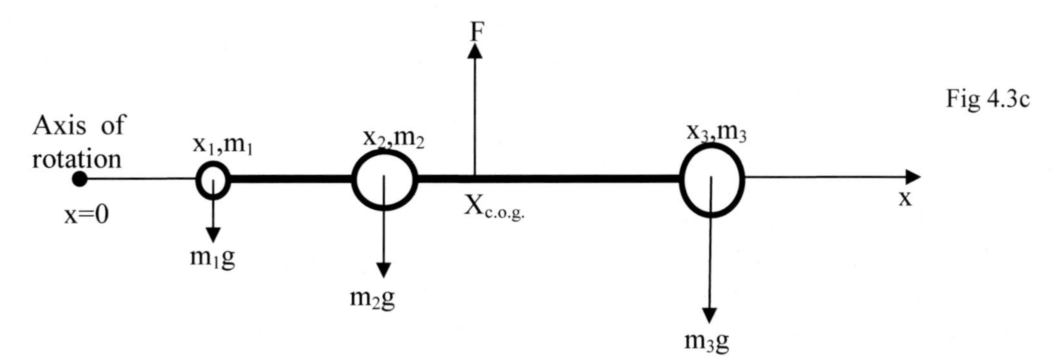

Fig 4.3c

$$X_{c.o.g.} = \frac{m_1x_1 + m_2x_2 + m_3x_3}{m_1 + m_2 + m_3} \qquad [4.3]$$

This formula can be extended to any number of masses:

$$X_{c.o.g.} = \frac{\sum_i m_ix_i}{\sum_i m_i} \qquad [4.3a]$$

The center of gravity calculated by these formulas is with respect to the origin of the arbitrarily chosen x-axis. If any point of the object is chosen as the origin, the calculated c.o.g. is with respect to that chosen point.

Example
An uneven barbell has a left mass of 20 kg, and a right mass of 25 kg. The bar itself is uniform. Its mass is 8 kg and its length is 1.2 m. Find the c.o.g. of the system.

Fig 4.3d

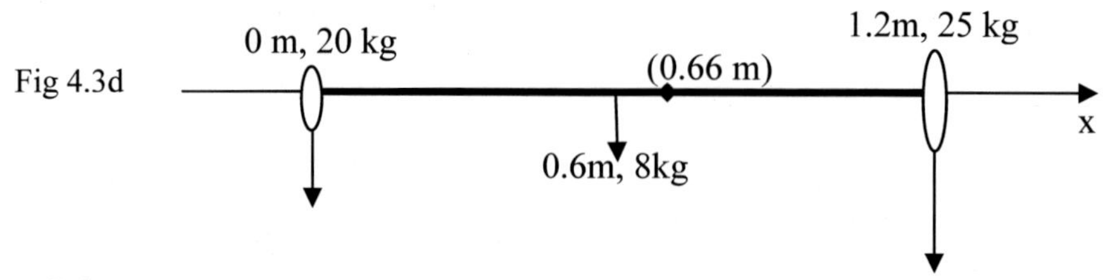

Solution

The weights are considered point masses, and the bar is represented by a point of mass 8 kg, located at the bar's c.o.g, at the middle of the bar. The origin of the x-axis is chosen at the left mass. Figure 4.3d illustrates the situation.

Substituting in [4.3] yields: $X_{c.o.g.} = \frac{20 \cdot 0 + 8 \cdot 0.6 + 25 \cdot 1.2}{20 + 8 + 25} = 0.66\text{m}$. So, the c.o.g. is 0.66 m to the right of the left weight.

The center of gravity of a two dimensional object, such as a plate, is expressed by two coordinates $X_{c.o.g.}$ and $Y_{c.o.g.}$. Similarly, three coordinates: $X_{c.o.g.}$, $Y_{c.o.g}$, and $Z_{c.o.g.}$ express the c.o.g. of a three dimensional object. Following [4.3], those coordinates can be calculated as follows:

$$X_{c.o.g.} = \frac{\sum_i m_i x_i}{\sum_i m_i}$$

$$Y_{c.o.g.} = \frac{\sum_i m_i y_i}{\sum_i m_i} \qquad [4.4]$$

And

$$Z_{c.o.g.} = \frac{\sum_i m_i z_i}{\sum_i m_i}$$

Example

Find the c.o.g. of an L shaped plate, illustrated in figure 4.3e. The plate is made of a uniform material of mass density σ kg/m^2.

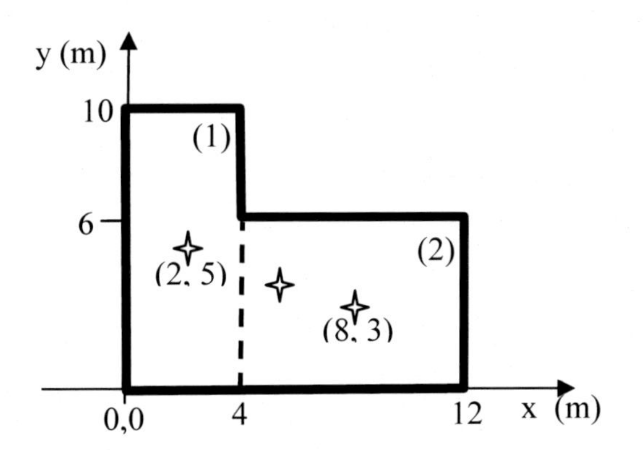

Figure 4.3e

Solution
The object is assigned an x-y coordinate system. The measures of the object are indicated on the coordinates. The L shape is now divided into two rectangles. Rectangle (1) is 4m x 10m = 40m^2, and rectangle (2) is 8m x 6m =48m^2. The c.o.g. of (1) is at x_1=2, y_1=5. The c.o.g. of (2) is at x_2=8, y_2=3. The mass of (1) is m_1=40σ, and the mass of (2) is m_2=48σ. Substituting in [4.4] yields:

$$X_{c.o.g.} = \frac{40\sigma \cdot 2 + 48\sigma \cdot 8}{40\sigma + 48\sigma} = 5.3 \text{ m}$$

$$Y_{c.o.g.} = \frac{40\sigma \cdot 5 + 48\sigma \cdot 3}{40\sigma + 48\sigma} = 3.9 \text{ m}$$

4.3 Statics of basic structural elements

Structure designers have to know the forces that would act in the structure that they design. The external forces, such as gravity, that would act on the structure are usually known. The main problem is to evaluate the internal reaction forces that are evoked by the known external forces. A general approach is to divide the structure into elements, and to analyze each element separately. The following are some common elements and situations that are found in many structures.

4.3.1 Three co-planar forces acting on a point.

When three forces that act on a point are found in the same plane, we have a 2-D situation. Let's use an x-y coordinate system to describe these situations. According to the first equilibrium condition, the sum of the x components of the three forces and the sum of the y-components of the three forces have to be equal to zero:

$$\sum F_x = 0$$
$$\sum F_y = 0$$

If the forces are F^1, F^2, and F^3 (figure4.4), the first equilibrium condition becomes:

$$F_x^1 + F_x^2 + F_x^3 = 0$$
$$F_y^1 + F_y^2 + F_y^3 = 0 \qquad [4.5]$$

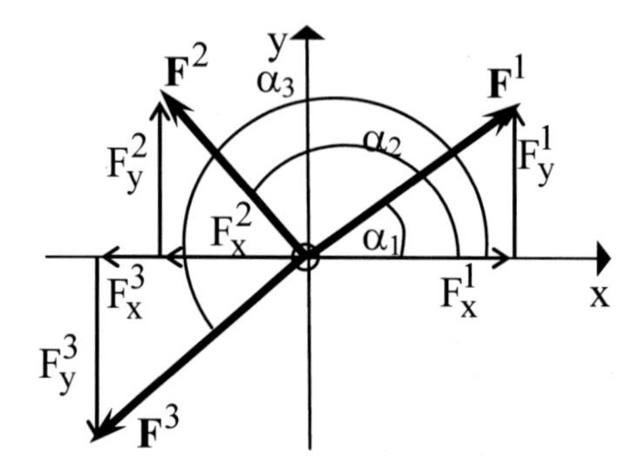

Figure 4.4: Notation used in the analysis of three co-planar forces that act on a point object.

Equations [66] use the components of the three forces. If any of the forces is given by magnitude and direction, or if we need to know the magnitude and direction of a force whose components were found, we use equations [3.1] and [3.2]. These equations give the relationships between the two representations of any vector, including force vectors:

$$V_x = V \cdot \cos\alpha \qquad [3.1a] \quad \text{(same as [1.2b])}$$

$$V_y = V \cdot \sin\alpha \qquad [3.1b] \quad \text{(same as [1.2a])}$$

$$V = \sqrt{V_x^2 + V_y^2} \qquad [3.2a] \quad \text{(same as [1.3b])}$$

$$\alpha = \tan^{-1}\frac{V_y}{V_x} \qquad [3.2b] \quad \text{(same as [1.4])}$$

We now have to make sure that the second equilibrium condition is also satisfied by the torques of the three forces that act on the point. These conditions are satisfied automatically in this case, because the torque of each force is zero. The point of attachment of each force is also the axis of rotation of the point. Therefore, the arm of the force is zero, and the torque is zero (equation [3.6]).

Example

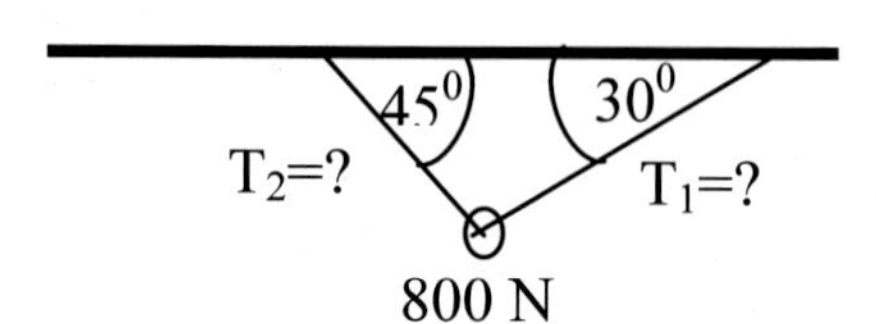

Figure 4.5a

A chandelier of weight 800 Newton hangs from the ceiling on two cables, as shown in figure 4.5a. Find the tension T_1 and T_2 in the cables.

Solution

The tension in a cable is equal to the force that the cable exerts on each of the objects at its ends (section 2.5). The cables exert forces on the chandelier. Therefore the tension T_1 is equal to the force that cable 1 exerts on the chandelier, and the tension T_2 is equal to the force that cable 2 exerts on the chandelier. Each of these forces is directed in the direction of its cable. The force of gravity on the chandelier (W=800 Newton) points straight down. Figure 4.5b shows these three forces in an x-y coordinate system. Their angles are with respect to the positive direction of the x-axis.

We now need to organize the information that we have so that we can substitute it in equations [4.5], solve the equations, and get the answer to the problem. The information is organized in table 4.1.

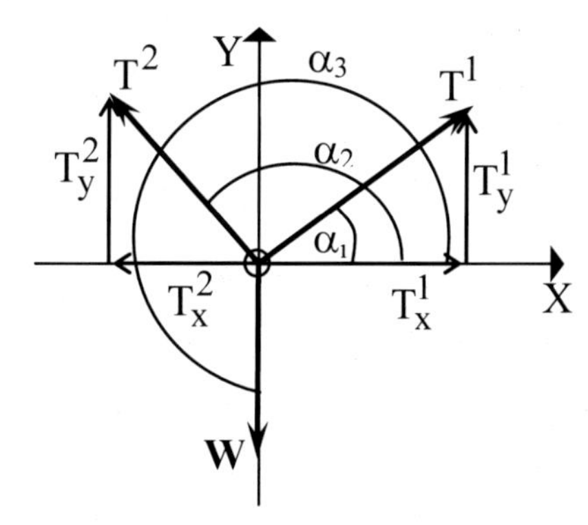

Figure 4.5b

Force	*Magnitude*	*Angle*	*x-component [3.1a]*	*y-component [3.1b]*
F^1	T_1=?	$\square_1=30^0$	$T_1 \cdot \cos 30^0=0.8660T_1$	$T_1 \cdot \sin 30^0=0.5T_1$
F^2	T_2=?	$\square_2=135^0$	$T_2 \cdot \cos 135^0= - 0.7071T_2$	$T_2 \cdot \sin 135^0=0.7071T_2$
F^3	W=800N	$\square_3=270^0$	$W \cdot \cos 270^0=0$	$W \cdot \sin 270^0= -800N$

Table 4.1

We now add all the x-components and equate their sum to zero, and add all the y-components and equate their sum to zero (according to [4.5]) to get:

$0.8660T_1-0.7071T_2+0=0$

$0.5T_1+0.7071T_2-800N=0$

This is a set of two equations with two unknowns: T_1 and T_2. We solve them (check it out) and get: T_1=585.65N, and T_2= 717.26. These are the tensions that we had to find.

4.3.2 A beam and posts

A horizontal 100-lb. beam of length 8 ft is supported at its ends on two posts (figure 4.6). The center of gravity of the beam is at its middle.

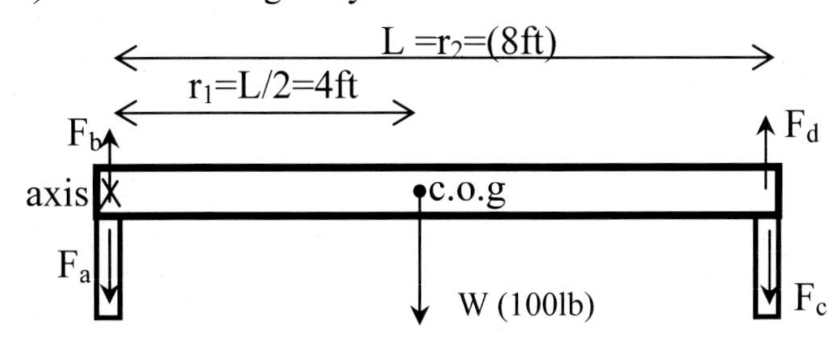

Figure 4.6: A beam supported by two posts

(a) Find the force that the beam exerts on each post.

The force that the beam exerts on the left post ($\mathbf{F}_a$) is equal in magnitude and opposite in direction to the force that that post exerts on the beam ($\mathbf{F}_b$). So, $|\mathbf{F}_b|=|\mathbf{F}_a|$. Similarly, for the right post, $|\mathbf{F}_c|=|\mathbf{F}_d|$. We will find the forces on the posts by figuring out their reaction forces on the beam.

Since the beam is in equilibrium, the forces that act on it satisfy the equilibrium conditions [4.5] and [4.2]. If the chosen x-axis points to the right, and the y-axis points up, all the forces point in the y direction. Therefore, the sum of the y-components of three forces that act on the beam according to [4.5] is in general

I: $F_b-W+F_d=0$

and in this case: $F_b-100+F_d=0$

Now we have to add the torques of the three forces that act on the beam. First, we choose an axis of rotation. The axis can be anywhere on the beam, and we choose it to be at the contact point with the left post.

The torque τ_b of the force $\mathbf{F}_b$ is equal to zero, because it acts on the axis of rotation.

The arm of the torque of the weight τ_w is $r_1=L/2=4$ft. This torque is negative, because it would create CW rotation. The sine of the angle between the force and the arm is $\sin 90^0=1$. Putting it all together we get, according to [3.6], $\tau_w=-W \cdot r_1 \cdot \sin 90^0=-W \cdot r_1=-400$lb·ft.

The arm of $\mathbf{F}_d$, is r_2. The torque τ_d of this force is positive, because it would cause a CCW rotation. Putting it all together gives

$\tau_d=F_d \cdot r_2 \cdot \sin 90^0=F_d \cdot 8$.

Based on [4.2], the sum of these torques is zero $\tau_b+\tau_w+\tau_d=0$. In general, that means:

II: $0-W \cdot r_1+F_d \cdot r_2=0$

and in this case: $-400+F_d \cdot 8=0$

By solving equations I and II we get in general: $F_d=W \cdot r_1/r_2$ and $F_b=W \cdot (r_2-r_1)/r_2$, and in this case $F_d=F_b=50$lb. The total weight, which acts as if it were concentrated at the center of the beam, is distributed evenly between the two posts.

(b) A load of weight G of 80 lb is placed on the beam, 7 ft from its left end (figure 4.6a). Find the forces that act now on the posts.

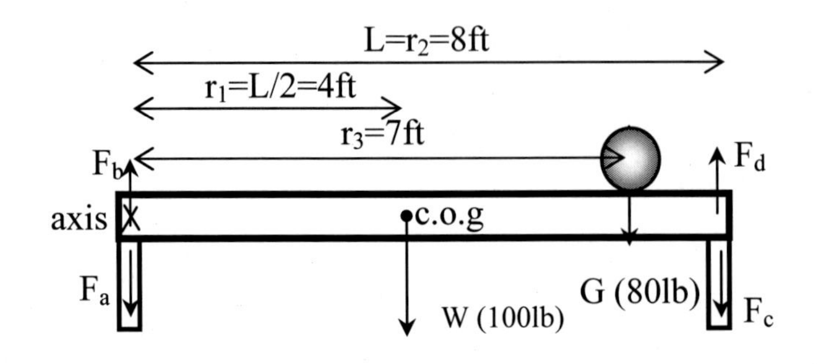

Figure 4.6a: A beam and an object on it, supported by two posts.

The forces that acted on the posts in figure 4.6 and their torques remain the same also in this case. In addition, the new force G=80lb acts in the negative direction of the y-axis, and its torque around the axis of rotation is $\tau_G=-G\cdot r_3\cdot\sin 90^0=-80\cdot 7=560$ lb·ft. The force is added to I and the torque is added to II to give:

I': $F_b-W-G+F_d=0$ in general

and in this case: $F_b-100-70+F_d=0$

and

II': $0-W\cdot r_1-G\cdot r_3+F_d\cdot r_2=0$ in general

and in this case: $-400-560+F_d\cdot 8=0$

The solution of equations I' and II' in this case is $F_d=120$lb, $F_b=60$lb, and in general: $F_d=(W\cdot r_1+G\cdot r_3)/r_2$, $F_b=W+G-(W\cdot r_1+G\cdot r_3)/r_2$.

The total weight is distributed unevenly between the two posts. The post closer to the additional load (the right post) supports more weight than that the furthest post (the left one).

(c) The same as (b), but using [4.2] twice

As stated earlier, it is possible to express the equilibrium conditions by two torque equations [4.2] around two different axes of rotation. Let's apply this method to the last example. Let one axis be as in (b), at the top of the left post, resulting in torque equation: $-400-560+F_d\cdot 8=0$. Let the other axis be at a different x coordinate, at the top of the right post, resulting in torque equation: $80\cdot 1+100\cdot 4-F_b\cdot 8=0$. The solution of these two equations is $F_d=120$ lbs, and $F_b=60$ lbs, the same as in (b).

(d) Any object on two posts

When an object of any shape, size and weight is at rest on two posts, the downward forces that the object exerts on the posts are the same as those that would have been exerted by a beam of the same weight, whose c.o.g. is right underneath the c.o.g. of that object (figure 4.6b).

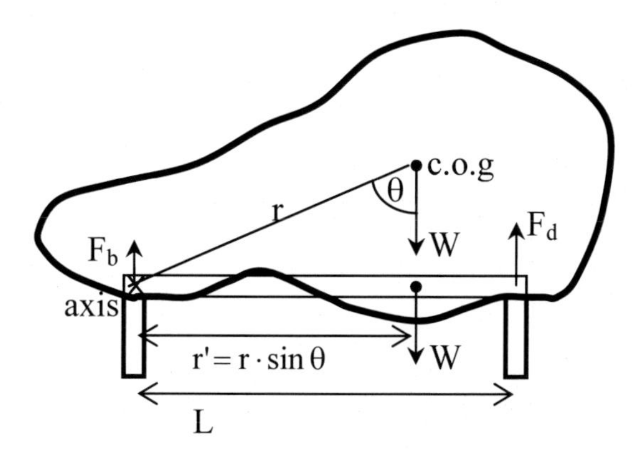

Figure 4.6b

Using the same arguments as in the previous two cases, the forces that the object exerts on the posts have the same magnitudes as the reaction forces from the posts: F_b and F_d. The vertical forces that act on the object are the same as those that act on the imaginary beam: F_b, W, and F_d. The torques of the object's forces around the chosen axis are the same as the torques of the beam's forces: The torque of F_b is zero for both.

The torque of the c.o.g is $-W \cdot r \cdot \sin\theta$ for the object, if we use $|\tau| = |F \cdot r \cdot \sin\theta|$ [3.6]. If we use $\tau = x \cdot F_y - y \cdot F_x$, and the origin of the x-y coordinate system is at the axis of rotation we have: $F_y = -W$, $F_x = 0$, $x = r \cdot \sin\theta$, yielding the same $-W \cdot r \cdot \sin\theta$, which is also the same as the torque of a beam whose c.o.g. is directly underneath that of the object. (This is a special case of formula [3.6a].)

Therefore, F_b and F_d are the same for the object and for the beam. (Write these equations and verify that $F_d = W\dfrac{r \cdot \sin\theta}{L}$ and $F_b = W\dfrac{L - r \cdot \sin\theta}{L}$).

(e) A horizontal beam on slanted posts

Figure 4.6c shows a horizontal symmetric beam, resting on two posts with slanted contact surfaces. The slant angle is δ in both posts, and $\phi = 90^0 - \delta$. There are no friction forces between the beam and the posts.

In this case, the reaction forces $\mathbf{R}^1$ and $\mathbf{R}^2$ between the beam and the posts are normal forces, which are perpendicular to the contact surfaces. These forces have x and y-components, as shown in figure 4.6c.

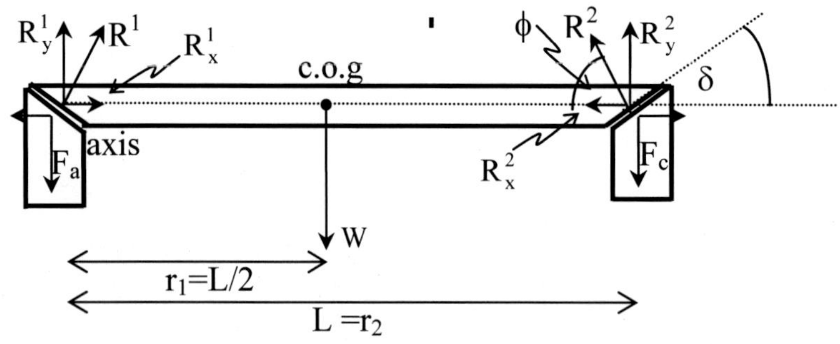

Figure 4.6c: A horizontal symmetric beam on slanted posts.

The torques on the beam due to the x-components of the reaction forces are zero, because they are in the direction of the arm (figure 4.6c).

The y-components of the reaction forces match the reaction forces in the case of the regular beam (figure 4.6): $R_y^1 = F_b$ and $R_y^2 = F_d$. The equilibrium conditions for the forces in the y-direction and for their torques are the same as for the regular beam and post system (figure 4.6). Therefore, as in figure 4.6, we get for the y-components of the forces:

$$R_y^1 - W + R_y^2 = 0$$

and for the torques:

$$-W \cdot \frac{L}{2} + R_y^2 \cdot L = 0$$

By solving these two equations we get:

$R_y^1 = R_y^2 = \frac{W}{2}$, which means that the downward force of the beam's weight is distributed evenly between the two posts.

In order to find the x-components of the reaction forces, we first find the magnitudes of R^1 and R^2. Based on equations [3.1] we get:

$R^1 = \frac{R_y^1}{\sin\phi} = \frac{0.5W}{\sin\phi}$, and using [3.1] again we get:

$$R_x^1 = R^1 \cdot \cos\phi = \frac{0.5W}{\sin\phi} \cdot \cos\phi = \frac{0.5W}{\tan\phi} = 0.5W \cdot \tan\delta$$

Similarly, for the second post we'll get also $R_x^2 = 0.5W\tan\delta$.

R_x^1 and R_x^2 are the horizontal forces with which the beam pushes the slanted posts outwards.

These formulas express also the horizontal forces that a symmetrical object of any shape would exert on two supporting, frictionless, slanted posts. For example, figure 4.6d shows an arch made of trapezoidal blocks. The horizontal force (thrust) that any segment of the arch (dark area in figure 4.6d) exerts on its supporting block depends only on the weight of that segment (W) and the slant angle at the basis of the segment (δ in figure 4.6d), as in the previous case of a beam on slanted posts.

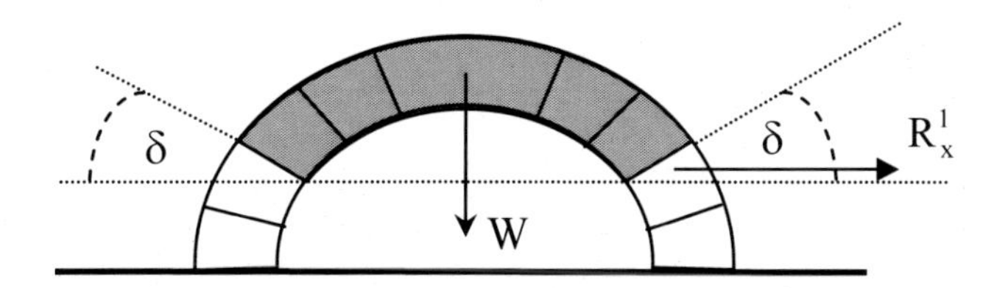

Figure 4.6d: Horizontal force due to a segment of an arch.

4.3.3 A ladder against a wall

A ladder leaning on a wall may serve as a model for similar elements in other structures. The ladder has its own weight, attached to its center of gravity, and an additional load, such as the weight of a person standing on it (figure 4.7). In addition to these forces, one interaction force from the wall and another one from the floor also affect the ladder's state.

In order to analyze the forces that act on the ladder, we first choose an x-y coordinate system, where x is along the floor and y is along the wall. The wall's and floor's interaction forces can be expressed by their components: $\mathbf{R}^w = (R_x^w, R_y^w)$ and $\mathbf{R}^f = (R_x^f, R_y^f)$.

The component R_x^w is responsible for the ladder not penetrating the wall. The component R_y^f is responsible for the ladder not sinking into the floor. The component R_y^w represents forces that oppose the sliding of the ladder on the wall. The component R_x^f represents the forces that oppose the sliding of the ladder on the floor. The latter component happens to point in the negative direction of the x-axis. R_y^w and R_x^f may be due to various causes such as friction, tension of cables that are attached to the ladder, nails, etc. W is the weight of the ladder, and G is the weight of an external load. The ladder makes an angle α with the floor. Another angle is $\beta=90^0-\alpha$. Figure 4.7 shows where these angles play a role.

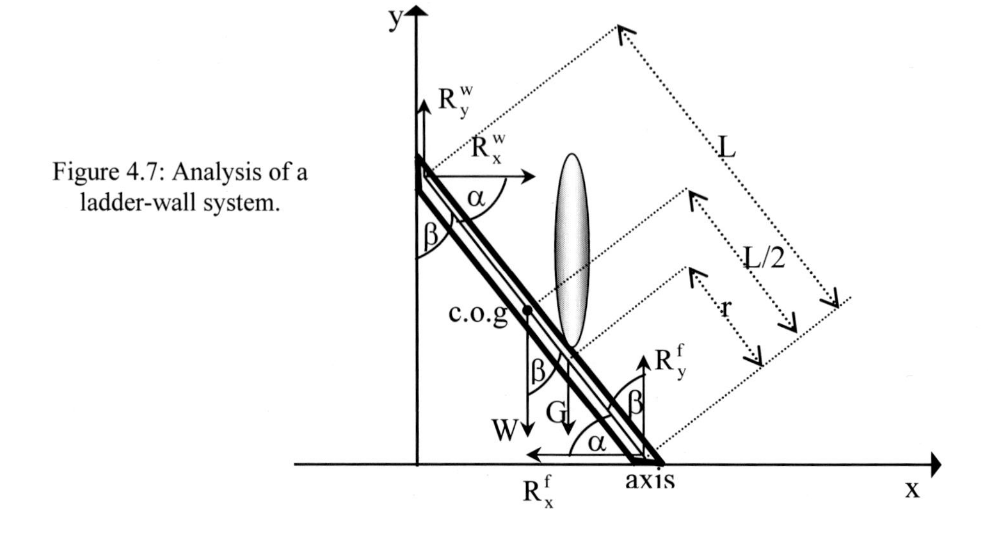

Figure 4.7: Analysis of a ladder-wall system.

When the ladder is at rest, it satisfies the equilibrium conditions [4.5] and [4.2]. The force components for [4.5] are already shown in figure 4.7. In order to use [4.2] we need to know the torque that is generated by each of the forces. These torques may be around any axis of rotation. If the axis of rotation is chosen at the bottom of the ladder, the torques are:

$$\tau_{R_x^f} = 0 \quad \text{(force on axis)}$$

$$\tau_{R_y^f} = 0 \quad \text{(force on axis)}$$

$$\tau_G = G \cdot r \cdot \sin\beta$$

$$\tau_W = W \cdot \frac{L}{2} \cdot \sin\beta$$

$$\tau_{R_x^W} = -R_x^W \cdot L \cdot \sin\alpha$$

$$\tau_{R_y^W} = -R_y^W \cdot L \cdot \sin\beta$$

When we substitute in the equilibrium conditions we get:

For the x-components of the forces: $$R_x^W - R_x^f = 0 \qquad [4.6]$$

For the y-components of the forces: $$R_y^f - G - W + R_y^W = 0 \qquad [4.7]$$

and for the torques:

$$G \cdot r \cdot \sin\beta + \frac{1}{2} \cdot W \cdot L \cdot \sin\beta - R_x^W \cdot L \cdot \sin\alpha - R_y^W \cdot L \cdot \sin\beta = 0 \quad [4.8]$$

Discussion

Equations [4.2abc] contain active forces (G and W) and four reaction forces: R_x^f, R_y^f, R_x^W, and R_y^W. In many practical applications, the active forces are known, and it is needed to find the four reaction forces. However, there are only three equations in [4.2abc] for the four reaction forces. That means that many combinations of the four reaction forces could keep the ladder in equilibrium. In some cases, the nature of the reaction forces provide an additional equation, making it possible to have a unique set of four reaction forces that keep the ladder in equilibrium. One such case is when one of the four reaction forces is zero, e.g. $R_y^W = 0$. This condition means that the ladder leans on a frictionless wall. In this case, R_x^f, R_y^f, and R_x^W are determined in a unique way.

Sometimes, reaction forces can adjust themselves to active forces up to a certain limit. For example, if R_x^f is due to the friction force between the ladder and the floor, its maximum value could be $\mu \cdot NF$ ([2.15]), where μ is the friction coefficient and $NF = R_y^f$. So, if $R_x^f > \mu \cdot R_y^f$ (and $R_y^W = 0$), the floor would not be able to provide the necessary friction force to prevent the ladder from sliding.

Example 1: A ladder of length 4m and mass 30 kg is leaning on a wall. It makes an angle of $\alpha=30^0$ degrees with the floor. There is no friction force with the wall, but the friction force with the floor keeps it from sliding. Find the friction force with the floor.

According to the notations of [4.2abc], $R_x^f = ?$, $R_y^W = 0$, W=30g=294N, L=4m, $\alpha=30^0$, $\beta=90^0-\alpha=60^0$.

When we substitute this information in [4.2abc] we get:

I $\quad R_x^W - R_x^f = 0$

II $\quad R_y^f - 0 - 294 + 0 = 0$

III $\quad 0 + \frac{1}{2} \cdot 294 \cdot 4 \cdot \sin 60^0 - R_x^W \cdot 4 \cdot \sin 30^0 - 0 = 0$

From III we get: $\quad R_x^W = 254.6N$

From I we get: $\quad 254.6 - R_x^f = 0$, or $R_x^f = 254.6N$, which is the answer.

Example 2: Show that when the friction between the wall and the ladder is negligible, the horizontal force at the basis of the ladder, which is needed to prevent the ladder from sliding, increases as the angle between the ladder and the floor decreases. (You know it already from experience. A ladder whose feet are too far from the wall is not safe.)

If the friction between the ladder and the wall is negligible, R_y^W=0. From [4.8] we get:

$$(G \cdot r + \frac{1}{2} \cdot W \cdot L) \cdot \sin\beta - R_x^W \cdot L \cdot \sin\alpha = 0$$

or

$$R_x^W = \frac{(G \cdot r + \frac{1}{2} \cdot W \cdot L) \cdot \sin\beta}{L \cdot \sin\alpha} = \frac{(G \cdot r + \frac{1}{2} \cdot W \cdot L)}{L} \frac{\cos\alpha}{\sin\alpha}$$

$$R_x^W = (G \cdot \frac{r}{L} + \frac{W}{2}) / \tan\alpha \qquad [4.9]$$

and based on I we get also [4.10]

$$R_x^f = (G \cdot \frac{r}{L} + \frac{W}{2}) / \tan\alpha$$

(We have used here the identities $\sin\beta = \cos(90 - \beta) = \cos\alpha$, and $\frac{\cos\alpha}{\sin\alpha} = \frac{1}{\tan\alpha}$)

G, r, W, and L do not change as the angle of the ladder changes. As α increases from 0 to 90^0, $\tan\alpha$ keeps increasing, and therefore R_x^f keeps decreasing. In other words, as the ladder gets closer to the wall, the horizontal frictional force that needs to prevent its sliding on the floor becomes smaller.

4.3.4 Triangular arch

A triangular arch can be viewed as two ladders leaning against each other (figure 4.8). The entire system is held in place by bulges at the basis, which prevent the ladders from sliding.

In order to calculate all the forces that act in a triangular arch, the equilibrium conditions are applied to each of the legs and to the system as a whole. The individual forces are then found by solving the equations.

Let's start with the bulges. The right bulge prevents the right leg from sliding to the right. It exerts a leftward force on the right leg at its bottom. Similarly, the left bulge exerts a rightward force on the left leg. The force that a bulge exerts on its leg is equal in magnitude and opposite in direction to the force that the leg exerts on the bulge. Overall, the bulges are pushed outward by the legs of the arch.

Let's turn now to the upper tip of the system. According to Newton's Third Law, the force that the right leg exerts on the left leg is equal in magnitude and opposite in direction to the force that the left leg exerts on the right leg. This is indicated in figure 4.8 by the components of the forces that act at the top of each leg (R_x^w and R_y^w). Since each leg is in equilibrium, the sum of the forces that act on it in the x- direction must be zero, yielding $R_x^w = R_x^f$ for both legs.

We now want to focus on the vertical forces that act at the contacts between the arch and the floor: R_y^f at the right leg, and R'^f_y at the left leg. These two forces may be different from each other because the right leg has a weight G on it, whereas the left leg doesn't. We now look at the entire arch as one unit in equilibrium. The sum of the torques on that unit is equal to zero. We choose our axis at the bottom of the right leg.

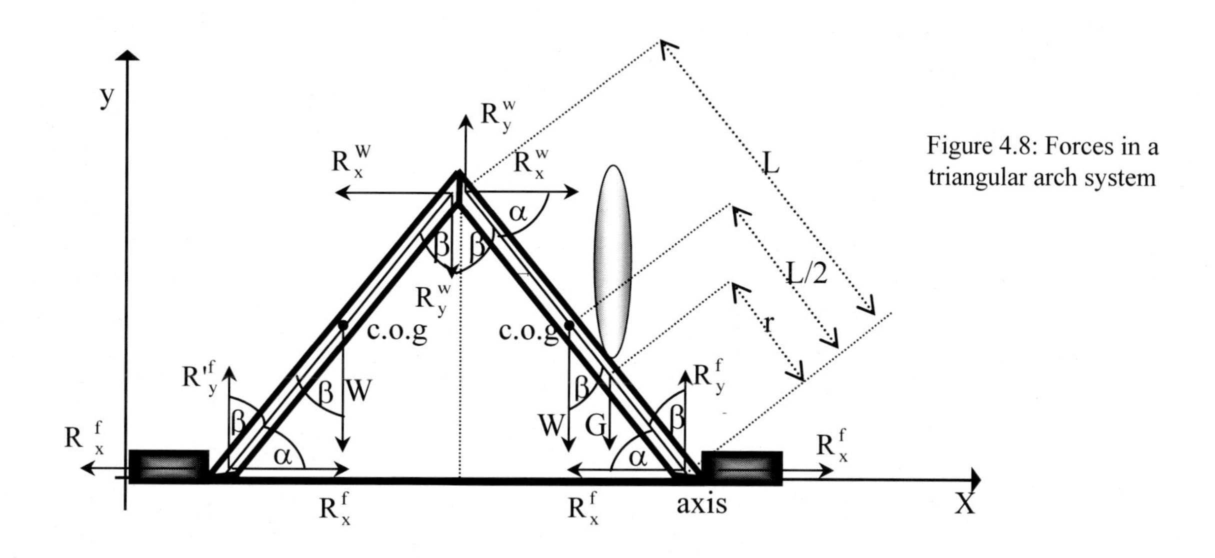

Figure 4.8: Forces in a triangular arch system

The torques of R_x^f, R_y^f, W, and G, which are due to forces on the right leg, are the same as in the case of a ladder against a wall:

$$\tau_{R_x^f} = 0 \quad \text{(force on axis)}$$

$$\tau_{R_y^f} = 0 \quad \text{(force on axis)}$$

$$\tau_G = G \cdot r \cdot \sin\beta$$

$$\tau_W = W \cdot \frac{L}{2} \cdot \sin\beta$$

The sum of the four forces at the top is zero. Since they act on the same point, they share the same arm and therefore, the sum of their torques is zero.

There are three more forces in the left leg that we have to consider (figure 4.8a).

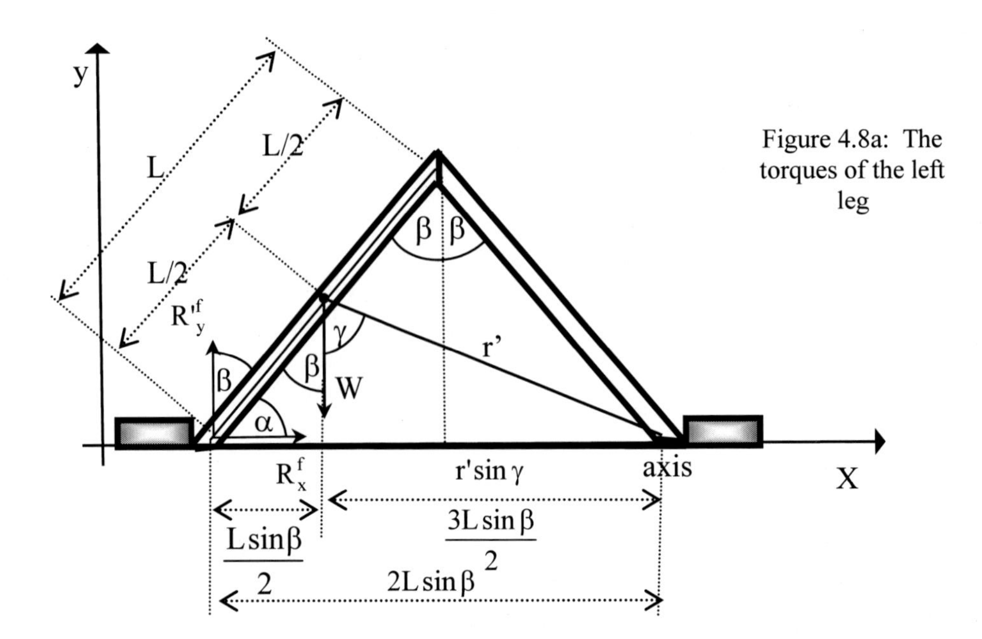

Figure 4.8a: The torques of the left leg

The torque of the weight of the left leg is $\tau_W^{left} = W \cdot r' \cdot \sin\gamma$. However, $r' \cdot \sin\gamma = \frac{3L\sin\beta}{2}$ because of the geometry of the arch. Therefore, $\tau_W^{left} = \frac{3}{2} W \cdot L \cdot \sin\beta$.

The torque of R'^f_y, whose arm is $(2L\sin\beta)$, is:

$\tau_{R'} = -R'^f_y \cdot (2L\sin\beta) \cdot \sin 90^0 = -R'^f_y \cdot (2L\sin\beta)$.

The torque of R^f_x is zero because the force points along the arm.
We now add all these torques and equate their sum to zero:

$$G \cdot r \cdot \sin\beta + \frac{1}{2} \cdot W \cdot L \cdot \sin\beta + \frac{3}{2} W \cdot L \cdot \sin\beta - R'^f_y \cdot (2L\sin\beta) = 0$$

The factor $\sin\beta$ cancels from all the terms and we get:

$$R'^f_y = G\frac{r}{2L} + W \qquad [4.11]$$

The sum of all the y-components of the forces is zero:

$$R^f_y - G - W + R^W_y - R^W_y - W + R'^f_y = 0$$

Substituting R'^f_y from [4.11] and solving we get:

$$R^f_y = W + G\frac{2L - r}{2L} \qquad [4.12]$$

In order to find R_y^W, we use the condition that the sum of the y-components of the force on any element is zero, and choose the right leg. We get:
$R_y^f - W - G + R_y^W = 0$, and based on [4.12] we get:

$$R_y^W = G\frac{r}{2L} \quad [4.13]$$

In order to find R_x^f, we consider the torques of the forces on the left leg, as shown in figure 4.8a. This time, we choose the axis of rotation at the top of the leg. We do it because we don't want the forces at this point to appear in the equation. We get:

$$R_x^f \cdot L \cdot \sin\alpha - R'^f_y \cdot L \cdot \sin\beta + W \cdot \frac{L}{2} \cdot \sin\beta = 0$$

Which gives?

$R_x^f = (R'^f_y - \frac{W}{2})\frac{\sin\beta}{\sin\alpha}$, and based on [4.11] we get:

$$R_x^f = (G\frac{r}{2L} + \frac{W}{2})\frac{1}{\tan\alpha} \quad [4.14]$$

Discussion

People already knew how to build magnificent arches more than fifteen hundred years before Newton's days. Knowledge was accumulated and refined by trial and error and through ingenious observations generation after generation. Equations [4.3], which are consequences of Newton's Laws, enable us to understand some of the fundamental principles of the working of arches by sheer thinking power. Let's see what we can learn from these equations.

Let's start with completely symmetric arches. In such cases there is no external load on any of the legs of the arch: G=0. From [4.11] we get that if G=0 then $R_y^W = 0$. No vertical force is needed at the top of the arch for it to stay in place. However, if $G \neq 0$ then $R_y^W \neq 0$, meaning that for the arch to stay in place, a vertical force must be present at the top. But what may be the source of such a vertical force? One source may be friction. However, if the friction force will not be strong enough, the arch would collapse. Instead of relying only on friction, the two legs could be glued or nailed to each other at the top. The arch designer has to ascertain that whatever means is taken to couple the two legs at the top, the vertical force that the coupling provides should not be less than the required $G\frac{r}{2L}$ of formula [4.11]. This problem does not arise in symmetric arches that are not intended to carry external asymmetric loads.

Similar to equation [4.9] for a ladder against the wall, the horizontal force at the base of the arch, as given by [4.14], is inversely proportional to $\tan\alpha$. That means that the "flatter" the arch, the harder it pushes against whatever is there to prevent it from sliding sidewise. This sidewise push depends both on W, the weight of the arch, and on G, the weight of the external load. Because of the $\tan\alpha$ in [4.14], the outside push may be much greater than the total weight of the leg of the arch and the external weight. The "bulges", or any other means that provide the horizontal forces that prevent the sliding of the legs at the base, have to be strong enough to provide that force.

At the top of the arch, the horizontal force compresses the two legs against each other. If the only coupling force at the top is friction, this horizontal force is beneficial to the stability of the arch. A friction force is proportional to the normal force (section 2.4.4). In this case, the horizontal force is the normal force.

When the two legs are rigid and they are coupled by a rigid connection at the top, for example metal legs that are welded to each other, this connection provides the horizontal force needed to keep the legs in place. In such cases, there will not be horizontal forces at the base that would try to push the legs outwards

4.3.5 From arches to domes

Arches use short structural elements to span longer distances. They take advantage of the force of gravity and use it to compress the structural elements against each other, thus holding them together as one big piece. Additional bindings of the elements, such as by cementing or by riveting, add to the overall strength of the structure.

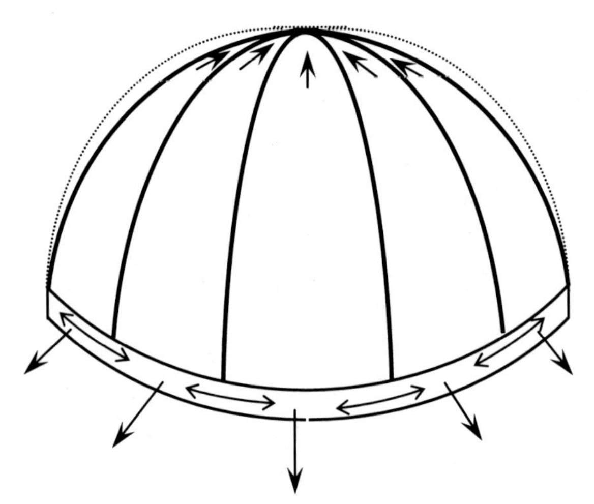

Figure 4.9: A dome and a circular retaining ring at its base. Thick arrows: horizontal forces exerted by each leg. Thin arrows: tension in the retaining ring.

Various combinations of arches have been used as roofs for structures of all sizes, from small shacks to large public buildings. In the simplest combination, two arches that are orthogonal to each other provide a supporting skeleton to the roof. In others, the entire dome-shaped roof can be viewed as a continuum of arches, arranged around a circle. The main characteristics of the triangular arch are present also in all such structures. At the base, each leg of an arch has to be held in place against its outward sliding force (thick arrows at the bottom of figure 4.9). At the top, two opposing legs of an arch are pressing against each other, thus exerting compression forces on the center of the dome. At the base, it is possible to buttress each leg of an arch by a horizontal force. However, it is also possible to buttress the entire base by a solid ring (figure 4.9). The outwards forces of the legs will be neutralized by the tension of the ring (thin arrows, figure 4.9). Similarly at the top, it is possible to replace the top of a dome by an open ring, whose compression neutralizes the inward forces created by the legs. The opening at the top of the Pantheon (the oculus) is based on this principle.

4.3.6 Buttressing

Walls in buildings do not topple mainly because they are connected directly and indirectly to other walls, to the roof, to the floors, and to the foundations. It may be said that the entire building holds a wall in place. If the support of the rest of the structure is not sufficient, buttressing is one possible way of strengthening a wall. Figure 4.10a shows typical forces that act on a buttressed wall. Consider first a rigid wall. The weight of the structure and reactions from the ground are two of the forces that exert restoring torques, which oppose that of P'. If the torque of P' overcomes those restoring torques, the buttress enters the picture. The wall will start tilting to the right and a force P will be exerted on the buttress. The buttress will respond with its reaction force f_x that will stabilize the wall. In general, P will be different from P', and it will depend on the torque of P' minus the torques of the restoring forces.

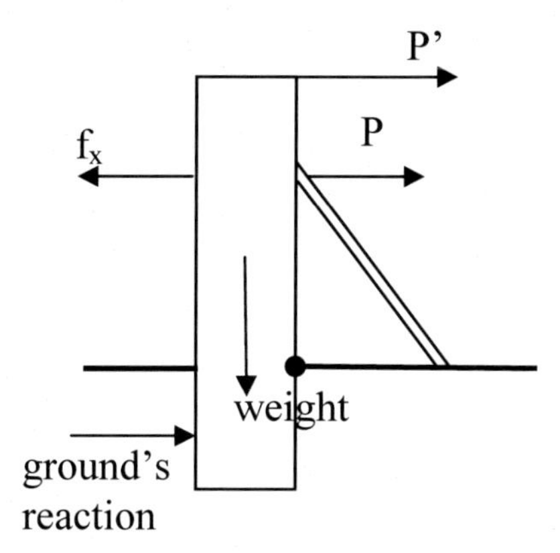

Figure 4.10a: Rigid wall. If the torques of the restoring forces (weight and ground's reaction) are greater than the torque of the thrust P' at the top of the wall, the buttress won't feel any thrust. The reaction of the buttress (f_x) depends on the net thrust (P) at the top of the buttress.

If the wall is bendable and flexible, and if the buttress is supposed to prevent the wall from bending, the buttress will have to provide a force f_x that equals in magnitude to P. If some wall bending is allowed, the elastic forces of the wall and of other building elements connected to it combine forces with the buttress to oppose P'.

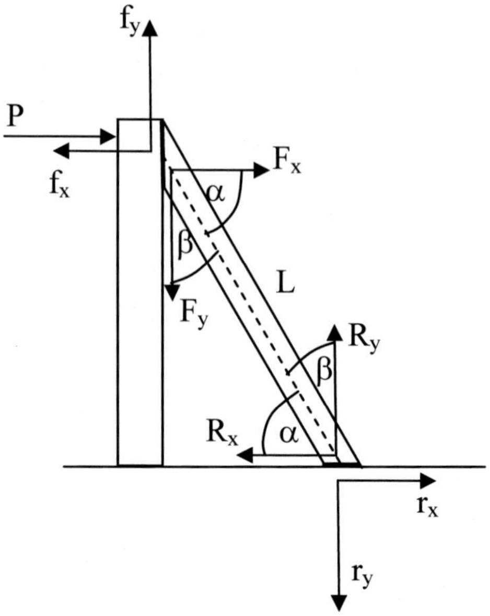

Figure 4.10b: Forces induced in a buttress system by a horizontal thrust P of the wall.

We now want to analyze the internal loads that P induces in the wall-buttress system, and the forces that the buttress relays to the ground. Figure 4.10b shows the main elements of a wall and its diagonal buttress. The wall and the buttress are in static equilibrium. The vertical reaction forces of the ground are balancing the weights of the wall and of the buttress. The figure shows the forces that are induced in the wall-buttress system by a horizontal thrust P that acts on the top of the buttress. Other forces, that have been acting in the system, and which are already in a static equilibrium, are not shown. The thrust P induces two forces at the top of the buttress: F_x and F_y. These forces are induced because the buttress is attached to the wall (by cement, nails, friction, etc.), and as it leans to the right, the wall pushes the buttress both down and to the right. According to Newton's Third Law, the buttress exerts on the wall the forces $f_x=F_x$ and $f_y=F_y$. The buttress exerts on the ground the forces r_x and r_y, and the ground exerts on the buttress the reaction forces $R_x=r_x$ and $R_y=r_y$.

From the static equilibrium equations as applied to the wall we get $f_x=P$. Since $f_x=F_x$ we get

$$F_x=P \qquad [i]$$

The vertical force on the wall, f_y, is usually balanced by the greater weight of the wall. (If the wall is too light, P and the resistance of the buttress may pull the wall up.)

From the static equilibrium conditions as applied to the buttress we get:

From the sum of the x-components: $F_x-R_x=0$ [ii]

And based on [i] we get $R_x=P$ [iii]

For the sum of the torques we get:

$$-F_x \cdot L \cdot \sin\alpha + F_y \cdot L \cdot \sin\beta = 0 \qquad [iv]$$

since $\beta=90^0-\alpha$ we get

$$-F_x \cdot L \cdot \sin\alpha + F_y \cdot L \cdot \cos\alpha = 0 \qquad [v]$$

where the length of the buttress is L, and the axis of rotation of the buttress is at the ground.

Combining [i] and [v] we get

$$F_y = F_x \cdot \sin\alpha / \cos\alpha = P \cdot \tan\alpha \qquad [vi]$$

Based on Newton's Third Law we get:

$$r_x = P \quad \text{and} \quad r_y = P \cdot \tan\alpha \qquad [vii]$$

Conclusion: Equations [vii] tell us that the buttress neutralizes the thrust P at the top of the wall, and relays it to the ground (as a horizontal thrust r_x). In addition, the buttress induces a vertical force r_y into the ground, in the amount of $P \cdot \tan\alpha$.

Flying Buttresses

Figure 10.c: Flying buttresses, National Cathedral, Washington, DC

The external walls of Gothic cathedrals have many tall and narrow windows with magnificent stained glass artwork. Those external walls have to support a heavy roof that may exert horizontal thrusts at their tops. Due to their slenderness and the large area of their windows, the walls may be in a precarious situation, caused by the outward thrust of the roof. Supporting the walls by simple diagonal buttresses, as those described in figure 4.10a, is not feasible, because of the size of the cathedral. The architects of the cathedrals came up with an ingenious solution: the flying buttress. The essence of a flying buttress is a diagonal buttress that connects the wall to a column, instead of to the ground. The column is much thicker than the wall of the cathedral. It is lower than the wall and stands away from it. The masonry buttress is supported by its own arch (or a part of an arch). Several diagonal buttresses, originating from different points of the wall, may lean on the same column. In some cathedrals, one high buttressing system is supported by a second lower buttressing system.

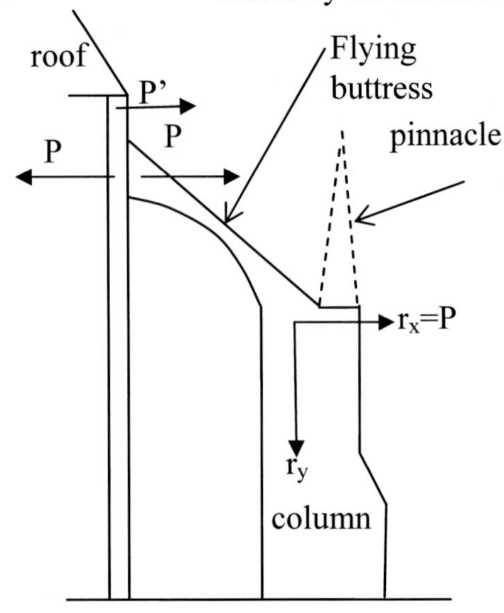

Figure 4.10d: The thrust P that acts on the flying buttress induces in the column two forces: r_x and r_y. The flying buttress neutralizes the outward thrust of the roof on the wall.

Figure 4.10d shows the forces that are relayed to the top of the column by the flying buttress. Basically, these are the same forces that are relayed to the ground by a simple diagonal buttress (figure 4.10b). The thrust P on the cathedral's wall is neutralized by the flying buttress. The wider and shorter column, to which

Figure 10.e: Flying Buttresses, National Cathedral, Washington, DC.

The National Cathedral was built from 1907 to 1990. It is a Gothic style cathedral, 102'6" high at the nave (inside). "The Cathedral has been built in the traditional "stone-on-stone" method. The Cathedral contains no structural steel, rather stones are placed one on top of the other and held in place by the force of the flying buttresses against the stone walls and the downward thrust of the vaulting bosses."

the thrust P is relayed, is much more resistant to the torques of P than the slenderer and taller cathedral's walls. The vertical force r_y, which is induced by the flying buttress due to the thrust P, further adds to the stability of the column. In order to increase the downward force on the column even more, and thus increase its stability, a pinnacle was usually built on the top of the column (dashed line).

4.3.7 Multi-member structures.

Forces that act on any structural element may be divided into two groups: external and internal forces. The sources of external forces are outside of the structural element, and the sources of internal forces are from within. Equations [4.1] and [4.2] provide the conditions that are satisfied by the forces that act on any structural element at equilibrium. Both external and internal forces have to be included in these equations. However, due to Newton's Third Law, the sum of the internal forces and the sum of their torques are zero, and therefore they may be omitted from these equations. Nonetheless, internal forces, which are usually not given explicitly, are important, because they cause stresses in the structural element.

In order to find those internal and reaction forces, a structure is divided into virtual substructures. By so doing, internal forces of the entire structure become external forces to some of its substructures. As such, they contribute to Equations [4.1] and [4.2] of the substructure. By solving these equations for the substructure, internal and reaction forces can be found. To be able to find them, the number of independent equations for that substructure has to be equal to the number of unknown forces. There are cases that this cannot be done without considering elastic and similar properties of the substructures, as will be discussed in section 4.3.8. These are called **indeterminate** cases. This section deals only with **determinate** structures that have a large number of internal and reaction forces. The following two examples illustrate how these ideas are implemented in structures that have a large number of elements.

Example 1: Trusses

Consider the truss in figure 4.11a. It is made out of beams connected in frictionless hinges. Horizontal beams are 3 m long, vertical beams are 4m, and diagonal beams are 5m. The truss is supported at two points (A and E), and an external load of P Newton is acting at point B. All the beams are assumed to be weightless. (If we want to consider the weights of the beams, the weight of each beam could be attached to the center of the beam, and treated similarly to P).

Figure 4.11a

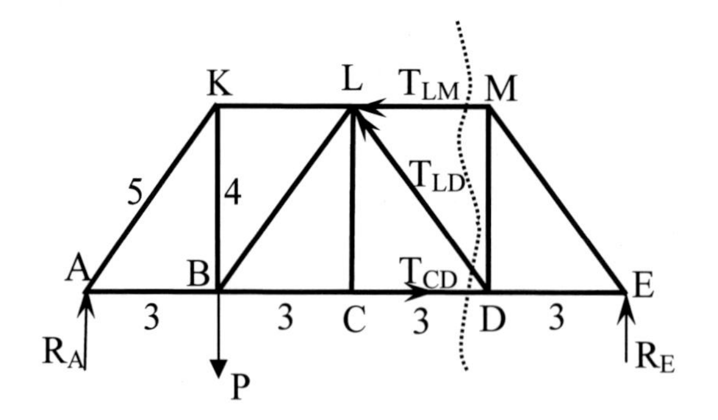

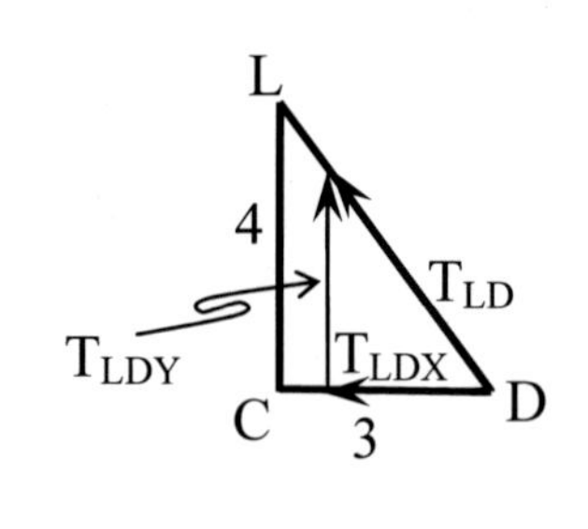

The external forces on the truss are P, and the reaction forces at the support points: R_A and R_E. Assume that we want to find out the tensions T_{LM} in beam LM, T_{LD} in beam LD, and T_{CD} in beam CD. If we consider the entire truss, these are internal forces.

First, let's find R_A and R_B. Together with P they act on the entire truss in the (vertical) y direction. These are the only forces that act in the y direction. Therefore, their sum must be zero, and their torque about any point, say E, must also be zero. We get the following two equilibrium equations for the entire truss:

$R_A + R_E - P = 0$ for the forces [*i*]

and

$R_A \cdot 12 - P \cdot 9 = 0$ for the torques. [*ii*]

These are two equations with two unknowns R_A and R_E. By solving them we get: $R_A = \frac{3}{4}P$ and $R_E = \frac{1}{4}P$. [*iii*]

We now divide the entire truss into two subsystems, as indicated by the dotted line in figure 4.11a. The tensions T_{LM}, T_{LD}, and T_{CD} are now external forces to the subsystem ABCLK. We assume that beams LM and LD are compressed, and beam CD is stretched. The arrows on those beams indicate the directions of the forces that they exert on subsystem ABCLK. (If we guessed wrong the type of any of those forces, we will get a minus sign for it in our final results). The forces P and R_A are also external forces to this subsystem. While T_{LM} and T_{CD} act in the (horizontal) x direction, T_{LD} acts diagonally. The components of T_{LD} are T_{LDX} and T_{LDY}. If the x-axis points to the right and the y axis points upward we get (as can be seen in the right part of figure 4.11a):

$T_{LDX} = -\frac{3}{5}T_{LD}$ and $T_{LDY} = \frac{4}{5}T_{LD}$ [*iv*]

We now apply the equilibrium conditions to subsystem ABCLK:
The sum of the y forces on subsystem ABCLK gives:

$$R_A - P + T_{LDY} = 0$$

Combining this with [*iii*] we get

$T_{LDY} = \frac{1}{4}P$ [*v*]

$T_{LDX} = \frac{-3}{16}P$ [*vi*]

and

$$T_{LD} = \frac{5}{16}P \qquad [vii]$$

The sum of the torques around point L gives:

$-R_A \cdot 6 + T_{CD} \cdot 4 = 0$ or (based on [*iii*])

$$T_{CD} = \frac{3}{8}P \qquad [viii]$$

The sum of the forces in the x-direction that act on subsystem ABCLK gives:

$-T_{LM} + T_{CD} + T_{LDX} = 0$

and based on [*iii*] and [*viii*] we get

$$T_{LM} = \frac{3}{16}P \qquad [ix].$$

Example 2: Suspension Bridges

The second example is of a section of a suspension bridge. The section starts at the horizontal part of the main suspending cable (usually the center of the bridge, marked as A in figure 4.11b), and includes n vertical cables. In our example n=7. The separation between consecutive vertical cables is d, and the distance between the two outer vertical cables is L=(n-1)·d (6d in the figure). The length of the segment is n.d. The vertical cables are connected to the main cable at the joints, and each joint supports the same share w of the segment's total weight, which is W=n·w (7d in the figure). The segment's center of gravity is L/2 away from the first joint (3d in the figure). The weight of the cables is small compared to the

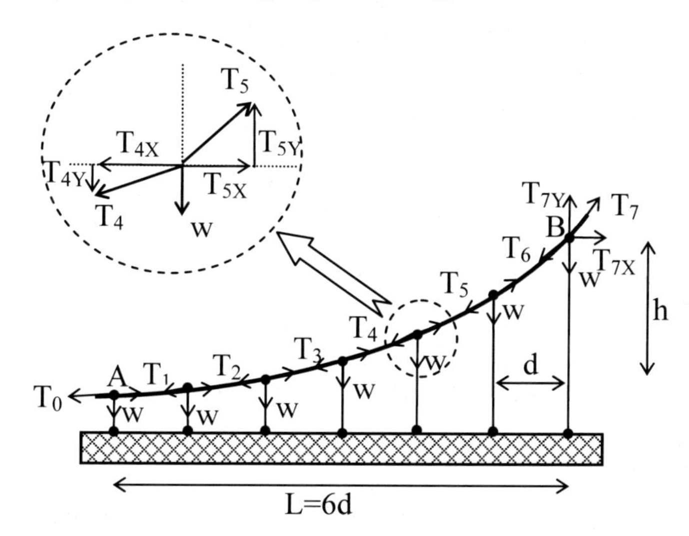

Figure 4.11b: A segment of a suspension bridge.

weight of the bridge, and therefore is neglected.

Between two consecutive joints, the segments of the main cable are assumed to be straight lines. The right end of the main cable (B in the figure) is h meters higher than the left end (A in the figure). T_1 to T_6 are the internal tensions in the corresponding segments. T_0 is the external horizontal tension at the left end, and T_7 is the external force at the right end of the main cable.

We now find the tensions T_0 to T_7. The sum of the x components and the sum of the y components of the three forces that act at each joint must be equal to zero. That gives the following relationships between the force components that act at each joint:

Joint	x components	y components
1 (A)	$T_0=T_{1X}$	$T_{1Y}=w$
2	$T_{1X}=T_{2X}$	$T_{2Y}=w+T_{1Y}$
3	$T_{2X}=T_{3X}$	$T_{3Y}=w+T_{2Y}$
4	$T_{3X}=T_{4X}$	$T_{4Y}=w+T_{3Y}$
5	$T_{4X}=T_{5X}$	$T_{5Y}=w+T_{4Y}$
6	$T_{5X}=T_{6X}$	$T_{6Y}=w+T_{5Y}$
7 (B)	$T_{6X}=T_{7X}$	$T_{7Y}=w+T_{6Y}$

From the equations for the x components we get that all the x components are equal to each other. From the equations of the y components we get that:

$T_{1Y}=w$; $T_{2Y}=2w$; $T_{3Y}=3w$; $T_{4Y}=4w$; $T_{5Y}=5w$;
$T_{6Y}=6w$; and $T_{7Y}=7w$. (or $T_{nY}=n\,w$)

In order to find the value of T_0, (which is also the value of all the other x components of the tensions), we consider now the main cable as a whole. We choose point B as an axis of rotation for the entire cable. The torque of T_0 around B is $-T_0 \cdot h$. The torque of the weights is $nw \cdot (L/2)$ (in this case $7w \cdot L/2$). Since the sum of these two torques must be zero we get:

$$T_0 = \frac{W \cdot L}{2 \cdot h} = \frac{n \cdot w \cdot ((n-1) \cdot d)}{2 \cdot h} = \frac{w \cdot d \cdot (n^2 - n)}{2 \cdot h} \qquad [4.15]$$

or in our case $T_0 = \dfrac{7w \cdot L}{2h}$.

The magnitude of the tension in the n'th cable is

$$T_n = \sqrt{T_{nX}^2 + T_{nY}^2} = \sqrt{T_0^2 + (n \cdot w)^2} \qquad n = 0,1,2,... \qquad [4.16]$$

Now we can find the geometrical shape of the cable. By rearranging equation [4.15] we get:

$$h = \frac{w \cdot d}{2T_0}(n^2 - n) \qquad n = 0,1,2,... \qquad [4.17]$$

where n is the sequential number of the joint on the main cable, and h is the height of that joint with respect to the left end of the main cable (point A for which n=1). Equation [4.17] describes a parabola ($y=ax^2-bx$, where $y=h$, $x=n \cdot d$,

$a = \frac{w}{2 \cdot T_0 \cdot d}$; $b = \frac{w}{2T_0}$). So, the main suspension cable has the shape of a parabola, whose parameters depend on the span of the bridge, its weight, and the tension in the main cable.

Note: the parabolic shape of the suspension cable was obtained because we assumed that the weight of the cable can be neglected with respect to the weight of the bridge. That resulted in all the segments of the bridge having the same weight. A suspended cable that supports only its own weight has a shape called catenary.

Example 3 The Catenary and Stone Arches.

It is possible to carve an entire arch from one big rock. The arch will stay as one strong piece due to the bonding forces that exist between atoms inside a solid. However, when an arch has to be built from smaller stones, other forces have to hold the stones together. Since ancient times, people have been building arches from smaller stones, relying on gravity, friction, and normal forces to hold the arch together. The builders knew how to do it based on trials and errors and keen intuition. In this chapter, we will explore some of the physics underlying those structures. We will start with idealized arches, made of one dimensional beams. Then, we will move to realistic arches, made of two dimensional beams (the depth of the arch being the third dimension).

Figure 4.12: Landscape Arch, Arches National Park, Utah. For thousands of years, natural erosion forces have been carving the arch, which is now 108 m long and about 1.8m thick at its top.

Arches of one dimensional beams

Figure 4.13 describes half of a symmetric arch, made of one-dimensional beams. There are three kinds of beams, each assigned its own index: the top horizontal beam is indexed as 0; the bottom beam is indexed as N; the rest are indexed as i,. $0 < i < N$. The forces that act on a beam are the unspecified action-reaction pairs, H and V, between the beam and its neighbors, and W, the beam's own weight. H indicates the horizontal component of the action-reaction force, and V the vertical component. L is the length of the beam, and d is the distance of its center-of-gravity from its lower end. Each force of an action-reaction pair acts on a different beam. Forces marked by thick arrows act on beams marked by thick lines, and similarly for thin arrows and lines.

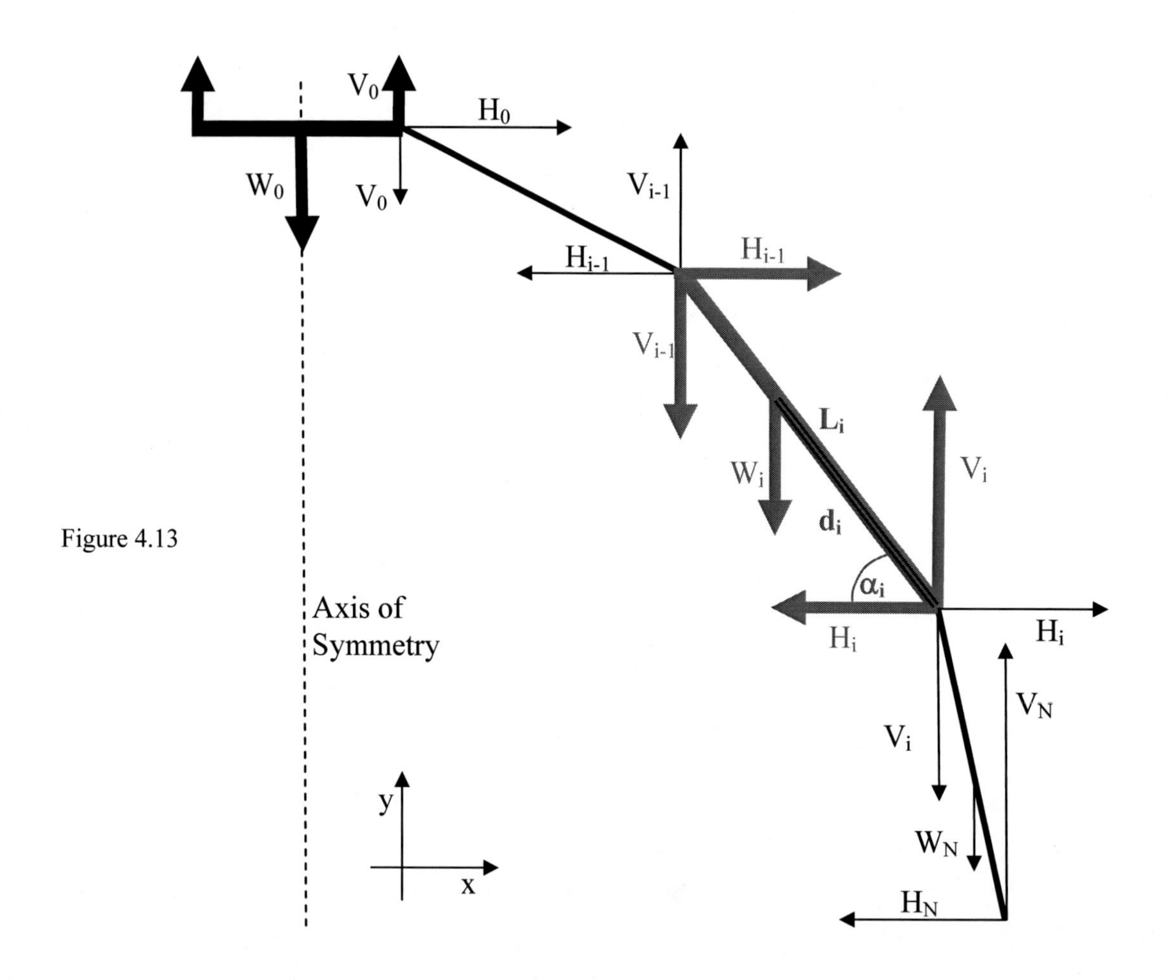

Figure 4.13

The static equilibrium conditions for a beam yield:

For the vertical forces at the top beam:

$$V_0 = \frac{W_0}{2} \quad [4.18]$$

For the other beams:

For the horizontal forces:

$$H_i = H_{i-1} \quad [4.19]$$

Therefore

$$H_0 = H_1 =,...,H_i,...,= H_N$$

For the vertical forces:

$$V_i = V_{i-1} + W_i \qquad [4.20]$$

Therefore

$$V_i = \frac{1}{2} W_0 + \sum_{j=1}^{i} W_j$$

For the torques:

$$V_{i-1} L_i \cos\alpha_i + W_i d_i \cos\alpha_i - H_{i-1} L_i \sin\alpha_i = 0$$

which yields

$$\tan\alpha_i = \frac{V_{i-1} + \frac{d_i}{L_i} W_i}{H_0} \qquad [4.21]$$

From [4.20 and 4.21] we get

$$H_0 = \frac{V_0 + \frac{d_1}{L_1} W_1}{\tan\alpha_1} \qquad [4.22]$$

For any given set of beams, one of the variables H_i or α_i (i=1,…,N) could be chosen arbitrarily and that will determine the shape of the entire arch. If, for example, α_1 is chosen arbitrarily, equations [4.18-21] uniquely determine the entire arch. For the arch to be at equilibrium, the foundation has to provide the horizontal force $H_N = H_0$, and the joints between the beams have to provide action-reaction forces according to formulas [4.18-21].

Equation [4.21] gives the relationship between the slope of beam i, its weight, the weights of the beams above it, and the horizontal forces that act in the arch.

If the weight Wi of a typical beam is much smaller than the total weight of all the beams above it, equation [4.21] could be approximated by

$$\tan \alpha_i = \frac{V_i}{H_0} \quad i=1,\dots,N \qquad [4.23]$$

The weight of the beam, W_i, is proportional to its length, L_i,

$$W_i = \rho A_i L_i \qquad [4.24]$$

where ρ is the beam's weight density and Ai is the cross sectional area.

Based on equations [4.20 and 4.24], equation [4.23] tells us that $\tan\alpha_i$, the slope of the arch at any point, is proportional to $\sum_{j=1}^{i} L_j$, the length of the arch from that point to the top of the arch. This property defines a catenary. That proves that the shape of the arch is a catenary.

(The generic formula of a catenary is $y(x) = \frac{1}{2}a(e^{x/a} + e^{-x/a})$, where a is a parameter that determines the opening of the catenary. This generic shape can be rotated and shifted to match any catenary structure. Figure 4.14 shows four catenaries. The general equation of these shapes is $Y(x) = C - \frac{1}{2}a(e^{x/a} + e^{-x/a})$, where a determines the opening of the catenary and C its vertical shift. Using calculus, one gets that the ratio between L(X), the length of the catenary from x=0 to x=X, to the slope of the catenary at x=X, $y'(X) = \frac{dy(X)}{dx}$, is constant: $\frac{L(X)}{y'(X)} = 2a$.)

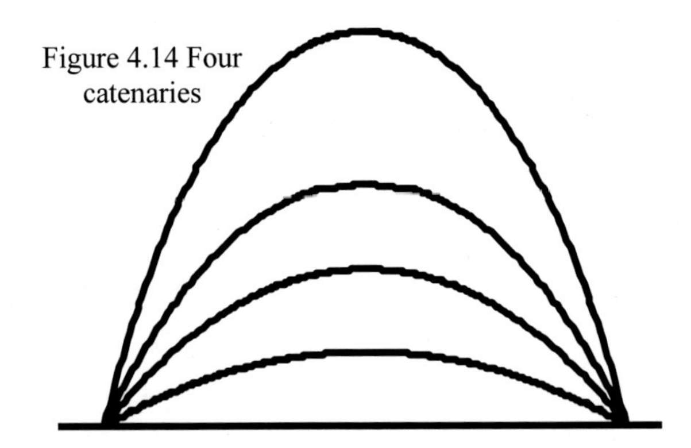
Figure 4.14 Four catenaries

Hinge-joints between the beams

The action-reaction forces Hi and Vi depend on the details of the junctions between the beams. If the junctions are frictionless hinges, and the beams are arranged according to formulas [4.18-21], the arch will be at a static equilibrium.

For example, a bicycle's drive-chain, like those that connect the front and rear sprockets, is an example of beams connected by hinges. It is possible to shape the chain as a catenary arch, according to formula [4.21] and figure 4.13. The hinges would provide the appropriate H's and V's. However, that structure will be extremely unstable. Any slight shift from the equilibrium state would collapse the entire structure.

On the other hand, if such a chain is hung from its two ends, the sagging chain will form an inverted arch. That sagging chain is at a stable equilibrium. If some links are shifted and then let go, the chain will return to its initial shape.

The schematics of a sagging chain and its forces are shown in figure 4.15. A closer look reveals that the orientations of the beams and the forces are very similar to those of Figure 4.13. The only differences are in the directions of some of the forces with respect to their affected beams. However, when the conditions of static equilibrium are spelled out, the resulting equations are exactly equations [4.18-22]. Therefore, the shape of the sagging chain is the same catenary as the catenary arch of figure 4.13.

Voussoirs

The pieces that make up an arch are called voussoirs. They have the trapezoidal shape of a truncated wedge. The tip of the wedge faces the inside of the arch, and the wider base faces out, see figure 4.16. An external force that acts on the wide base of a voussoir would push it towards the inside of the arch. That would create stress around the arch, much like the sidewise forces that are created by a wedge

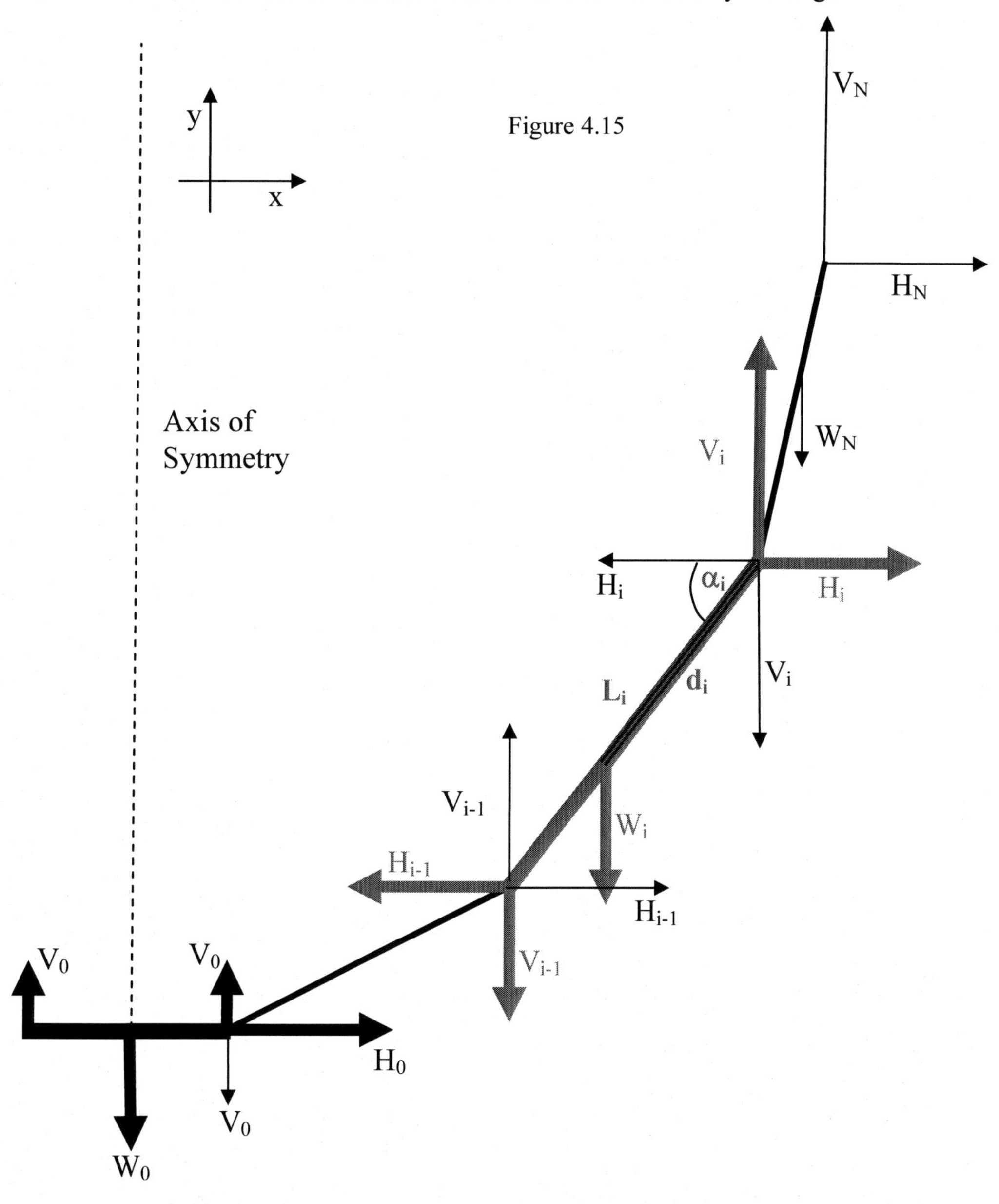

Figure 4.15

that is being pushed into a piece of wood. If the rest of the arch could withstand those sidewise forces, the pushed voussoir won't be able to move inwards, and the entire arch would remain intact.

Voussoirs exert contact forces on their neighbors. Those action-reaction forces may be oriented in any direction, according to the nature of the voussoirs' surfaces. Any such force could be represented in two coordinate systems; one, the Horizontal-Vertical with respect to the ground. We have used that coordinate system in formulas [4.18-21]. The horizontal component was denoted as H, and the vertical as V (Figures 4.13, 4.15). It is also possible to represent the action-reaction force between voussoirs i and i+1in a coordinate system that is defined by their common face. The axes of that coordinate system would be in the tangential direction and the normal direction to that contact surface. The components of the force in that coordinate system would be denoted as T_i and N_i; N_i in the normal direction, and T_i in the tangential direction to that surface, see Figure 4.16.

Since the pair H_i and V_i represent the same force as the pair N_i and T_i, the relationship between them (see Figure 4.16) is:

$$H_i=T_i\cos\theta_i+N_i\sin\theta_i \qquad i=0,\ldots,N \qquad [4.25]$$
$$V_i=-T_i\sin\theta_i+N_i\cos\theta_i$$

Or the inverse

$$T_i=H_i\cos\theta_i-V_i\sin\theta_i$$
$$N_i=H_i\sin\theta_i+V_i\cos\theta_i$$

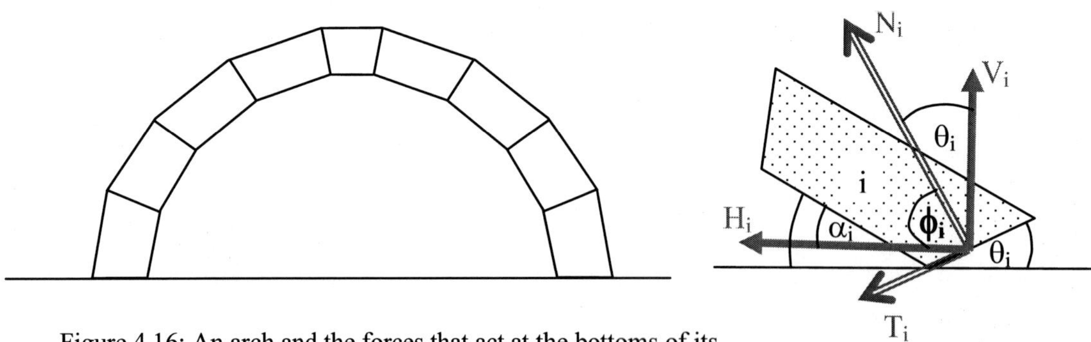

Figure 4.16: An arch and the forces that act at the bottoms of its voussoirs.

In stone arches that do not use cement, T is the friction force between the two voussoirs, and N in the normal force. Friction between voussoirs adds to the stability of the arch. An arch that is stable without friction forces would be more stable with friction. Of the two forces, friction and normal, friction is by far the weakest, and it would be the first to fail.

The normal and friction forces that participate in equations [4.18-25] are reaction forces that respond to the gravitational force that acts on the voussoirs. In a sense, the gravitational force "wants" to bring the arch down, and the normal and friction forces “turn the gravitation force against itself.” If cement is used in the structure, a different kind of reaction force, adhesion, enters the game. Unlike the normal force that prevents voussoirs from entering into each other’s space, adhesion opposes the separation of the voussoirs. If an arch is at a static equilibrium without cement, a thin layer of strong cement would strengthen the structure. If a thick layer is applied, the overall effect would depend on the properties of the cement. Sometimes, cement deteriorates with time faster than stone. If that happens, detrimental gaps may be formed between the voussoirs. On the other hand, if the cement is strong and resilient, arches may be constructed even from stones that are not shaped as wedges. The bonding force of the cement would act like the inter-atomic forces inside a solid, which pull the solid together. The cement will hold the stones together as one big chunk. The equations derived in this chapter do not apply to situations where such adhesion forces are active.

Frictionless voussoirs

If there is no friction (and no adhesion) between the voussoirs, T=0.

In such cases [4.25] yields : $H_i = N_i \sin\theta_i$ and $V_i = N_i \cos\theta_i$ or

$$\tan\theta_i = \frac{H_i}{V_i} \qquad i=0,\ldots,N$$

$$\tan\phi_i = \frac{V_i}{H_i} \qquad [4.26]$$

Equations [4.22 and 4.26] give all the angles with respect to the horizontal direction. In order to cut the voussoirs, one needs to know their base angles, γ_i and δi. Figure 4.17 shows the relationships between the various angles:

$$\gamma_i = \alpha_i + \theta_i \qquad 4.27]$$

and

$$\delta_i = 180^0 - (\theta_{i-1} + \alpha_i)$$

Figure 4.17

When voussoirs that are cut according to equations [4.21, 26, 26] (Figure 4.17) are assembled as an arch, the resulting shape is a catenary. The arch does not rely on friction forces to hold it together.

If, on the other hand, equation [4.26] is ignored, the resulting catenary arch would require both friction and normal forces, T_i and N_i, to hold it together [4.25]. The Ti’s are the required friction forces. For the voussoirs to stay in place, the conditions $T_i \leq \mu N_i$ must hold, where μ is the friction coefficient between the voussoirs.

Example

Design the voussoirs of a frictionless arch, consisting of 7 stones of equal central-lengths, L=1m each, (thick lines in figure 4.17a). The voussoirs are numbered according to formulas [4.18-4.27]. Voussoir 0 is the keystone, voussoirs 1 makes an angle of 30^0 with the horizontal. Assume that the center of gravity of each voussoir is at L/2 and at the middle of its width, as indicated in the right voussoir #1.

Figure 4.17a

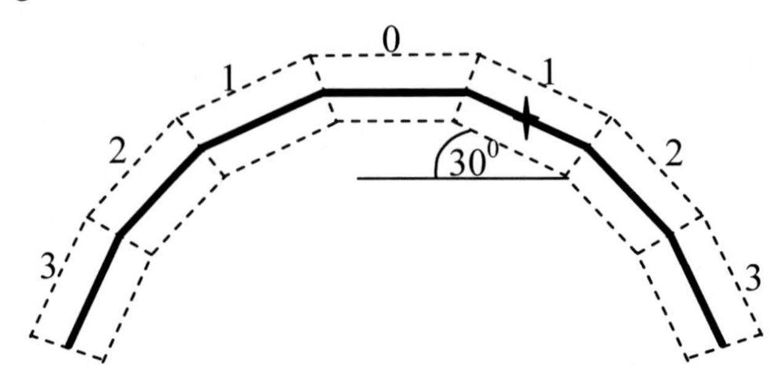

Solution

We have to find the α, θ, δ and γ of each voussoir, according to formulas [4.18-4.27]. $\alpha_1=30^0$. Based on [4.24] we may replace W_i in those formulas by L=1m. Based on [4.22] we get $H_0=1.731$. The results of the calculations are provided in the table.

Voussoir # i	V_i [4.20]	α_i [4.21]	θ_i [4.26]	γ_i [4.27]	δ_i [4.27]
0	0.5	0.00	73.90	73.90	73.90
1	1.5	30.00	49.11	79.11	76.10
2	2.5	49.11	34.72	83.82	81.79
3	3.5	60.00	26.33	86.33	85.28

Stability of Arches

Equations [4.18-26] deal with arches that carry their own weight. If an additional load that was not included in [4.18-26] is applied to any part of the arch, the equilibrium conditions may be violated, and the arch might fail. For an arch to collapse, at least one of its voussoirs has to move outwards. That would make room for other parts of the arch to fall down. As long as no voussoir can move outwards, the arch is safe. Friction forces, which always exist between voussoirs, add to the stability of the arch. Careful additions of loads and abutments that press the voussoirs inwards and oppose outwards slippage add to the stability of the arch, beyond formulas [4.18-26].

Semi-Circular Arches

The Romans used arches extensively in public and private structures. Many of those arches stand till today, attesting to their stability. A typical Roman arch had a semicircular shape. Figure 4.18 illustrates a semicircular arch whose width is about one tenth of its radius, a typical value of many Roman arches. A catenary curve is drawn inside the arch.

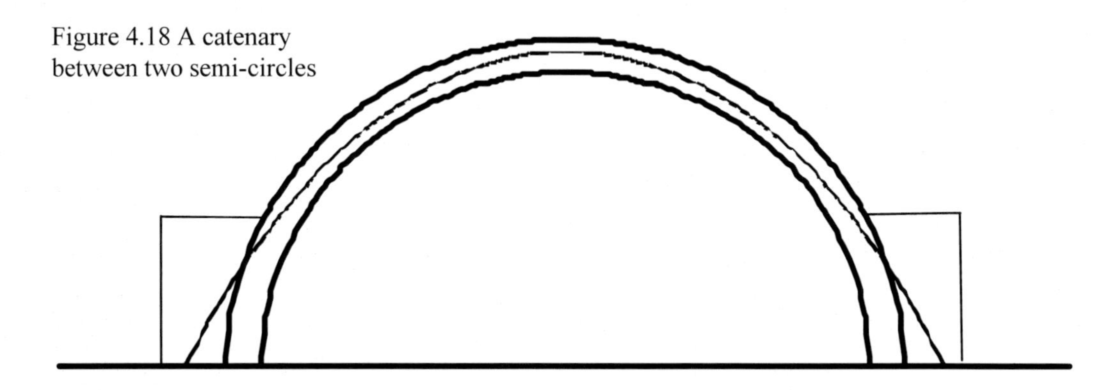

Figure 4.18 A catenary between two semi-circles

Figure 4.19 Remnants of the aqueduct near Caesarea, Israel. Notice the beginning of an arch at the left, and its abutment. It is still standing on its own to about half of the height of the arch (compare Figure 4.18). The aqueduct system was built by King Herod in the first century BC, and then expanded and maintained for about 1200 years, before it was abandoned. It carried water from springs at the feet of Mount Carmel to the thousands of people who lived in the port city Caesarea

It can be seen that the catenary fits quite nicely into the upper 60-70 percent of that arch. If the lower parts of the arch are broadened, as was really the case in Roman arches, the catenary would fit into the entire length of the arch. So, basically, semi-circular Roman arches are built around a catenary core. If their voussoirs were cut according to equation [4.21, 26, 27], their normal forces would be enough to hold them together. Abutments, friction forces, and sometimes cementing, provided safety margins that held the structures together for generations.

The Gateway Arch

The Gateway Arch in Saint Louis is a tribute to St. Louis' role in the westward expansion of the United States during the nineteenth century. The arch has a catenary shape, whose height is 630 feet above the ground. Its cross sections are equilateral triangles of 54 feet sides at its base, tapering to 17 feet at the top. The three faces of the arch are double-walls of carbon-steel inside and stainless steel outside. The separation between the walls is 3 feet at the base of the arch and 7.75 inches at the height of 400 feet. Up to the 300 feet level, the separation is filled with reinforced concrete, and above that, steel stiffeners are used. The walls were welded on-site from sections of 8-12 feet in height. The outside welds were polished to give the arch a uniform, sparkling shine

Figure 4.20: The Gateway Arch

Since a catenary supports its own weight, the backbone face of the arch could support itself, but that would have been an unstable equilibrium. The two side-faces of the arch add the necessary strength and permanency to the structure. Other than that, no trusses were needed, and the arch is hollow inside. A 40-passenger train carries visitors inside the arch to an observation deck inside its top.

4.3.8 Considering elasticity

All the structural elements that we have considered so far in this chapter were assumed to be rigid. A real beam that is supported by two posts will sag a little at its center due to gravity. The amount of sagging will depend, among other things, on the elastic properties of the beam. If this sagging is very small, the beam may be approximated as a rigid structural element. The same thing holds for the posts. When a load is put on it, a post would contract a little. The post may also compress the foundation on which it rests. If such deformations are negligible, the post and the foundation may be approximated as rigid elements. However, there are situation where the elastic deformations are very small, but their effects on the distribution of the loads between the elements of a structure are significant.

Consider, as an example, a horizontal stiff beam that rests on three posts (figure 4.21a). For the sake of the discussion we will assume that the middle post is right under the c.o.g of the beam, at its center. The weight of the beam (W) and the three reaction forces that act on it (R_1, R_2, and R_3) act in the vertical direction.

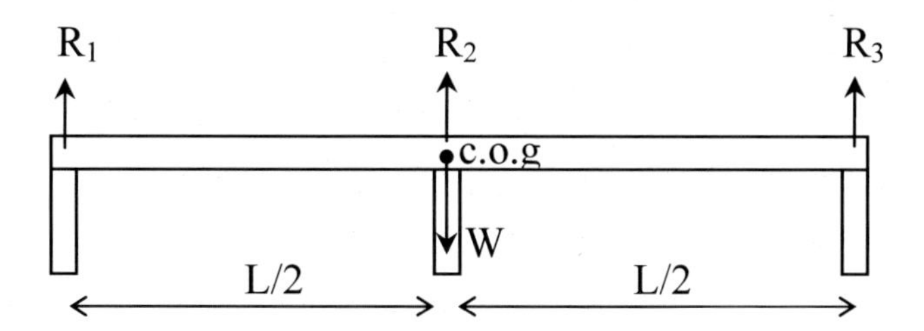

Figure 4.21a: Forces on a horizontal beam that rests on three posts.

Since the beam is in equilibrium, it satisfies the equilibrium conditions [4.1] and [4.2]. For the forces in the vertical direction we get:

$$R_1 + R_2 + R_3 - W = 0 \qquad [4.28]$$

and for the torques:

$$-R_1 \cdot \frac{L}{2} + 0 + 0 + R_3 \cdot \frac{L}{2} = 0 \qquad [4.29]$$

We have calculated the torques with respect to an axis of rotation at the center of the beam, in which case the torques of W and of R_2 were zero.

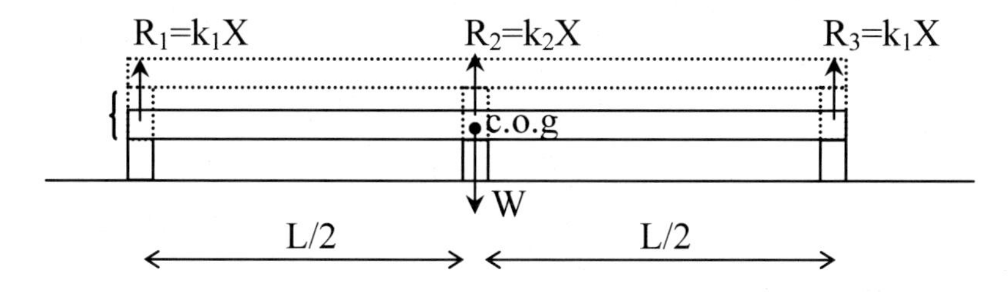

Figure 4.21b: A beam on three spring-like posts.

The weight of the beam and its length are known. Equations [4.5] are two equations for the three unknown reaction forces R_1, R_2, and R_3. Therefore, the reaction forces cannot be determined in a unique way. The elastic properties of the structural elements will determine the values of the reaction forces. We will consider the case where the beam is completely rigid, which means that it does not bend at all. The posts are elastic, and when compressed they contract by a very small amount. The three posts can be viewed as springs with very large spring constants k (equation [2.14]). That means that small contractions of the posts provide large reaction forces. For simplicity, we will assume that the two outside posts have the same k (k_1=k_3). The k of the middle post may be of any value.

Figure 4.21b shows the contraction of the posts (very exaggerated) after the beam has been placed on them. Because of the symmetry of the problem, all the posts have contracted by the same amount X.
When we substitute in [4.28] the expressions for R_1, R_2, and R_3 from figure 4.21b we get (based on [2.14]):
$k_1X + k_2X + k_1X = W$
which gives?
$X = \frac{W}{2k_1 + k_2}$
The reaction forces of the posts are therefore:

$$R_1 = R_3 = k_1 \frac{W}{2k_1 + k_2}$$
and
$$R_2 = k_2 \frac{W}{2k_1 + k_2} \qquad [4.30]$$

Any of these reaction forces is equal in magnitude to the downward force that the beam exerts on its post (the weight supported by the post). Equation [4.30] implies that a stiffer post, i.e. a post with a larger k, supports more beam-weight than a softer post.

4.4 Stress and strain

4.4.1 Types of deformations

External forces that act on an object may deform its shape. These deformations, which are sometimes noticeable even to the naked eye, can take a variety of forms, and they are the result of the atomic structure of all objects.

The atoms that make up solids are held in their places by the forces that they exert on each other. Figure 4.22 illustrates a part of a solid. The springs indicate forces that atoms (the dots) exert on their neighbors. In reality, no springs are present between atoms, and their spatial organization may be other than the cubical lattice shown here.

When external forces are applied on a solid, they act against these internal forces. As a result, atoms are shifted from their original equilibrium positions. In those shifts, "the springs" are stretched or compressed, causing internal forces. Three types of outcomes may happen due to these shifts: (1) if the shifts are small, the atoms will return to their original positions once the external forces are removed. This is called the elastic region. (2) If the applied forces are increased beyond the elastic region, the shifted atoms will not return to their original positions, but they will still be connected to each other. This is called the plastic region. (3) If the external forces are

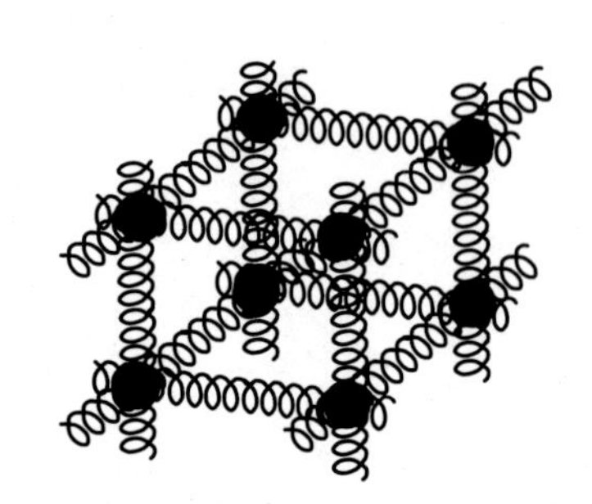

Figure 4.22: Atoms and forces in a cubic lattice

further increased beyond a critical point, the solid will break. Along the fracture, atoms will be separated from their previous neighbors to the extent that attraction forces between them don't exist any more.

In order to understand the relationships between the external forces and the reaction forces and deformations that they cause, it is helpful to first analyze simple situations. Since we are interested now only in internal processes, and not in the motion of the object as a whole, the sums of the external forces and torques are equal to zero. The simplest situation is when a pair of forces of equal magnitudes and opposite directions act along the same line, figure 4.23. There are two types of such pairs; (a) those that cause tension or stretching, and (b) those that cause compression.

When a pair of forces of equal magnitudes and opposite directions act off-line, their total torque is different from zero. Additional forces would be needed to hold the object in place. The overall deformation of the object would depend

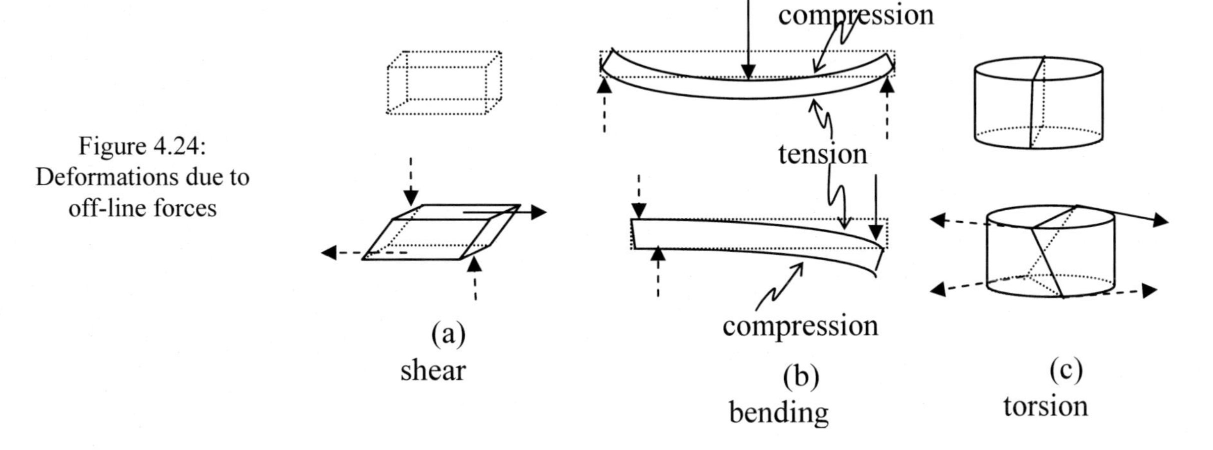

Figure 4.24: Deformations due to off-line forces

on the points of attachments of all the forces. Figure 4.24 illustrate a few situations in which off-line forces act on objects. One of the forces (solid arrow) is considered as the force responsible for the deformation, while the others (dotted arrows) are needed to hold the object in place. The figures illustrate the processes of shear, bending, and torsion. In shear, the main force is tangential to the plane of its attachment, as shown in figure 4.24a. The layers of the object 'slide' on each other, retaining their original parallel configuration and shape.

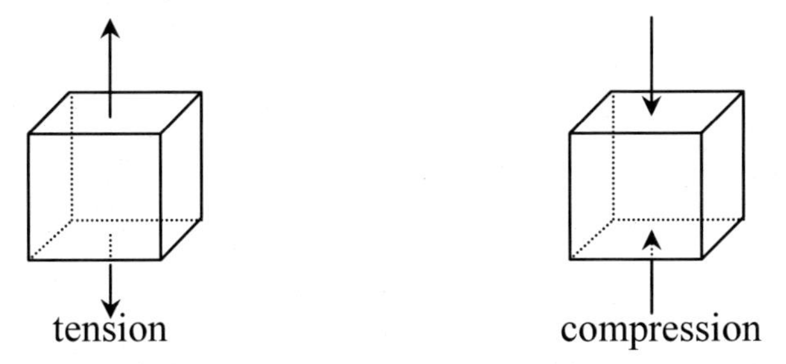

Figure 4.23: Tension and compression.

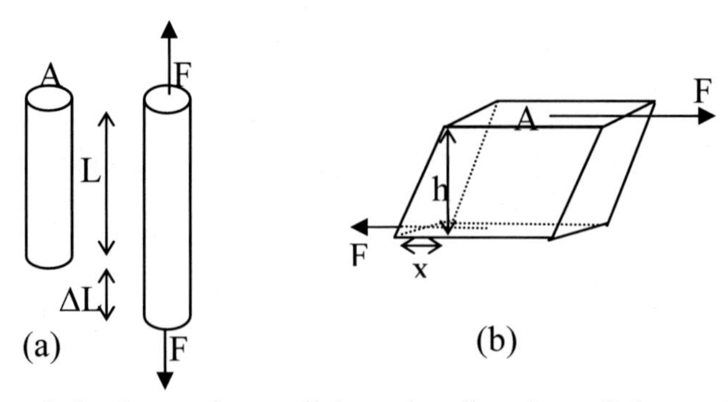

Figure 4.25: Parameters that participate in the calculation of stress and strain in tension and compression (left) and shear (right).

The motion of the layers is parallel to the direction of the applied force. In bending (figure 4.24b) layers perpendicular to the applied force are affected. "Inner" layers are compressed and "outer" layers are stretched. In torsion (figure 4.24c) layers are twisted around a central axis of the object.

4.4.2 The elastic moduli

Hooke's Law ($\mathbf{F} = -k \cdot \mathbf{x}$, [2.14]) relates the elastic force **F** exerted by a spring, when its length is stretched or compressed by the amount of **x**. Hooke's constant, k, describes the "stiffness" of the spring. The larger the k the more difficult it is to stretch or to compress the given spring. It was found that many objects, including steel, cement and wood columns, steel cables, etc., obey a similar law when they are stretched or compressed by external forces. Each such object has a constant k, similar to the spring's constant, which describes how stiff the object is. This k depends on the material and on the geometry of the object. For cylindrical objects of cross sectional area A and length L (figure 4.25a) it was found that

$$k = \frac{Y \cdot A}{L} \qquad [4.31]$$

where Y is a constant called Young's modulus, which is a property of the material. Equation [4.31] means that the stiffness, as expressed by k, is proportional to the cross sectional area A, and inversely proportional to the length L. When the k of [2.14] is replaced by [4.31] we get:

$$\mathbf{F} = -\frac{Y \cdot A}{L} \cdot \mathbf{x} \qquad [4.32]$$

When we consider only the magnitudes of **F** and **x**, rename **x** as ΔL, and rearrange the terms in [4.32] we get:

$$Y = \frac{(F/A)}{(\Delta L/L)} \qquad [4.33]$$

The quantity F/A is called tensile stress, the quantity $\Delta L/L$ is called tensile strain, and Y is Young's modulus. Equation [4.33], which reads: Young's modulus is equal to the tensile stress divided by the tensile strain, is just Hooke's Law, as applied to the object.

Young's modulus measures elastic properties of the material. Tensile stress combines the elastic force with a dimension of the material. Tensile strain measures an elastic deformation, ΔL, with respect to an original dimension, L.

In the case of compressing force, the quantity $\frac{F}{A}$ is the ratio between the force that compresses the object and the area on which it acts. This quantity is called pressure, and is denoted by p.

$$\text{PRESSURE} = p = \frac{\text{FORCE}}{\text{AREA}} = \frac{F}{A} \qquad [4.34]$$

(The unit of pressure in the SI system is the pascal (Pa). 1 pascal=$1 N/m^2$. The unit of pressure in the British system is pound/ft^2.) We can substitute p from [4.34] into [4.33], and after rearranging we get

$$\Delta L = \frac{L}{Y} \cdot p \qquad [4.35]$$

Thus, the change of the length of an object due to compression is proportional to the pressure that is acting on it. Let's look at some everyday's examples that illustrate this fact.

When you walk on the beach, your footprints in the sand are due to your weight compressing the sand under your feet. A skinny lady walking on pointed high-heel shoes would leave deeper footprints than a heavyweight wrestler walking barefoot. The skinny lady compresses the sand more than the heavier wrestler, even though she weighs less. Her weight acts on a much smaller area– her pointed heels. Therefore, she applies a larger pressure on the sand, and the compression of the sand, which is proportional to the pressure, is greater.

(Note: equations [4.6] deal with the compression or stretching of elastic materials. Sand is not elastic, but the general proportionality between pressure and compression holds, albeit not linearly.)

When designing buildings, the compression of the ground on which they stand due to the weight of the building is a factor that has to be considered very carefully. In general, in order to reduce the compression, the area of the foundation is made as large as possible. That reduces the pressure on the ground, thus reducing the compression and similar undesired effects. This is especially important when posts or columns are in direct contact with the ground. Metal plates or concrete slabs are put between the post and the ground, thus distributing on a larger area the weight that the posts transfer to the ground.

It was found that elastic shear forces also behave according to Hooke's Law. If F is the elastic shear force (figure 4.25.b), and the amount of shear deformation is measured by x, then F and x are related by $\mathbf{F} = -k \cdot \mathbf{x}$. In these cases, k measures the shear stiffness of the material. The larger the k the more difficult it is to shear the object. The k of the shear is usually different from the tensile k of any given object, so we'll denote it by k', and will call it **the elastic shear constant**. It was found that shear stiffness, as measured by k', is proportional to the cross section area A, and inversely proportional to the thickness h of the object (figure 4.25b).

$$k' = \frac{S \cdot A}{h} \qquad [4.36]$$

The constant S is called the shear modulus, and it measures the stiffness of the material with respect to shear forces. When the k' of [4.36] is substituted in Hooke's Law, [2.14] we get:

$$\mathbf{F} = -k' \cdot \mathbf{x} = -\frac{S \cdot A}{h} \cdot \mathbf{x} \qquad [4.37]$$

If we consider only absolute values, and rearrange terms we get:

$$S = \frac{(F/A)}{(x/h)} \qquad [4.38]$$

The quantity F/A, which combines elastic force and cross section, is called shear stress. The quantity x/h, which combines amount of deformation and an original size, is called shear strain. Altogether, [4.38], which is Hooke's Law for shear, states that the shear modulus is equal to the shear stress divided by the shear strain.

We have seen that tensile and shear elasticities can be represented by expressions of the form

$$\text{Modulus} = \frac{\text{Stress}}{\text{Strain}}$$

where

$$\text{stress} = \frac{\text{elastic force}}{\text{area on which force is acting}}$$

and strain measures the ratio between an elastic deformation and an original dimension of the object. This approach is used also for other types of elasticity. For example, another modulus is the bulk modulus. It deals with the change in volume of an object when it is compressed from all sides by an external pressure. If the external pressure (p) is the same all around the object, this pressure plays the role of stress. The strain is defined as the change in volume (ΔV) of the object due to the pressure divided by its original volume (V). The bulk modulus B, which is a property of the material, is defined as:

$$B = \frac{p}{(\Delta V/V)} \qquad [4.38]$$

The linear relationship between modulus, strain, and stress holds as long as the material is in its elastic region. Beyond that, the modulus starts to change. Soon, the material becomes plastic and eventually it breaks. Table 4.2 shows values of these constants for some common building materials. The strengths values are indication of the maximum stresses that the materials can take.

Strength=maximum value of (Force/Area) before the material fails.

Those stresses are outside the elastic region. There are other important types of deformations that have not been discussed here. For example, slender columns that are compressed by the loads that they support may fail because of buckling, which may happen before the compressive strength of the material, as given in table 4.2, has been reached.

Material	*Young's Modulus, Y (N/m²)*	*Shear Modulus, S (N/m²)*	*Bulk Modulus, B (N/m²)*	*Tensile Strength (N/m²)*	*Compressive Strength (N/m²)*	*Shear Strength (N/m²)*
Aluminum	$7.0x10^{10}$	$2.5x10^{10}$	$7.0x10^{10}$	$2.0x10^{8}$	$2.0x10^{8}$	$2.0x10^{8}$
Steel	$2.0x10^{11}$	$8.0x10^{10}$	$1.4x10^{11}$	$5.0x10^{8}$	$2.0x10^{8}$	$2.0x10^{8}$
Concrete	$2.0x10^{10}$			$2x10^{6}$	$2x10^{7}$	$2x10^{6}$
Wood	$1\text{-}10x10^{9}$			$4x10^{7}$	$3.5x10^{7}$	$5x10^{6}$
Water			$2x10^{9}$			

Table 4.2: Elastic moduli and ultimate strengths of materials

Example

A mass of 1,000 kg is supported by a steel cable of radius 0.5 cm, whose unstrained length was 3 m. (a) By how much was the cable stretched? (b) What is the elastic constant k of the cable? (c) What is the maximum weight that the cable can support safely?

Solution

Young's modulus of steel is $2x10^{11}$ N/m² (Table 4.2).

(a) Following the notations of [4.33] we get: Y= $2x10^{11}$ N/m², L=3m, $A=\pi\cdot 0.005^2=7.8x10^{-5}$ m², F=1,000·9.8=9,800 N, ΔL=?

Substituting in [4.33] yields ΔL=1.9 mm.

(b) Based on [4.31], and using the values from (a) yields $k=5.2x10^6$N/m.

(c) The tensile strength of steel is $5x10^8$ N/m² (Table 4.2). Since the cross section area is $7.8x10^{-5}$ m², the maximum weight F is given by $F/A=5x10^8$, yielding F=39,270N, which corresponds to 4,007kg.

4.4.3 Controlling deformations in structures

You have probably noticed that a beam with a narrow rectangular cross section will bend more if the weight that it supports rests on its wide edge (figure 4.26a) rather than on its narrow edge (figure 4.26b). Therefore, a beam that has to carry weight is usually positioned like 4.26b and not like 4.26a.

When supporting weight like in figure 4.26b, the most economical beam would be as narrow and as high as possible. However, if a beam is too narrow, it may buckle sidewise. Beams with different cross sectional profiles, such as I beams (figure 4.26c) or T beams, provide the needed directional strengths in the most economical way.

Many structures are made of frames that support surfaces such as walls and floors. The frames have to be strong enough to hold the structure, and economical in the amount of materials that they use. Since the beams that make up the frames and the joints of the frames are flexible, the structure engineer has to make sure that the internal and external forces that act on the structure not deform any element of the structure beyond certain acceptable limits.

Figure 4.26

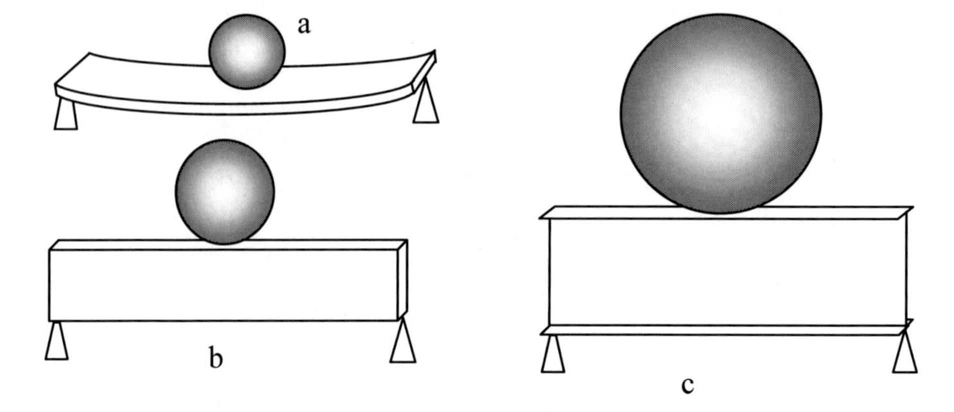

Frames are deformed by external forces similar to solid elements. Figure 4.27 shows some deformations modes of rectangular frames. While bending and shearing happen in two dimensions, twisting is a three-dimensional phenomenon.

In order to strengthen a frame, it could be braced by additional beams. There are many ways of bracing a frame. The lower parts of figure 4.27 show some ways of bracing rectangular frames, and the forces that act on the bracing beams. The bracing beams may respond in different ways to different types of forces that act on the frame.

An open frame can be strengthened also by enclosing it with attached

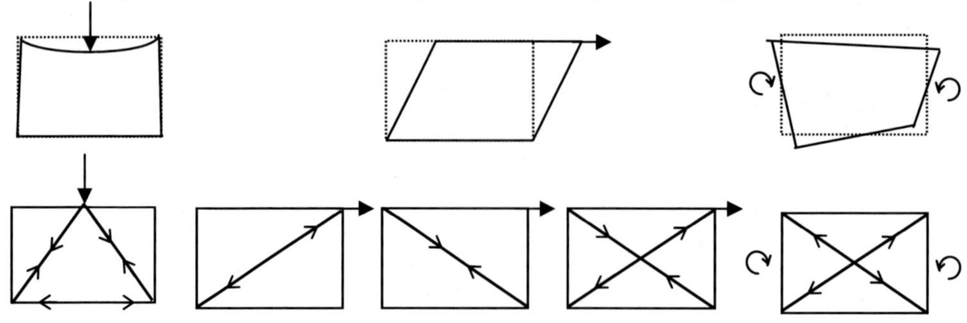

Figure 4.27: Effects of bracing of a rectangular frame.

surfaces. For example, a rectangular wood-frame becomes more resistant to shear forces when a sheet of plywood or dry-wall is nailed tightly to it. The combined frame and sheet become a shear wall. The sheet may be viewed as a continuous mesh of diagonal bracing beams, each of which resists shear forces. However, due to the two-dimensional nature of the sheet, the imaginary diagonal beams won't remain straight when the frame is deformed due to shear forces.

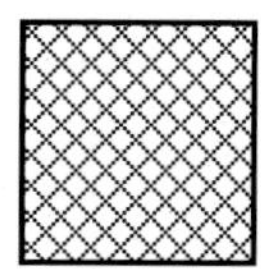

4.5 Hydrostatic pressure

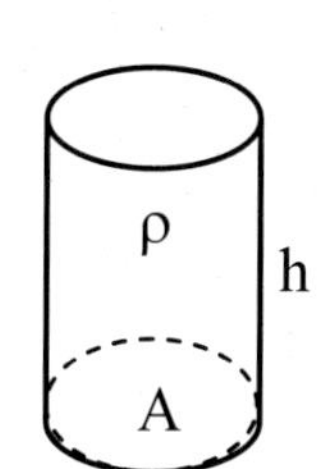

Figure 4.28

A solid column exerts pressure on the ground on which it stands. By definition, this pressure is equal to the weight of the column divided by the area of its base. Consider a cylindrical column (figure 4.28) whose base area is A and height is h. The density of the column's material is ρ. The volume of the column is V=A·h. The mass of the column is m=ρ·V=ρ·A·h (based on equation [2.12]). The force that the column exerts on its base, which is its weight w, is w=mg=ρ·A·h·g. Finally, the pressure that the column exerts on its base is:

$$P = \frac{F}{A} = \frac{w}{A} = \frac{\rho \cdot A \cdot h \cdot g}{A} = \rho \cdot g \cdot h\,. \qquad [4.39]$$

Consider now a barrel, of the same dimensions as the column (figure 4.28) filled with water to its rim. The water exerts pressure on the base of the barrel. The magnitude of this pressure would also be given by equations [4.39], where ρ would now be the water's density. This pressure is called hydrostatic pressure, because it is caused by static water. The term hydrostatic pressure is also used for the pressure caused by any static liquid.

We can appreciate the existence of the hydrostatic pressure if we make a small hole in the bottom of the barrel. Water will come out through that hole. In order to plug that hole we will have to push a plug into it, and apply an upward force on it so that it stays there. That upward force has to counter the hydrostatic force that is behind the hydrostatic pressure.

There is a big difference between hydrostatic pressure, caused by liquids, and pressure caused by columns of solid materials. The weight of the solid material, which is the source of the pressure, acts only downwards, while the hydrostatic force, which is the source of the hydrostatic pressure, acts in all directions. A solid column does not exert any side forces. On the other hand, if we make a hole in the side of the barrel, water will come out through it. In order to stop it, we will have to push a plug into that hole, and to apply a force on that plug to counter the sidewise hydrostatic force. Hydrostatic pressure acts also on the surfaces of an object that is submerged in the liquid, e.g. a tennis ball submerged in the barrel. When we make holes in that tennis ball, water will flow into it through those holes. Like in the case of water flowing out through holes in the wall of the barrel, this is an indication that water pressure acts perpendicularly to the surface of the tennis ball. All those familiar facts demonstrate that the hydrostatic force acts in all directions, both inside the liquid and on the walls of the container.

Measurements have shown that the hydrostatic pressures that acts on the walls of the container and on the surfaces of any object immersed in the liquid are given by the following formula:

$$P_{HS} = \rho \cdot g \cdot h \qquad [4.40]$$

Where P_{HS} is the hydrostatic pressure at a point which is at a depth h below the surface of the liquid. ρ is the density of the liquid.

Although [4.40] looks like [4.39], there are major differences between pressure due to solid objects ([4.39]) and hydrostatic pressure ([4.40]). In addition to the difference in the directionality ([4.39] is a downwards only pressure, [4.40] is pressure in all directions), equation [4.39] holds only for cylindrical objects. On the other hand, [4.40] holds for any shape of container. If a solid column has a base area A and a height h, but its volume is not given by $A \cdot h$ (e.g. if it is a cone of base A and height h), formula [4.39] does not apply to it. On the other hand, equation [4.40] applies to liquid in a container of any shape. For example, the hydrostatic pressure at the base of a conical container of base area A and height h, filled with a liquid of density ρ would be $P_{HS} = \rho \cdot g \cdot h$, even though the weight of the liquid in that cone is less than the weight of the liquid in a cylinder of the same A and h, correspondingly.

The formula for hydrostatic pressure ([4.40]) is applicable to pressure in water pipelines, when the water is standing still. Figure 4.29 illustrates the principles of a water supply system. A water tank at a high point, which serves as a reservoir, is connected by a network of water pipes to the consumers (the faucet in the figure symbolizes a consumer). The entire network of pipes and the reservoir constitute one big water container. The hydrostatic pressure at the faucet, which is one point in that big container, depends on the "depth" of that point with respect to the water surface at the water tank (h in the figure): $P_{HS} = \rho \cdot g \cdot h$. In practice, the water pressure in a main pipeline may be too high for domestic use. In such cases, pressure regulators are inserted between the main pipeline and the consumer.

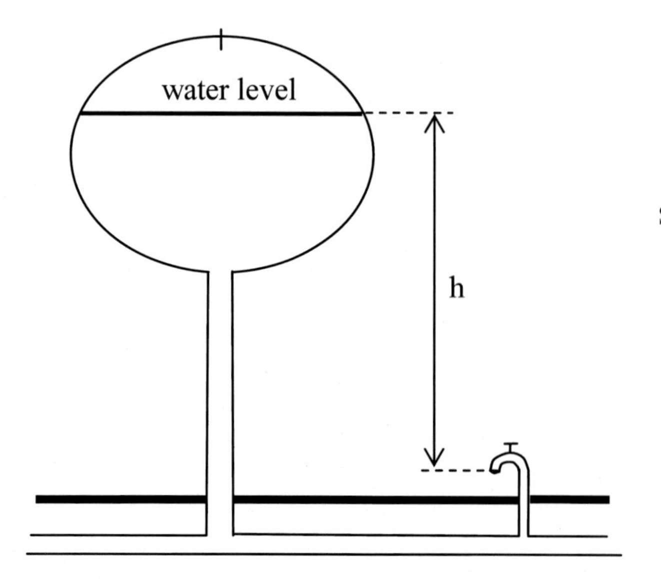

Figure 4.29: Schematics of a water supply system.

Example
What is the water pressure at a faucet which is 6.5 m above street level if the water level at the reservoir is 24 m above street level?

Solution
The hydrostatic pressure P_{HS} depends on h, the "depth" of the point where the pressure is given with respect to the surface of the water in the container. In this case, h=24-6.5=17.5m. The density of water is ρ=1,000 kg/m^3. Substituting in [4.40] yields P_{HS}=1,000·9.8·17.5=171,500 pascal.

Because hydrostatic pressure depends only on the height difference h between the liquid's surface and the point at which the pressure is assessed, and not on the shape of the container, pressure is sometimes expressed as a height of a column of liquid. A common unit is mm of mercury (Hg). For example, a pressure of 120 mm of mercury implies that a column of 120 mm=0.12 m of Hg, whose density is 13,600 kg/m^3, would generate, an hydrostatic pressure of P_{HS}=13,600·9.8·0.12=15,994Pa (based on [4.40].

The hydrostatic pressure given by [4.40] is a direct consequence of the force of gravity that acts on the liquid. If another force acts on the surface of a liquid, the pressure due to that external force will propagate in the same magnitude throughout the liquid. For example, gasses above the beer in a keg exert pressure on the beer. This pressure propagates throughout the beer. It can push the beer up through a spigot, which may be high above the surface level of the beer in the keg.

The air above us (the atmosphere) has weight. Air is relatively light (air density under normal conditions is approximately 1.3 kg/m^3, compared, e.g., to water's 1,000 kg/m^3). However, because of the height of the atmosphere, the pressure that it creates is significant: 1.013x10^5 Pa at sea level at 20^{0}C. A unit of pressure called "atmosphere" is equal to 1.013x10^5 Pa. Like hydrostatic pressure, atmospheric pressure acts in all directions, and it decreases as the assessed point moves up. However, atmospheric pressure does not obey equation [4.40].

PROBLEMS

Section 4.1

1. Two children are pushing a door with forces F_1= 75N and F_2= 60N, as shown in the figure. Their pushes are balanced, and the door does not move. What is the angle of F_2 with the door? The length of the door is 80cm, and F_2 is acting 15cm away from the door's edge.

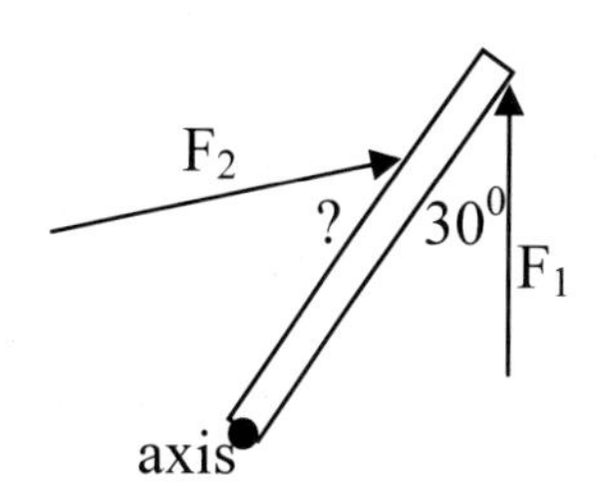

2. Three children, John (55 lbs), Jin (52 lbs.), and Jane (65lbs.) are sitting on a seesaw, as shown in the figure (distances are from the center of the seesaw).

Where should Abe (95 lbs.) sit so that the seesaw is balanced?

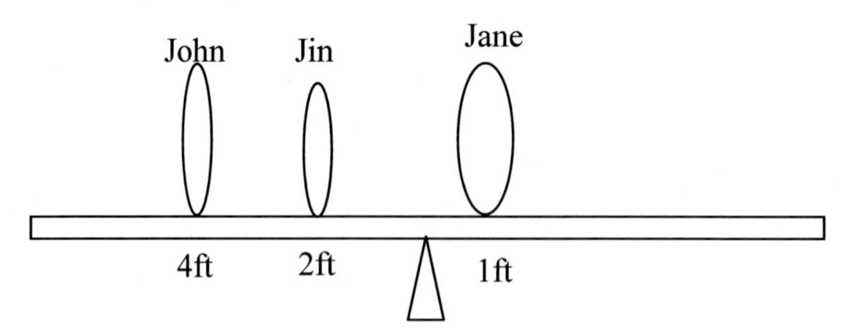

3. You want to design a mobile that hangs from the ceiling, whose dimensions are shown in the figure. What should be the weights W_1 and W_2?

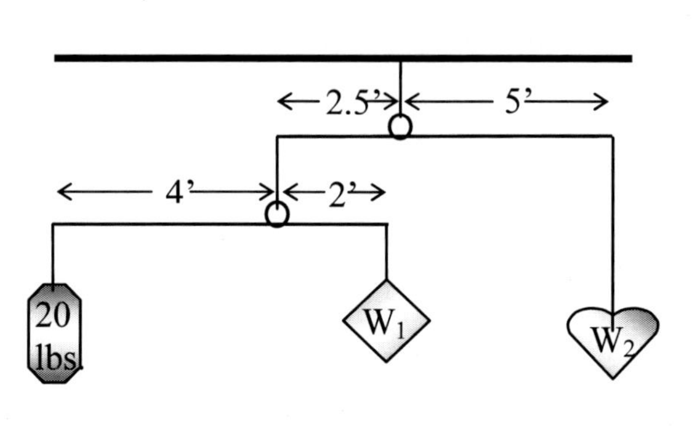

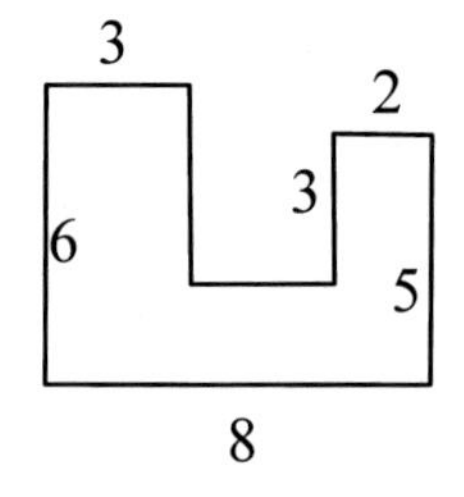

Section 4.2

4. Calculate the location of the c.o.g. of a concrete plate whose dimensions in meters are given in the figure:

Section 4.3.1-4.3.5

5. (a) Find the tension in each string in figures (a) through (f), and the compression in the beams of (c), (d) and (e). Assume that the strings and beams are weightless. Heavy lines indicate ceilings, walls, or floors.

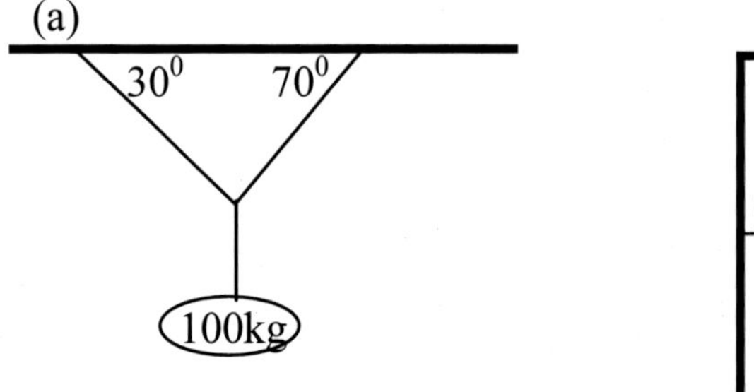

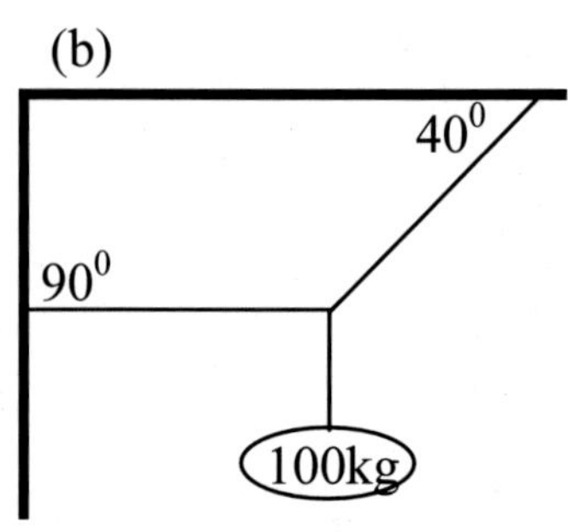

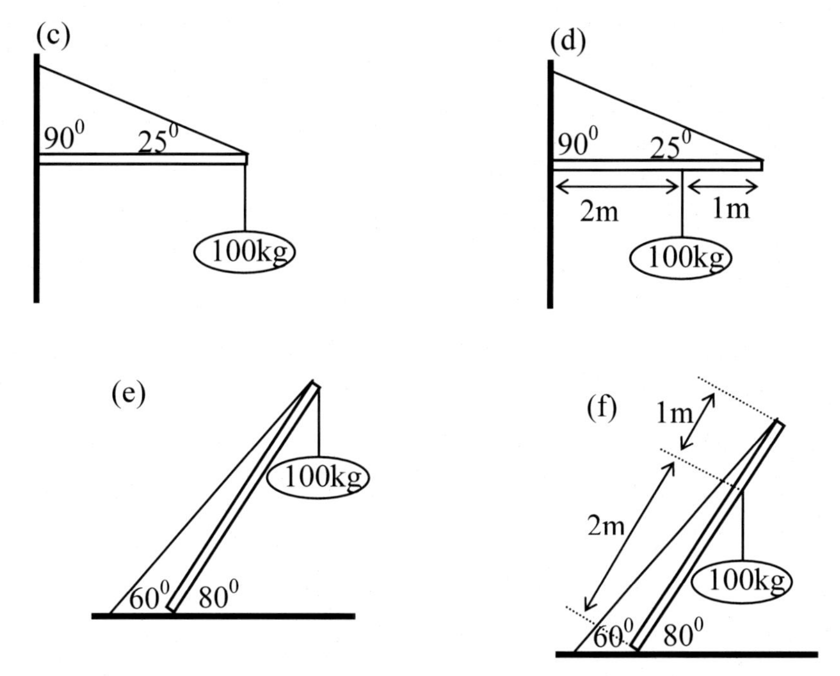

(b) Find the forces between the beam and the wall in (c) and (d), and between the beam and the floor in (e) and (f), in the figures above.

6. Solve problem 4 (e) through (f) above, if the mass of each beam is 80kg, and the beam's c.o.g. is at the beam's center.

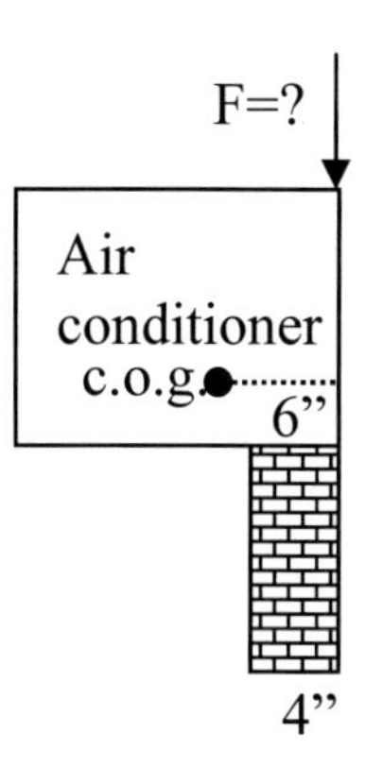

7. The figure to the right shows a side view of an air conditioner that is installed in a window. What is the smallest force F that a retaining bar at the top edge of the unit has to provide, so that the air conditioner does not fall out? The weight of the air conditioner is 85 lbs.

8. A horizontal truss, symmetric around its center, of mass 15,000 kg and total length 22 m, is placed on two posts as shown in the figure.
(a) Find the downward force that it exerts on each post.
(b) Find the additional downward force on each post when a load of 2,000 kg is placed at the center of the truss.

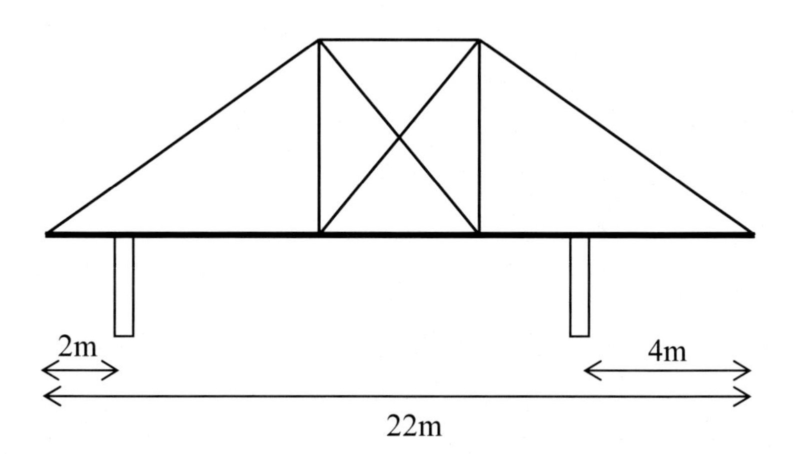

9. A symmetric arch, whose bases are slanted, is resting on vertical columns, as shown in the figure. The mass of the arch is 750kg, and the slant angle is 20^0 with the horizon. Find the sidewise force and the downward force that the arch exerts on each column.

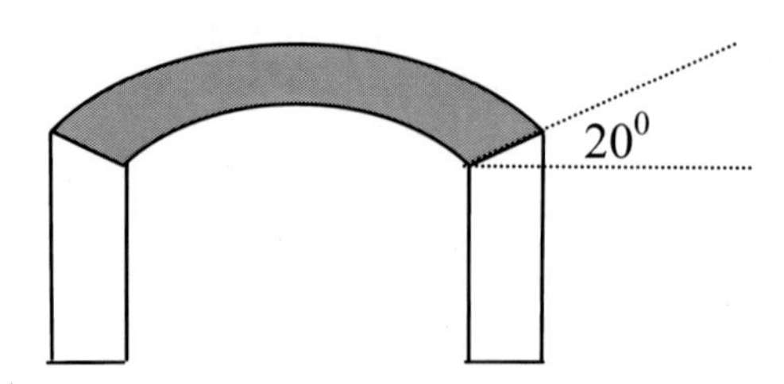

10. A plank of wood of mass 66 kg and length 3.30 m is cut in three. The length of one piece is 3 m, and the length of each of the two other pieces is 0.15m. You have to design a bench as shown in the figure such that the distance between the two legs would be as short as possible. However, it is required that if a 200 kg person sits on any of the edges, the bench would not flip over. What is the smallest distance between the legs of the bench that would comply with this requirement? Assume that the width of the legs is negligible.

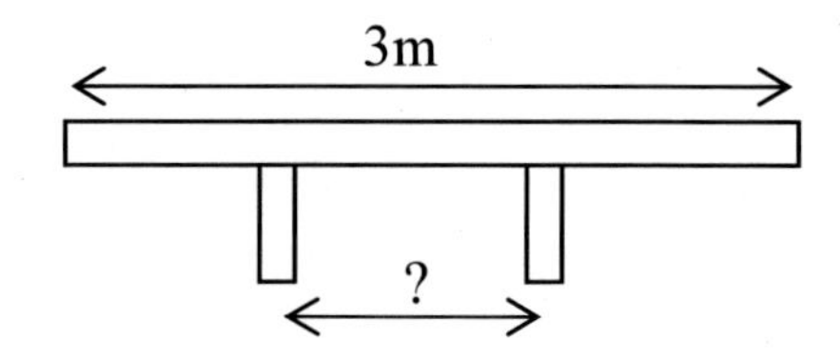

11. A 70 lbs. ladder of length 10ft leans against a frictionless wall.
Find what horizontal force is needed to hold the ladder in place when the angle that it makes with the floor is: (1) 80^0. (2) 45^0. (3) 15^0.
If the horizontal force between the ladder and the floor is due only to friction. What should be the smallest friction coefficient needed to hold the ladder in place in each of those three angles.

12. (a) Find the smallest horizontal force that has to be provided by the horizontal beam (thick line) in the cathedral ceiling of the figure, if there are no friction forces between the ceiling and the walls on which it rests. The two legs of the ceiling lean on each other at the top, without friction. Each leg is 2000 lbs.

(b) Find the smallest horizontal force that the walls would have to provide, if the horizontal beam is removed, and the legs are held in place only by their friction with the walls.

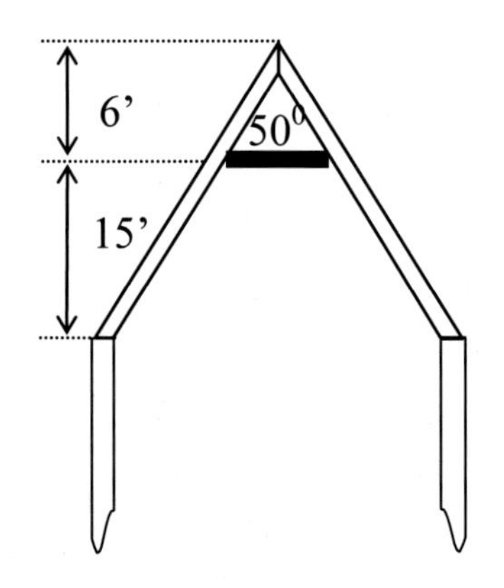

13. Two unconnected segments of a bridge are supported by three columns, as shown in the figure. The weight of each segment is 30,000 lbs. What is the force that each column exerts on its foundation?

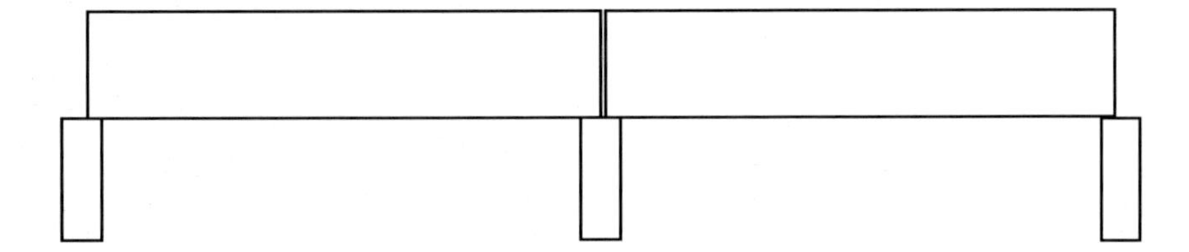

Figure 4.30: Stonehenge, a group of prehistoric ritual monuments of large stones near Salisbury, England. The site evolved in phases from 2550 to 1400 BC. It is suggested that the architect of the latest structures, parts of which are shown here, was familiar with buildings of the contemporary civilizations of the Mediterranean world.

14. Figure 4.32 is schematics of a segment of an arches bridge. Assume that all the parts (arches and beams) are rigid and are not tied or attached to each other. Under these conditions, how would the external weight of 10,000 lbs. at the center be distributed at the foundations of the two large arches (the arrows in the figure)?

Figure 4.31: Pont du Gard, Nimes, France. This Roman bridge is about 50 m high. (The small specks on the top are people walking on it). It was part of a 50 km long aqueduct that brought fresh water to the city of Nimes. The entire structure was built approximately 2,000 years ago without the use of cement.

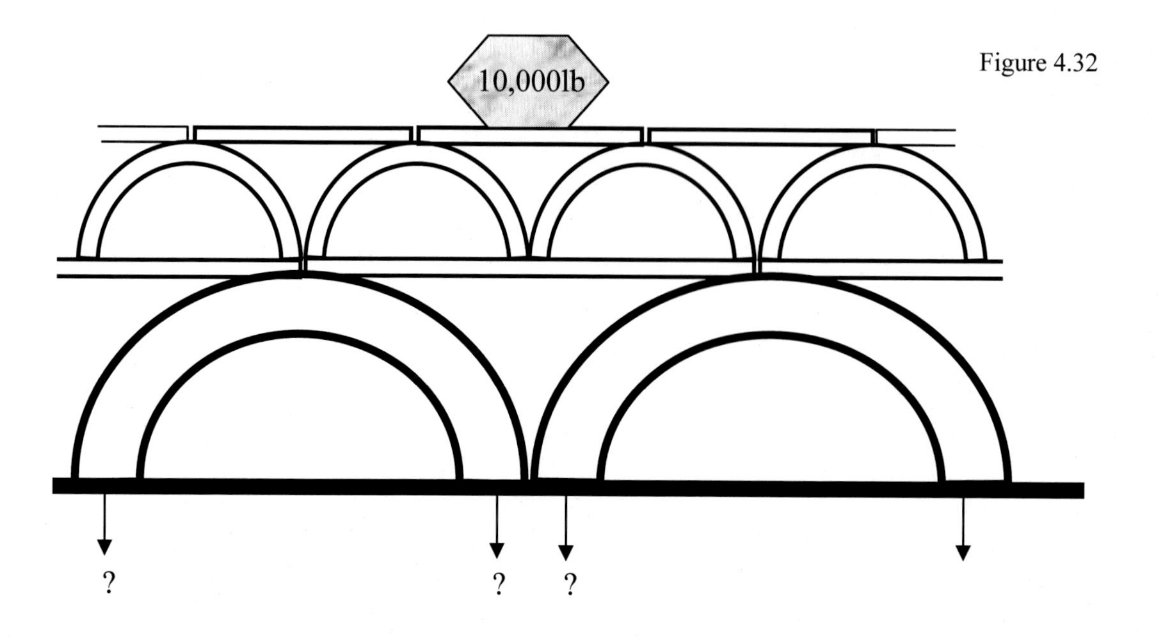

Figure 4.32

Section 4.3.6-4.3.7

15. The span between the two towers of a suspension bridge is 240 m. The towers, of equal heights, are 42m above the pavement of the bridge. The lowest points of the two main supporting cables are 2 m above the pavement. There are equally spaced 12 vertical suspension cables on each main cable. The distance between two vertical suspension cables (that hang from the same main cable) is 20m. The first and last vertical suspension cables of each main cable are 10m from their edge of the pavement. The bridge is uniform and its mass is $2.4x10^6$kg. The mass of the cables is small compared to the mass of the bridge. Find the tension in each main cable at its lowest point and in its segment that lies on the tower.

16. Calculate the tension (in multiples of P) in each of the beams of the truss in figure 4.11a. Indicate if each beam is compressed or stretched.

17. Design the voussoirs of a frictionless arch, consisting of 7 stones of equal central-lengths, L=1m each, (thick lines in figure 4.17a). The voussoirs are numbered according to formulas [4.18-4.27]. Voussoir 0 is the keystone, voussoir 1 makes an angle of 60^0 with the horizontal. Assume that the center of gravity of each voussoir is at L/2 and at the middle of its width, as indicated in the right voussoir #1 of figure 4.17a.

Section 4.3.8-4.4

18. A plank of wood is placed on three posts, as shown in the figure. It is fastened to the two outer posts, and just rests on the middle one. A load is then placed on the plank as shown in the figure. Which regions of the plank will experience tension and which will experience compression due to that weight?

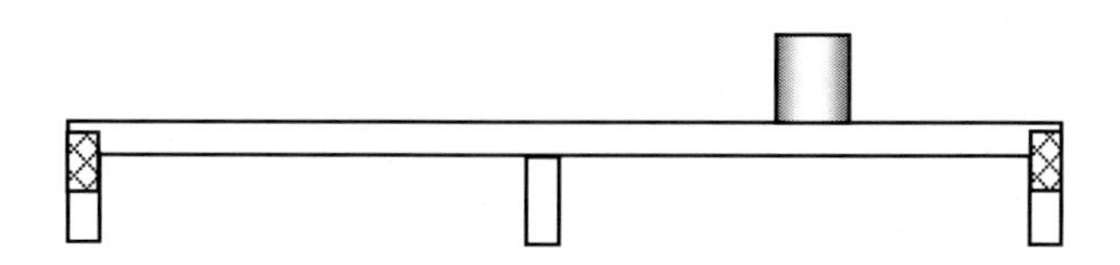

19. A beam whose weight is 10,000N is supported on three posts, as shown in figures 4.21. Find the weight distribution on each post when:
(a) k_1= $2x10^7$ N/m, k_2=$2x10^8$ N/m
(b) k_1= $2x10^8$ N/m, k_2=$2x10^7$ N/m
(c) k_1= $2x10^7$ N/m, k_2=$2x10^7$ N/m

20. A box of dimensions 1.5m x 2m x 0.8m weighs 2000N. It can rest on any of its sides. What is the largest pressure that it can exert on the floor?

21. A cylindrical concrete column is 3m high, and its diameter is 80cm. What pressure does it exert on the floor? (You'll find some relevant data in table 2.2).

22. When it was cast, the height of a concrete column was 10'. After the building above it was completed, the pressure on that column was 30,000 pounds per square inch (psi). By how much did the column contract due to that pressure?

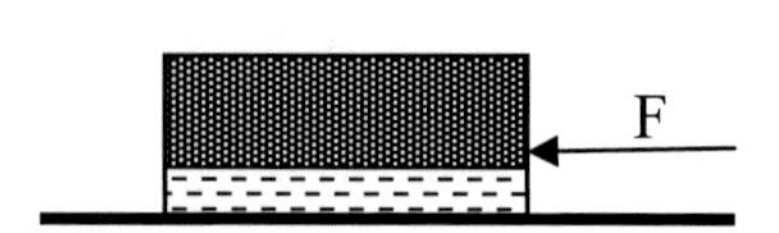

23. A brick of base area 4"x8"is cemented by concrete to the floor. What horizontal force F is needed in order to break the concrete? Assume that both the brick and the bonding between the brick and the concrete are stronger than the concrete.

24. A steel pipe, 12' long, of inside diameter 2" and outside diameter 2.2" is filled with concrete, and then used as a column in a structure. It carries a load of 10,000 pounds. By how much did it contract because of that load? Ignore lateral deformations of the column.

25. An elevator of gross weight of 2,000 lbs hangs on a steel cable. The building code requires a safety factor of 10 for such a cable. (The weakest point of the cable should not break when loaded with 10 times the maximal operational load.)
(a) What should be the smallest diameter of the required cable? The weight of the cable itself can be neglected.
(b) What should be the smallest diameter of the cable if the elevator accelerates upwards with an acceleration of 0.2g?

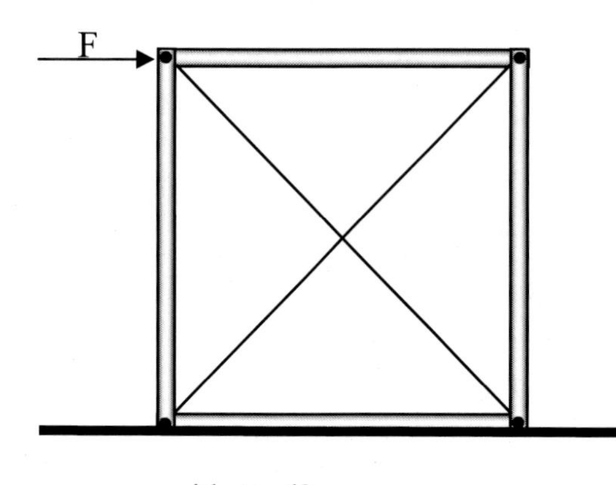

26. A square frame is made of solid steel beams, connected by pins that cannot prevent its collapse when pushed by even a small shearing force. The bottom beam is anchored to the ground. To make it shear resistant, the frame is braced by diagonal steel cables of diameter 1.6 mm (approximately 1/16") (see figure). The cables' cross section is much smaller than that of the beams.
(a) What would now be the maximum shearing force F that the braced frame can withstand?
(b) What would be the maximum F that the frame can withstand if instead of cables it is braced by steel rods of diameter 1.6 mm?
(Hints: Ignore buckling effects. When F is applied, the stretch of one diagonal rod is equal to the compression of the other. Those stretch and compression are so small, that it is ok to assume that the frame itself remains almost a perfect square.)

Section 4.5

27. In a zoo, a big aquarium has a side wall with small glass windows, through which visitors can watch the fish. A circular glass window of radius 5 cm is located in that wall 3.5m below the water's surface. (a) What is the water pressure on that window? (b) What is the total force that the water exerts on that window?

28. A reservoir at the top of a high-rise building is directly connected to faucets in floors below. The water level in the reservoir is 95 m above street level. What would be the pressure at a closed faucet (a) 82 m above street level? (b) 6 m above street level?

29. What is the pressure in pascal of (a) 80 mm of mercury? (b) 760 mm of mercury? (c) of 10m of water? (d) 0.8m of water?

30. What should be the minimum pressure P inside a beer keg, so that the beer gets to the mouth of the spigot, 0.8 m above the beer level? (Assume that the beer's density is 980 kg/m^3).
(Usually, that pressure is provided by gasses that the beer releases. If that is not enough, pumping in external air provides additional pressure).

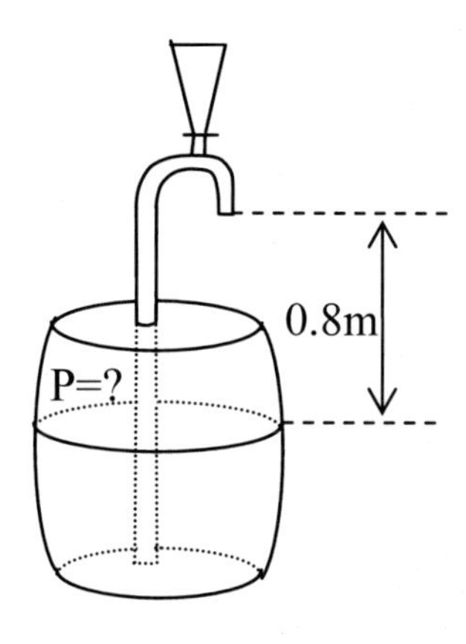

5. LATERAL FORCES

5.1 Wind effects

5.1.1 Wind flow

Wind, which is flowing air, is one of the factors that have to be considered when designing buildings. Buildings affect the flow of air around them, and at the same time winds exert forces on the buildings. To study the effects that a building might have on the flow of air around it, reduced models of the building and its surroundings are often placed in wind tunnels. There, tiny particles, such as smoke, are injected into streams of air that simulate a variety of wind conditions. The patterns of the smoke in these streams make it possible to visualize patterns of wind flow around the building.

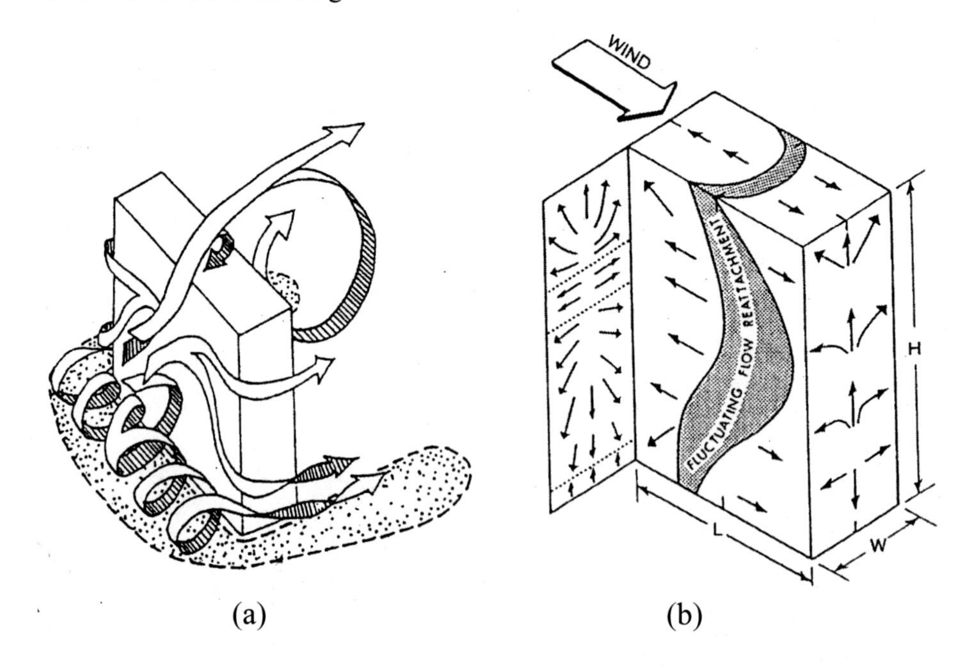

Figure 5.1: Schematics of air flow around buildings. (a) Patterns of air flow around rectangular buildings. Dotted area indicates strong surface wind. (b) Flow patterns on the surface of a building. Dotted area indicates fluctuating flow.©ASHRAE

Figure 5.1 illustrates patterns of air flow around a rectangular building. The wind approaches the front of the building in an orderly laminar flow, parallel to the ground. Usually, wind speed increases with elevation. The building diverts that flow (figure 5.1 (a)). The part that is diverted down forms turbulence on the ground. That turbulence flows around the building to its downwind side. The part that is diverted up forms turbulence around the top edge of the building before it moves to the downwind side. Usually, there are also regions of relatively calmer air, at the back of the building and close to the roof. The exact flow patterns depend on the shape of the building and on the wind profile. Air flow patterns have to be addressed when designing chimneys and vents, because their exhausts should not be blown back to the building or to the street. Surface winds at the street level that are formed by high-rise buildings may become annoying or even dangerous to pedestrians. Special awning-like elements may have to be included in the design in order to eliminate them. There are practical design standards and codes that address those issues for common situations. In more unique cases, wind-tunnel experiments are needed to find the appropriate solution.

5.1.2 Wind forces.

'Fluid' is the general term used for gasses and liquids. An object that is moving with respect to a fluid experiences forces due to that motion. These forces depend on the relative motion between the object and the fluid. For example, when we stand in the wind we feel its force on our face. In this case, we are at rest, and the fluid around us is moving. We also feel the force of the wind when we ride on a bike in calm air. In this case, the force that we feel is due to our motion with respect to a stationary fluid.

Two kinds of forces act on an object that moves with respect to a fluid. The first force, the drag, is acting in the direction of the motion. An example of air drag is the force that opposes the motion of any object in air. Moving cars, bicycles, airplanes, and any object that moves in air has to overcome that drag force in order to keep moving. The second force acts in the transverse direction to the direction of the motion. This force is also called 'lift', because it is the force that acts on the wings of airplanes, and lifts them up. Figure 5.2 shows the forces that act on a stationary object due to the flow of air (wind).

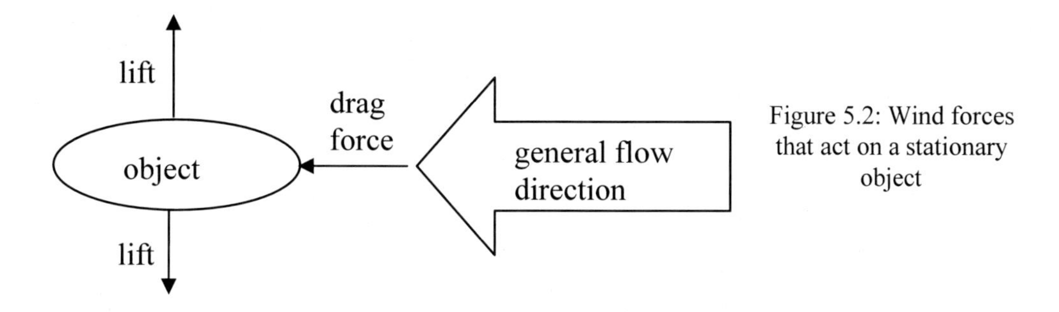

Figure 5.2: Wind forces that act on a stationary object

Drag forces

The drag force is similar to friction force, because both oppose the motion. However, there are some significant differences between the two. Unlike friction, the drag force depends on the relative velocity between the object and the fluid. The drag force increases with that velocity. For small velocities, the drag force

(D.F) is proportional to the velocity (v): $D.F. \propto v$. As the velocity increases beyond a critical point, the drag force becomes proportional to the velocity squared: $D.F. \propto v^2$. The transition between the two regions occurs when the flow of the fluid around the object changes its nature from laminar flow to turbulent flow. In both regions the drag force is proportional to the cross sectional area of the object that faces the flow (e.g. the wind). Unlike friction, the drag force depends on the shape of the entire object. So, two objects that have the same cross section but different shapes may be subject to different drag forces. The drag force depends also on the smoothness of the object's surfaces and on the nature of the fluid.

A number of strategies are used in order to reduce the drag force. One strategy is to reduce the effective cross section of the object. Skiers and bike riders know that they have to keep a 'low profile' in order to reduce the resistance of the air to their motion. A low profile reduces the object's cross-section. Another strategy is to keep the drag force in the linear region, where it is proportional only to v and not to v^2. In order to keep the drag in the linear region, the object has to be aerodynamically streamlined. The flow of air around it should be turbulence free. Turbulence is the main cause of the v^2 dependency of the drag force. Those strategies are sometimes employed when designing bridges and high-rise buildings, where it is essential to reduce drag forces due to winds. The profiles of these structures are optimized using wind tunnel simulations. They retain their intended functionality, and at the same time they create the smallest possible drag forces.

Figure 5.3: The top of the cab is equipped with a curved deflector. It reduces the amount of turbulence at the front of the trailer, by streamlining the air flow in this region, thus reducing drag forces.

Lift forces

A stream of air that flows tangentially over the surface of an object tends to suck that object. This is due to the lift force of the air stream. You can verify this fact by a simple experiment, illustrated in Figure 5.4. Hold a paper in two hands below your lips, as shown in Figure 5.4 (a). The paper will dangle down. Now blow air from your mouth straight forward. The paper will be sucked up to the air stream (part (b)). Please note that the air that you blew did not drag the paper. Dragging, or pushing, would happen when you blow the air directly on the paper, e.g. if you hold the paper at your eyes level, and blow straight on the dangling paper.

A simple formula, which is a special case of the more general Bernoulli's Law, gives the relationship between P, the lift (suction) pressure, and v, the tangential velocity of the air over the surface. The formula is:

$$P = \frac{1}{2}\rho v^2 \qquad [5.1]$$

where ρ is the density of air, which at sea level is approximately 1.2 kg/m^3. This relationship holds for non-turbulent flows.

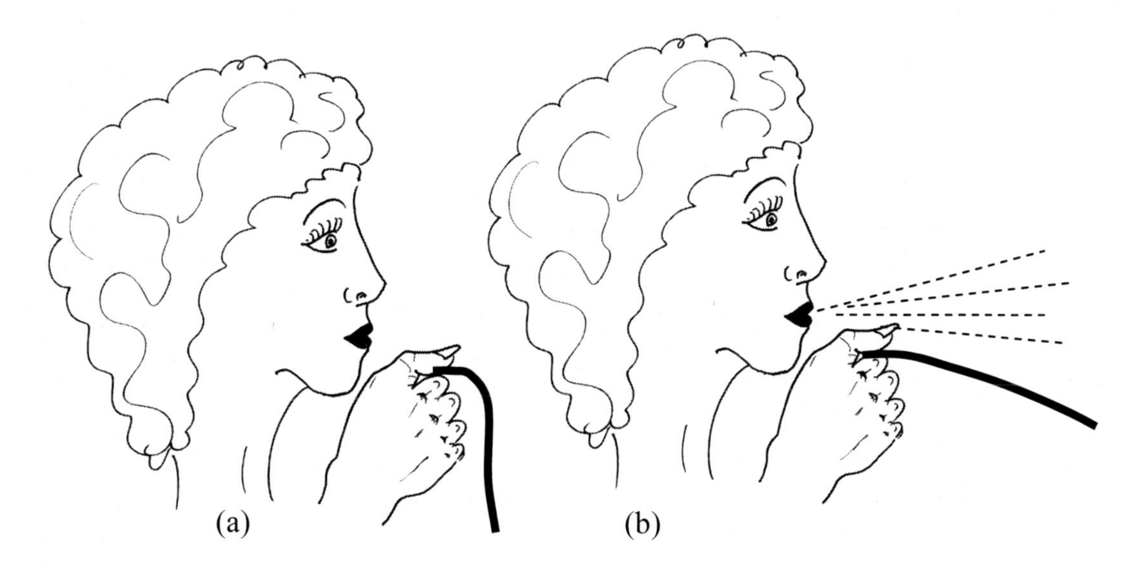

Figure 5.4: (a) Before blowing air. (b) Blowing air above a sheet of paper causes it to lift, due the lift forces of the air stream.

Example

A small warehouse has a flat roof whose area is $100m^2$. A wind blows tangentially to the roof at 80 miles per hour (hurricane). Assuming laminar flow, calculate the lifting force exerted by the wind on the roof.

Solution

The wind's speed is 80 x 1610 /3600=35.8m/sec.

According to [5.1], the lift (suction) pressure would be $P=0.5 \times 1.2 \times 35.8^2 =$ 769 pascal.

The lifting force F is the lifting pressure times the area of the roof:

F=769 x 100 =76,900N. This is in the range of the weight of a truck.

A roof is usually designed to withstand downward forces. If a roof like that in the example above is not designed to withstand lift forces in the **upward** direction, it may fall apart in strong winds. If it stays intact, but its weight and anchors are not strong enough, it will be blown away.

Drag and lift forces act also on the walls. Drag forces push walls in, while lift forces suck them out. Turbulence effects on the walls and the roof may be greater than those predicted by [5.1]. To survive a hurricane, a structure will have to withstand all these forces.

Shingles may be blown away by lift forces of winds that are much slower than hurricane. Shingles have to be adequately fastened to the roof to prevent this from happening.

While lift forces in steady flows can be calculated by a simple formula [5.1], drag forces in steady and turbulent air flows do not usually have simple formulas. They are still important in determining wind loads of buildings. They are determined case by case based on empirical information and wind tunnel measurements.

Lift forces of tangential winds have also beneficial effects on building. When such winds blow across openings of vents and eaves they suck out stale air from enclosures inside the house, such as bathrooms and attics. Fresh air enters these enclosures through other narrow openings and replaces the exhausted stale air.

When a stream of air hits an object, the stream may split into two parts that move around the object, and then rejoin after passing it. This happens when a stream of air hits a wing of an airplane. Part of the stream flows above the wing, and another part flows below the wing, as shown in Figure 5.5. The air flow above the wing has greater velocity than the air flow under the wing, because of the special profile of the wing. The upper flow has to cover a larger distance than the lower flow, and therefore it is moving faster. Based on equation [5.1], the upward lift force above the wing is stronger than the downward lift force below the wing. The net lift force on the wing is the difference between the two lifts, and it points upwards. This net lift force is the force that lifts the airplane, and keeps it up.

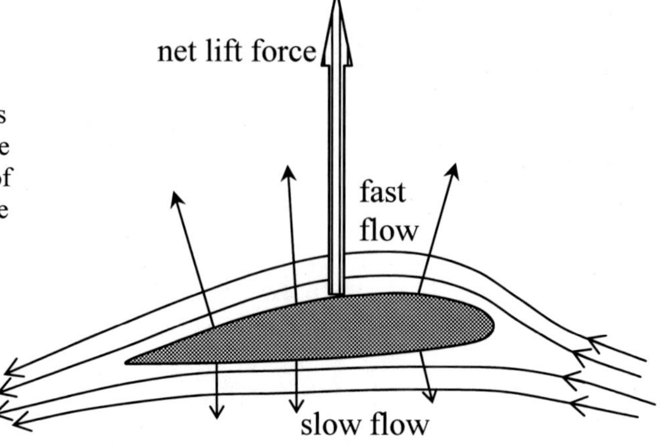

Figure 5.5: Air flow around an airplane's wing (cut view). Air flow above the wing is faster than the flow below it. The sum of the upwards lift forces is greater than the sum of the downwards lift forces (thin arrows). The result is an upwards net lift force.

High-rise buildings can be viewed as vertical wings. They split an oncoming wind into streams that flow around the building. Like in an airplane's wing, each flow will have its lift and drag forces. Those lifts, which are in opposite directions to each other, would have horizontal components. If they have different magnitudes due to asymmetry in the building or to the way that the wind hits the building, a net horizontal force will result. Torques may develop due to drag and lift forces, and depending on the particular circumstances, they may cause shear and torsion strains in the building. Due to the elasticity of the building, gusts of wind may trigger oscillations that continue even after the gust has passed.

Turbulences are often generated in the wake of wind-flows around buildings. Such turbulences may have their own frequencies, even if the wind is steady. If the frequency of a turbulence is in resonance with a natural frequency of the building, the building may enter into undesired resonance.

High-rise buildings are designed to withstand these various lateral effects of winds. They have to be stiffened against wind-generated lateral shear and torsion forces. In general, the amount of steel needed to attain such stiffening increases in a non-linear way with the height of the building. In high-rise buildings, as the height of the building increases, more steel is needed to stiffen the structure than to support its weight.

Tops of high-rise buildings may sway in the wind for several feet, with time periods of a few seconds. To make living and working there possible, these sways must not exceed a certain level. In addition to stiffening the structure, elements that damp oscillations may be included in the structure. These elements may be passive or dynamic. Dynamic elements adjust their response to sway based upon real time data that are sensed by sensors, processed by a computer, and activate feedback mechanisms.

One common way of strengthening high rise buildings against earthquakes and wind shear forces is by bracing their exterior tube with diagonals. The braces at each face of the tube form a two dimensional truss whose major role is to withstand the shear forces. Sometimes, those two dimensional trusses share also in supporting the gravitational loads. Various patterns of braces are being used, depending on the specific circumstances of each structure. Quite often, the diagonal braces are hidden under the skin of the building, mostly for aesthetical considerations. For example, Citigroup Center (page 66) has such hidden braces.

5.2 Earthquakes

5.2.1 Plate tectonics

If you could drill deep at any point on Earth, eventually you would reach magma, which is hot molten rock. That would occur at depths of between 15 and 75 km. This may seem a lot because even oil drilling equipment cannot reach that deep. However, relative to the radius of the Earth, which is approximately 6,400 km, even 75 km is really just a small scratch on the surface. It can be said that the crust of the Earth is a thin layer floating on magma.

The Earth's crust is not an intact layer. Rather, it is made of a number of segments, called **tectonic plates**. Those plates, which float on the magma, are in tight contact with each other. Figure 5.6 shows the major tectonic plates of the Earth. The major plates have sub-regions that are separated by fault lines. In some cases, visible fault lines on the ground mark the underlying boundary between tectonic plates or sub plates. One example is the San Andreas Fault in California, which is found over the boundary between the North American and the Pacific plates.

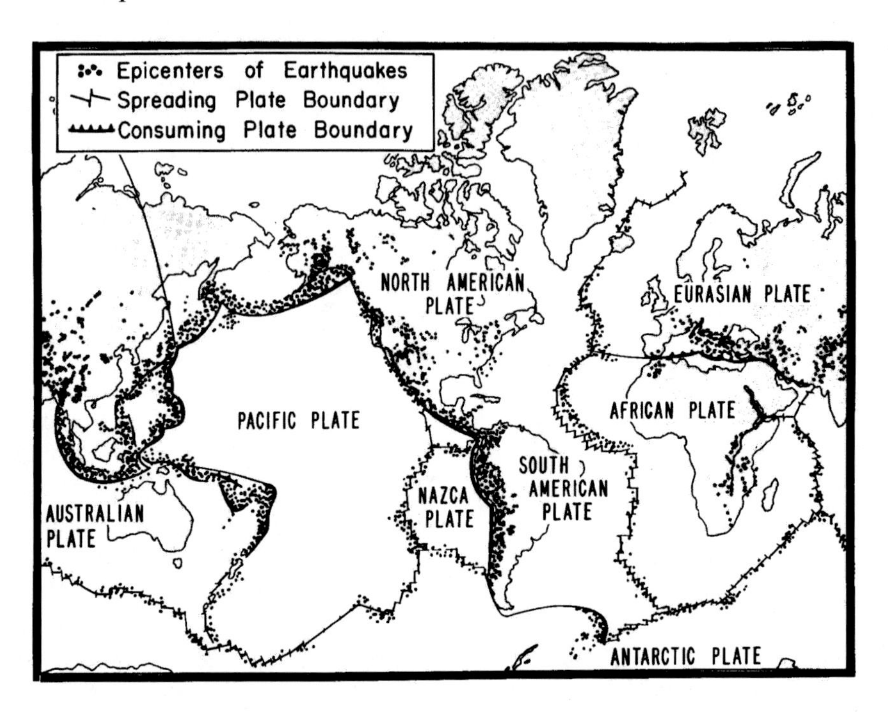

Figure 5.6: World map of major tectonic plates and seismic zones.
Epicenters of earthquakes are spread along boundaries of adjoining plates.
(USGS)

Each plate floats at its own velocity on the magma. The velocity of the North American Plate near San Francisco was measured to be approximately 5 cm per year in the North-Northwest direction.

Some plates are slightly riding over their neighbors. This happens at what is called consuming boundaries. In consuming boundaries, the edge of the lower plate is pushed down into the magma, where it melts. Other plates slide horizontally against their neighbors, and others are moving away from their neighbors. The latter are called spreading boundaries. The gaps that open at spreading boundaries, which occur frequently in plates at the bottom of the oceans, are filled from below by hot magma that solidifies into under-water mountain ridges. This is how the Mid-Atlantic and the Mid-Pacific ridges were formed.

Figure 5.7
Surface deformations.

Loma Prieta earthquake, 1989,

(USGS) Photograph by J.K. Nakata

The motion of tectonic plates against each other is not smooth. They exert forces on each other, and these forces strain the boundaries of the plates. Occasionally, the strain at a certain point would pass a critical value. Sudden violent motion of the plate at that point erupts, similar to the sudden release of a compressed spring. Because of the elasticity of the plates, this sudden release propagates in all directions as vibrations, which are called seismic waves. These vibrations reach the surface of the Earth, where they are felt as an earthquake.

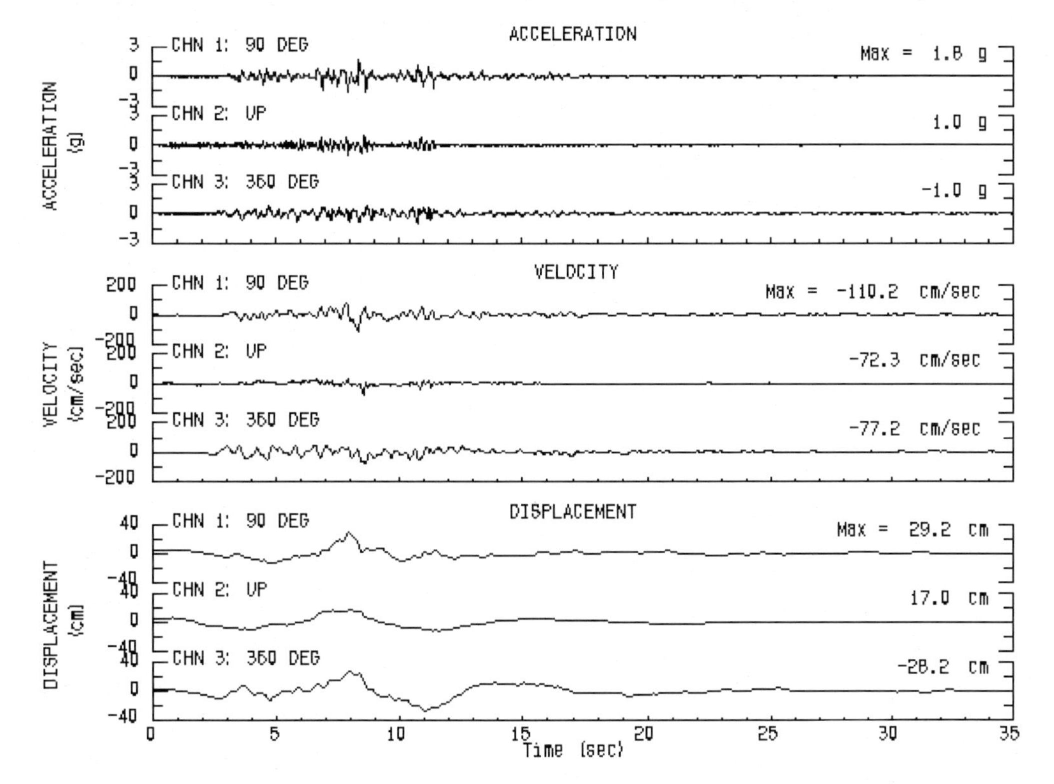

Figure 5.8: Seismographic data of Northridge 1994 earthquake, Tarzana station.
(USGS)

The point where the original release occurs is called the focus of the earthquake. The point on the surface of the earth right above that focus is called the epicenter.

Earthquakes often accompany volcanic eruptions, which are caused by upward flows of magma. These earthquakes are due to sudden movements of the crust in the vicinity of the volcano.

Earthquakes can last from a few seconds up to minutes. The amplitude of the ground's displacements may range from few millimeters up to a meter or more. After an earthquake subsides, the deformed landscape provides a vivid indication of the displacement of the underlying crust (Figure 5.7). Ground motion may have horizontal and vertical components. Figure 5.8 shows the recorded acceleration, velocity and displacement of the ground in the various directions of one earthquake.

5.2.2 Interaction of earthquakes with structures

While most earthquakes originate in solid rock, as their seismic waves propagate and reach the ground, they interact with various kinds of soil that support the foundations of structures. Soils consisting of sand with high concentrations of water may provide adequate support to structures under normal conditions. However, vibrations in such soils cause them to liquefy. This is similar to what happens when wet soil is turned into mud when stepped upon for several times. Liquefied soil cannot support structures, and they topple over, many times as a whole (Figure 5.9). Liquefaction may occur also in landfills. In hilly terrain, earthquakes may cause landslides, especially when the soil is saturated with water.

Figure 5.9 Liquefatction

Niigata earthquake

(USGS/ National Geophysical Data Center)

Masses of soil may enter into resonance with the original seismic waves. As a result, the vibrations of these masses will be stronger than the vibrations of the seismic waves that cause them. Such a mass of soil behaves like a pile of jelly on a plate. Small vibrations of the plate may cause larger vibrations of the pile of jelly. That would happen if resonance occurs between the frequency of the plate and a normal frequency of the jelly. Also, like the vibration of the jelly, the vibration of certain soils may continue even after the vibrations of the underlying rock (the plate) have ended. Unlike the rocks underneath them, the internal structure of these soils does not have strong damping mechanisms that attenuate the vibrations.

The forces of an earthquake act directly on the foundations of buildings. Computer simulations and experimentations with scaled models can help in the analysis of how these forces would affect the rest of the structure. Here, we will address only the main factors that play a role in these interactions. The resilience of a structure to earthquakes depends on a number of factors: the characteristics of the earth's vibrations (intensity, frequencies, and duration); inertia of the structure and its parts; the flexibility, damping capabilities, and breaking points of key components of the structure; and the stability of the design.

The impact of an earthquake on a structure depends not only on the vibrations of the ground, but also on the inertia and elastic properties of the structure. Once we know the displacement of the ground as a function of time, we can calculate its velocity and acceleration as functions of time (e.g. Figure 5.8). We represent the shaking ground as a vibrator, whose displacement, velocity, and acceleration as function of time are independent of the structure. We then figure out the effects of those vibrations on the structure, based upon the features of the structure.

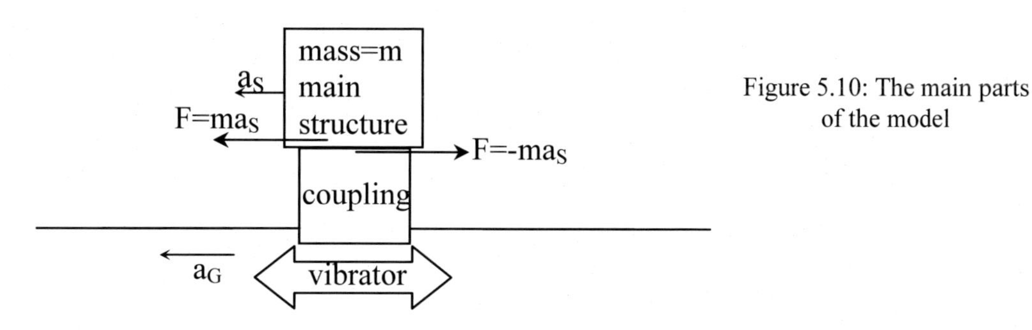

Figure 5.10: The main parts of the model

A simple model of the interaction between the earth's vibrations and a structure consists of a vibrator, the main structure, and a coupling element that connects the vibrator and the main structure (Figure 5.10). The vibrator represents the vibrations of the ground, which are completely independent of the structures on which they operate. The coupling is the part of the structure that relays the earth's vibrations to the rest of the structure. At any given time, the vibrator dictates the position of the end of the coupling to which it is connected. The other end of the coupling interacts with the main structure, and its position at any time depends on the vibration of the ground and on the response of the main structure.

A typical coupling consists of the columns of the ground floor in a multi-story building. They are anchored to the ground (the vibrator) on one side, and support the rest of the building on the other side. Quite often, the coupling is a critical part of the structure that might be the weakest link in the system (Figure 5.11).

At an arbitrary time during an earthquake, the ground would be moving with acceleration of $\mathbf{a}_G$ (Figure 5.10). The main part of the structure, whose mass is m, would be moving at an acceleration of $\mathbf{a}_S$. Based on Newton's Second Law, this is an indication that the force that the coupling has exerted on the main structure was $m\mathbf{a}_S$. According to Newton's Third Law, the reaction force that the main structure exerts on the coupling is

$$\mathbf{F}=-m\mathbf{a}_S. \quad [5.2]$$

From the 'point of view' of the coupling, this force, which acts on the coupling, is due to the inertia of the main structure. Therefore, this force is sometimes called the **force of inertia,** or the **inertial load,** of the main structure.

When the coupling consists of columns connected to a slab, the force of inertia pushes the slab sidewise, thus creating shear stresses that deflect the tops of the columns. If all the columns are identical and the weight is distributed evenly between them, these stresses will be distributed evenly between the columns. In certain situations, a slab is supported by columns of different heights. This happens, for example, when the foundation is on a slope of a hill, or if the columns originate at different floors. In such cases, the moving slab will deflect the tops of all columns by the same amount Δx. However, the shorter columns will experience more strain, because strain is $\Delta x/L$, where L is the length of the column. Therefore, the shorter columns will be the first to reach the critical strain and to break.

Figure 5.11: Shear Failure. The pictures show various degrees of failure of the coupling due to shear forces. The main of the structure is much less affected.

Loma Prieta earthquake, 1989, (USGS) Photographs by: top C.E. Meyer, bottom J.K. Nakata

We now discuss in more detail processes that take place in different kinds of couplings and main structures.

5.2.3 Rigid coupling and rigid main structure

In these situations, the coupling has no flexibility. Therefore, for the system to remain intact, the main structure and the coupling have to move in unison with the vibrating ground. If the acceleration of the ground is $\mathbf{a}_G$, the main structure has to move also with that same acceleration $\mathbf{a}_S=\mathbf{a}_G$. The force of inertia on the coupling would be $\mathbf{F}=-m\mathbf{a}_G$. If the junction between the coupling and the main structure can withstand this force, and if the coupling itself can withstand the shear and bending stresses, the main structure and the coupling would move in unison with the ground, and the structure will remain intact. Otherwise, the weakest link (the junction or the coupling) would break.

In reality, the coupling and the main structure will always have some flexibility. The "rigid" model is applicable to situations where the displacements of the ground are much larger than the elastic limit of the coupling. In such cases the main structure has to pretty much move in unison with the ground. For example, if a column will crack when its top is deflected by one inch with respect to its base, and the amplitude of the vibrations of the ground is one foot, the "rigid" approximation is justified. The motion of the main deviates only slightly from the motion of the ground. In such cases the column will fail should $m\mathbf{a}_G$ become greater than the critical force needed to shear it beyond its elastic limit.

5.2.4 Effects of structure's flexibility

All structures have some flexibility in them. In general, flexibility softens the effects of the earthquake's first jolt. After that, a flexible structure will wiggle in wavy shapes in response to the ground's vibrations. In some cases flexibility helps the structure, but in some cases it may work against it. Figure 5.12 illustrates how such wavy shapes are formed.

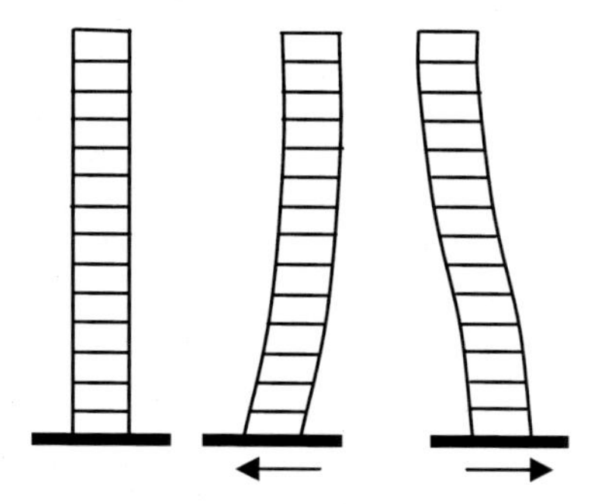

Figure 5.12: Swaying of a high-rise building in an earthquake (exaggerated). Arrows indicate direction of ground motion.

Any vibrating structure can be described by a superposition of fundamental waves, which are called the **normal modes** of the structure. At certain ground vibration frequencies, the building oscillates in only one of those normal modes. Those frequencies are called the **normal frequencies** of the structure. Each normal mode has its characteristic normal frequency. If the ground vibrates at several frequencies at the same time, or at a frequency that does not match any of the structure's normal frequencies, a superposition of several normal modes can describe how the structure vibrates.

When the ground vibrates at one of the normal frequencies, the structure enters into resonance: the amplitude of its vibration keeps growing with time. This is exactly the same situation that we discussed earlier in section 2.6.3, where Tiny Tina pushed Big Bob on a swing set. Here, the role of Tiny Tina is played by the vibrating ground, and the role of Big Bob by the structure. The structure may fail if the amplitude of its vibrations surpasses its tolerance limit. In the case of a swing set, friction forces act against the pushing of Tiny Tina. They set a limit on the amplitude of the swings of the swing set. In the case of earthquakes, internal damping forces, such as friction between moving elements of the structure, reduce the sway of the structure, and may keep it in the safe region.

The height of a building is one of the main factors that determine its normal frequencies, and through them its susceptibility to certain earthquake frequencies. Figure 5.13 illustrates the relationship between the number of stories in a building and the period T of its main normal mode. (Frequency f is related to period by f=1/T. Another useful entity, angular frequency, ω, is related to the period by $\omega = \frac{2\pi}{T}$)

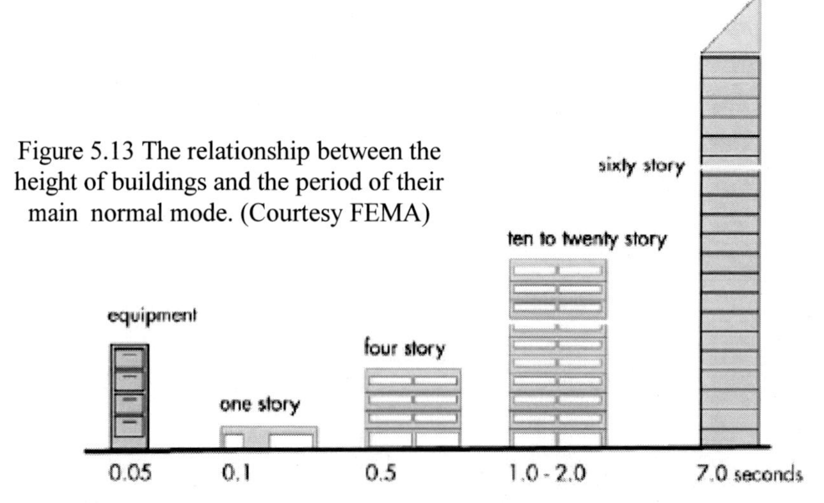

Figure 5.13 The relationship between the height of buildings and the period of their main normal mode. (Courtesy FEMA)

The ground vibrations of most earthquakes consist mainly of combinations of oscillations at periods of 0.5-2 seconds. That is why in certain earthquakes, such as the one that hit Mexico City in 1985, 4-16 story buildings collapsed, while much higher or lower buildings in the same neighborhood remained intact.

Earthquakes may destroy structures even if resonance conditions do not occur. Two obvious factors affect the severity of any earthquake: the amplitude of the ground's vibrations and their duration. If the amplitude is large enough, it may push the structure beyond its tolerance limit, regardless of the frequency. If the duration is too long, the materials of the structure may loose their strength and crumble due to fatigue and similar effects.

Although the frequency-response of complex structures can be found only by numerical computations and by experimentations with models, it is possible to use analytical methods in some idealized cases. One group of such cases is when the coupling of the structure is simple and elastic, and the main is rigid. (Figure 5.14 left.). Based on observations of damage to low-rise buildings (Figures 5.11) it appears that the ground floor behaved like an elastic coupling that eventually failed, and the main part of the buildings behaved like a rigid element that remained intact after all the vibrations. Modeling such buildings as a rigid main on an elastic coupling should give a reasonable description of the behavior of those structures. In addition to low-rise buildings, structures that consist of a rigid unit at the top of a long flexible "stem", such as certain water towers or one or two legged canopies at gas stations, may also be included in that category.

For the sake of the discussions, we will assume that the coupling undergoes shear deformation. The action-reaction shear forces that act on the top of the coupling and on the main are given by formula [4.37].

$$\mathbf{F} = -k' \cdot \Delta\mathbf{x} = -\frac{S \cdot A}{h} \cdot \Delta\mathbf{x} \qquad [4.37]$$

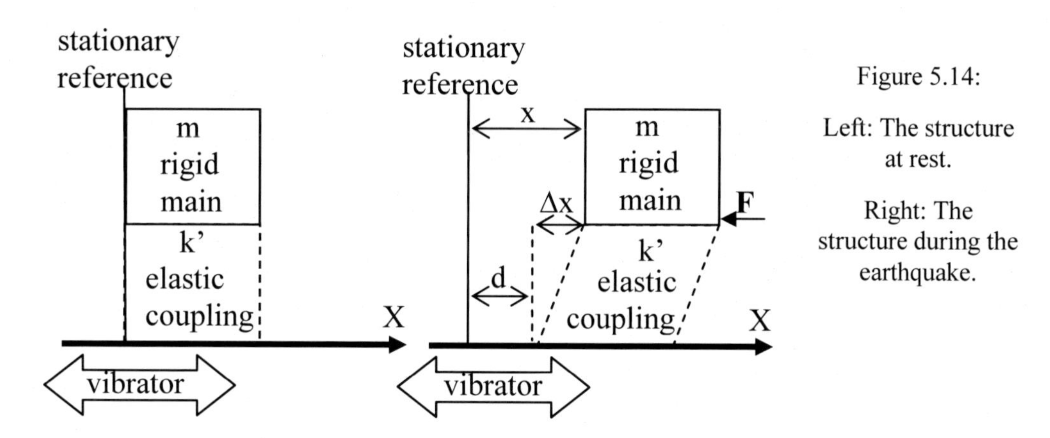

Figure 5.14:

Left: The structure at rest.

Right: The structure during the earthquake.

Here we use an x axis whose origin is at the bottom left corner of the structure, just before the earthquake starts. At any given time, d is the displacement of the ground, which is also that of the bottom of the coupling, x is the displacement of the top of the coupling, which is also the displacement of the main, and Δx is the shear displacement of the coupling: Δx=x-d (figure 5.14 right.). Δx is responsible for the shear force **F**. The shear elastic coefficient of the coupling is denoted by k'.

If the displacement of the ground is given by $d = -D\sin\omega t$, where D is the amplitude of the ground's oscillations, and $\omega = \frac{2\pi}{T}$, where T is the period of the ground's oscillations, then Newton's second law for the main becomes:

$$ma = -k'(x - D\sin\omega t)$$

or

$$a + \omega_0^2 x = \omega_0^2 D \sin\omega t \qquad [5.3]$$

where a is the acceleration of the main, and $\omega_0 = \sqrt{\frac{k'}{m}}$

Equation [5.3] can be expressed as a differential equation with the initial conditions that at t=0 the main is at rest and at the origin. The solution of this equation depends on the relationship between ω and ω_0:

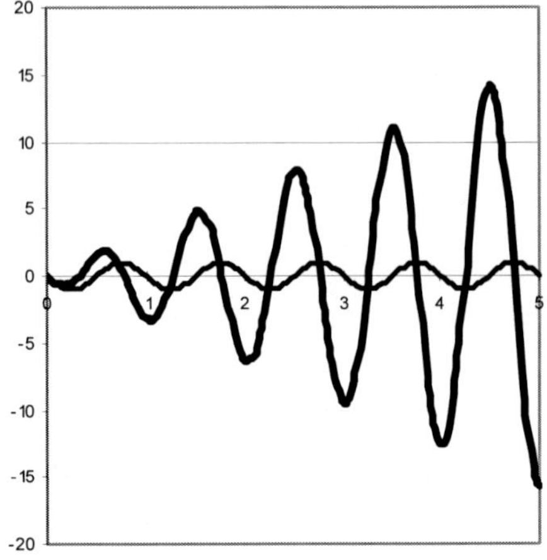

Figure 5.14a. The buildup of Δx as a function of time in resonance conditions [5.5]. The ground's oscillations have a constant amplitude of 1 (thin line). In 5 cycles, Δx (thick line) grows to more than 15 times the ground's amplitude.

(a) For $\omega \neq \omega_0$:

$$x = \frac{\omega_0^2 D}{\omega_0^2 - \omega^2}(\sin\omega t - \frac{\omega}{\omega_0}\sin\omega_0 t)$$

$$\Delta x = \frac{\omega_0^2 D}{\omega_0^2 - \omega^2}(\sin\omega t - \frac{\omega}{\omega_0}\sin\omega_0 t) - D\sin\omega t \qquad [5.4]$$

This describes oscillations within fixed boundaries. The closer are ω and ω_0 to each other, the larger is the amplitude of the oscillations.

(b) For $\omega = \omega_0$ (resonance conditions):

$$x = \frac{D}{2}(\sin\omega t - \omega t \cdot \cos\omega t)$$

$$\Delta x = \frac{D}{2}(-\sin\omega t - \omega t \cdot \cos\omega t) \qquad [5.5]$$

This describes oscillations with amplitude that increased linearly with time plus an oscillation with a fixed amplitude (figure 5.14a). When $|\cos\omega t|=1$, meaning for an integer number of half cycles (n=t/T=integer, or n=t/T=half integer, where T is the natural period of the structure) we get

$$|\Delta x| = \frac{\pi \cdot D \cdot t}{T} = n\pi \cdot D \qquad [5.6]$$

If the shear strength of the coupling or its critical distortion are known, equation [5.6] provides an estimate of the survivabilty of a flexible coupling with a rigid main structure in undamped resonance situations.

Example

The ground floor of a four stories building serves as a garage. It does not have any walls, only load-bearing columns. If the top of the columns move 5 cm with respect to their bottom, the garage would collapse. The overall elastic shear constant of the garage is 180,000 N/m, the mass of the floors above it is 12,000 kg. Will the garage collapse in a tremor that resonates with it, if the amplitude of the ground motion is 1.5 cm and it lasts 10 seconds?

Solution

The normal period of the structure is [2.17a] $T = 2\pi\sqrt{\dfrac{12{,}000}{180{,}000}} = 1.6\text{s}$.

Using [5.6]: t=10 s, D=0.015m, Δx=?, yields Δx=0.047 m. The garage will barely survive, because its critical distortion is 0.05 cm.

The duration of the vibrations is an important factor in the damage that they

Figure 5.15: Effects of shaking.

A brick wall disintegrated completeley due to shaking (above), while a reinforced highway support column (right) was not affected by it. The amount of the shaking of the column could be seen by the enlargement of the hole in the ground. The debris are of the roadbed that collapsed.

Loma Prieta earthquake, 1989

(USGS) Photographs by J.K.Nakata

may cause. Even if resonance conditions exist, it takes a number of cycles for the amplitude of the structure's vibration to reach its breaking point. If the earthquake ends before that happens, the structure will survive. Prolonged shaking may cause crumbling even without resonance. The reinforcement of the structural element may determine its survivability (Figure 5.15).

In some cases, an earthquake may shift elements of a structure, even without breaking them. That may take away the support and stability that these shifted elements provide to other parts of the structure, and cause their collapse (Figure 5.16).

Figure 5.16: Collapsed portion of San Francisco-Oakland Bay Bridge.

Loma Prieta earthquake, 1989

(USGS) Photograph by C.E. Meyer

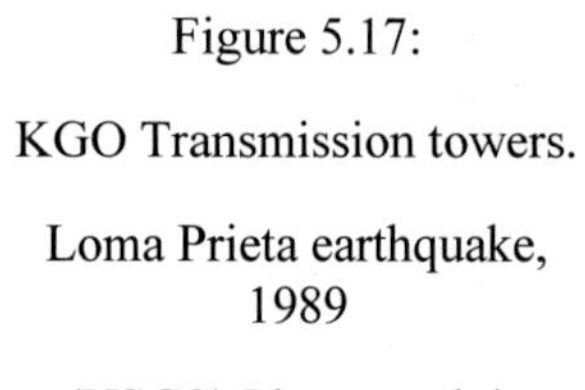

Figure 5.17:

KGO Transmission towers.

Loma Prieta earthquake, 1989

(USGS) Photograph by H.G. Wilshire

5.2.5 *Toppling of structures*

A structure will topple as a whole should the torques due to the "inertia force" caused by the earthquake become stronger than the reaction torques that the structure can provide. All those torques may be calculated with respect to any arbitrary axis of rotation. For the structure to topple, the dominance of the "inertia torque" over the reaction torques has to be sustained long enough, so that the structure's center of gravity (c.o.g.) would move beyond the base area of the structure. The structure has to be internally strong enough to withstand all the stresses caused by the earthquake. Otherwise, it might crumble before it topples.

When we discussed the static equilibrium of structures, we dealt with the torque due to the force of gravity. The force of gravity acts on all the elements of the structure. However, to calculate the torque of the force of gravity (mg), we may assume that it is attached to the c.o.g. Similarly, we may assume that in cases of earthquakes the torque of the "inertia force" is also attached to the structure at its c.o.g. According to Newton's Second Law, that "inertia force", which acts at the c.o.g, would be $m \cdot a_S$, and if the structure is rigid ($\mathbf{a}_S=\mathbf{a}_G$), it would be equal to $m \cdot a_G$.

To illustrate the interplay between the different torques, consider a rigid ($m\mathbf{a}_S=m\mathbf{a}_G$), high, and narrow structure (e.g. a tower or a chimney), Figure 5.18. Both the force of gravity **mg** and the "inertia force" $m\mathbf{a}_S$ are attached to the structure at the c.o.g. If the solid ground is accelerating to the left, we can choose the arbitrary axis of rotation at the structure's lower right corner. The two torques are acting against each other. As long as the torque due to the "force of inertia" would be less than the torque due to the gravity, the structure will not rotate. With respect to the chosen axis of rotation, the force of gravity has an arm of $\mathbf{r}_1$ and the "force of inertia" has an arm of $\mathbf{r}_2$. The critical earthquake's acceleration, at which the inertia torque equals the static-weight torque, is given by

$$m \cdot g \cdot \mathbf{r}_1 = m \cdot a_S \cdot \mathbf{r}_2 \qquad \text{or} \qquad a_S = g\frac{r_1}{r_2} \qquad [5.7].$$

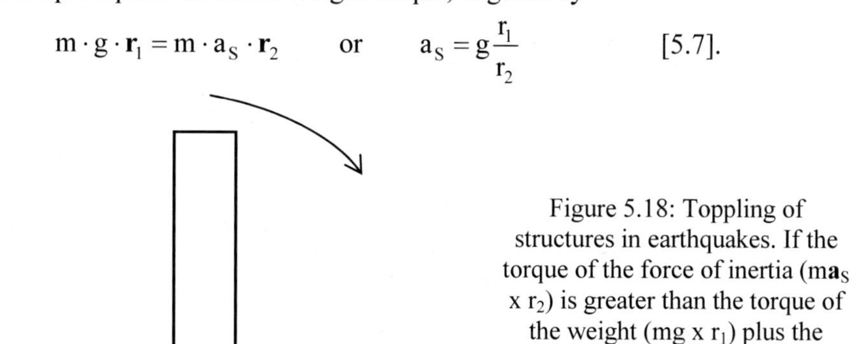

Figure 5.18: Toppling of structures in earthquakes. If the torque of the force of inertia ($m\mathbf{a}_S$ x r_2) is greater than the torque of the weight (mg x r_1) plus the torque of the tie-down forces (T x r_1), the structure will start to turn around the axis. If $\mathbf{a}_S$ lasts long enough, and if the structure does not crumble, it will topple.

If the earthquake's acceleration $\mathbf{a}_G$ surpasses that value of $\mathbf{a}_S$, the structure will start to turn around the axis. If there are additional tie down forces (denoted as T in Figure 5.18), their contribution should be added to the torque of the force of gravity.

Earthquakes are mighty. Nonetheless, understanding their nature has enabled engineers to design earthquake resistant structures, and to reinforce existing ones, as needed. The adverse effects of the horizontal inertia forces, that cause shear and bending stresses, can be countered by bracing frames with diagonal beams and by boxing them by sheets and creating shear walls. Strengthening the corners of open frames and the connections between columns and floor-slabs also helps to stiffen the structure. The height of the building plays an important role is choosing its appropriate earthquake-proofing strategy. Flexibility can be used to soften the impact of the vibrations. However, resonance conditions have to be considered very carefully.

There are modern designs in which the entire structure can slide on its foundation. In this way, it is disengaged from the vibrations of the earthquake.

Economic considerations are important in choosing the earthquake-proofing strategy of a structure. Quite often, non-critical parts are 'sacrificed' for the sake of the economical feasibility and the overall safety of the entire structure. These parts are designed to attenuate the vibrations by absorbing energy, even if they are being damaged in the process. Architects implement building codes that combine economic and communal considerations with risk assessments.

5.3 Lateral effects of hydrostatic forces

As was discussed earlier (section 4.5), a liquid exerts hydrostatic pressure on the walls of its container. That pressure is given by:

$$P_{HS} = \rho \cdot g \cdot h \qquad [4.40]$$

Where P_{HS} is the hydrostatic pressure at a point which is at a depth h below the surface of the liquid. ρ is the density of the liquid.

The hydrostatic pressure acts in the horizontal directions on dams. Figure 5.19 illustrates how that pressure changes linearly, from zero at the water surface to its maximum value at the base of the reservoir.

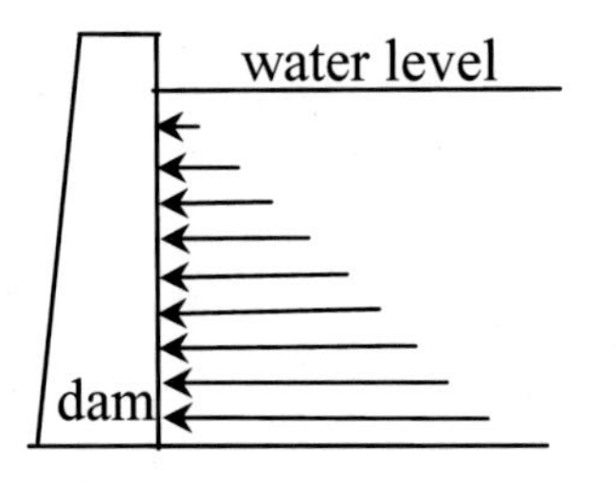

Figure 5.19: Variation of the hydrostatic pressure on a dam with water-depth (arbitrary scale).

Retaining walls are also affected by horizontal pressures exerted by the soil that they retain. Like hydrostatic pressure, soil pressure increases with the depth of the point below the surface. The exact relationship between the depth and the horizontal pressure is usually not linear. However, for many practical applications it is customary to approximate soil horizontal pressure by formula [4.40]. The density ρ, which is used in those cases, is a fraction of the density of the soil. The exact value of the fraction depends on the soil and its nature. In this section, both hydrostatic and soil pressure will be referred to as hydrostatic pressure.

After making sure that each small element of the dam (or the retaining wall) can withstand the stress exerted on it, the effects of the hydrostatic pressure on the structure as a whole have to be considered. Two main factors have to be figured out. First, the total translational horizontal force, which tends to push the dam horizontally. Second, the torque of the hydrostatic force, which tends to topple (rotate) the dam around its basis. This torque is affecting also the vertical pressure that the dam exerts on the soil underneath.

Since the hydrostatic force changes with the depth, surface elements of the dam at different depths will experience different forces. The total force and torque of the hydrostatic pressure will depend on the shape of the area of the dam. We will discuss in the following only rectangular dams, of length L and height H. (see Figure 5.20). The area A of such a dam is A=L·H. Please note that **H here is the height (depth) of the water behind the dam, and not the height of the wall**. The total force F that acts on the dam depends on the area of the dam (H·L) and the average hydrostatic pressure. This force is given by:

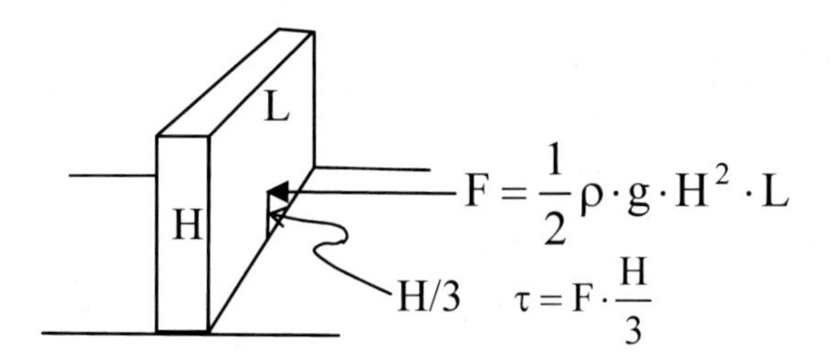

Figure 5.20: Total horizontal force and total torque around the basis, for a rectangular dam.

$$F=\frac{1}{2}\rho\cdot g\cdot H\cdot(H\cdot L)=\frac{1}{2}\rho\cdot g\cdot H^2\cdot L \qquad [5.8]$$

That means that the average pressure is equal to half the pressure at the toe of the dam. This average pressure acts on the whole area of the dam to generate the total force, as given by [5.8]

The total torque τ that acts on the dam, when the axis of rotation can be at any point of its basis (the bottom of the reservoir), is

$$\tau=F\cdot\frac{H}{3} \qquad [5.9]$$

where F is the average force given by [5.8]. That means that the arm on which the average force is acting is 1/3 of the height of the dam.

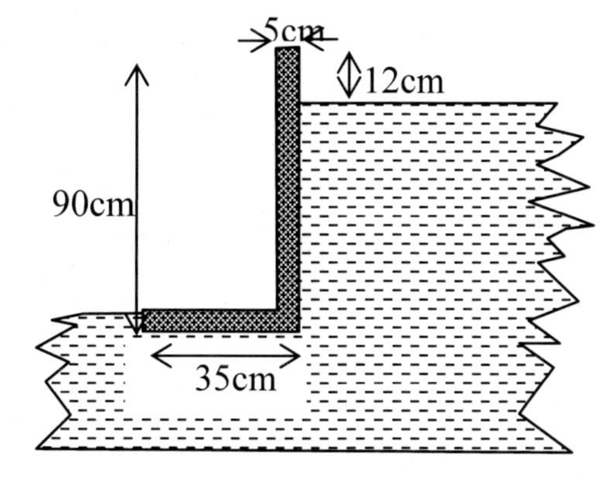

Figure 5.21

Example

Figure 5.21 shows a cross section of a retaining wall made of concrete. The thickness of the wall and its foot is 5 cm. The length of the wall is 3m. The density of the concrete is 2,300 kg/m^3. The density of the soil, which has to be taken in figuring the horizontal soil pressure is 800 kg/m^3. Find (a) the total horizontal force that act on the wall. (b) The total gravitational forces that act on the wall. (c) The torque of the horizontal forces. (d) The torque of the gravitational vertical forces. The particular building code requires that the total gravitational torque should be twice the horizontal torque. Is the building code met?

Solution

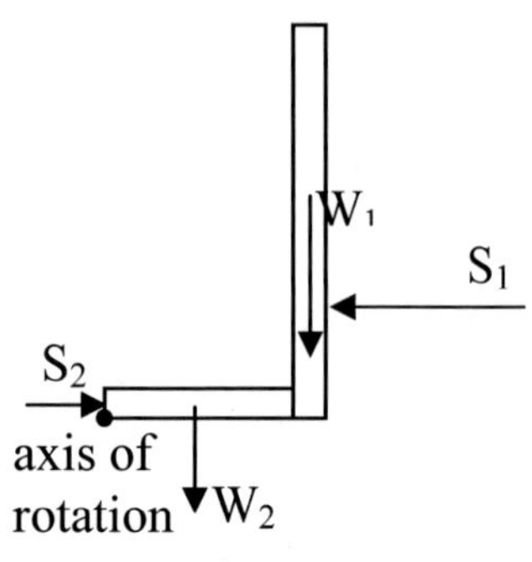

Figure 5.21a

In order to find the forces and torques that act on the wall we will divide it to two parts, as shown in Figure 5.21a. The weights of the parts are W_1 and W_2, and the side soil forces that act on the parts are S_1 and S_2 respectively. The axis of rotation is chosen at the bottom left end of the wall. Table 5.1 contains the details of the calculations of the weights and torques.

Based on the table we get:
The total horizontal force on the wall is 7,155-29=7,126N to the left.
The weight of the wall is 3,043+1,014=4,057N
The torque of the horizontal forces: 5,348-0.35=5,348 N·m counter clockwise.
The torque of the weights: -989-152=-1,141 N·m clockwise.
The torque of the horizontal forces is much greater than the opposing torque of the gravitational forces. This wall needs to be redesigned; it is at a risk of being toppled, or at least being rotated out of plumb, by the horizontal soil pressure.

Parameter \ part	1 (vertical wall)	2 (slab on the ground)
Volume (V), m^3	0.05x0.9x3=0.135	0.05x0.3x3=0.045
Weight (ρ.V.g), N	W_1=3,043	W_2=1,014
Torque of weight, N.m	-3,043x0.325=-989	-1,014x0.15=-152
Height of soil (H), m	0.90-0.12=0.78	0.05
Soil horizontal force ([5.8]), N	S_1= 7,155 to the left	S_2= 21 to the right
Soil torque ([5.9]), N.m	7,155x0.78/3=1,860	-29x0.05/3=-0.48~0

Table 5.1

5.4 Comments on lateral forces

We have discussed in this chapter three examples of lateral forces: wind forces, inertial loads that come up in earthquakes, and horizontal hydrostatic forces that act on dams and retaining walls. Each element of the structure has to be able to withstand those forces if and when they act on it. Any element should not fail under these forces. In addition, the structure as a whole has to be able to withstand them. When considering the structure as a whole, both the total magnitude of the lateral force and its torque have to be considered. The total inertial load of a structure does not depend on the shape of the structure. It depends only on the mass of the structure and on the acceleration of the ground. On the other hand, the total of the wind forces that acts on a structure depends on the shape of the structure. The same is true for lateral hydrostatic forces. Their total depends on the shape of the structure.

In order to find the torque of the total lateral force it is needed to know its point of attachment to the structure. The point of attachment of the inertial load is always at the center of gravity of the structure. That is because both the inertia and the weight of an object are proportional to its mass. The points of attachment of wind and horizontal-hydrostatic-forces depend on the shape of the object. The point of attachment in the latter cases may be other than the center of gravity.

PROBLEMS

Section 5.1

1. Winds of 50 mph happen frequently around high-rise buildings. What is the total lift force that such a wind exerts on a 2'x3' glass window? What is the weight of this pane, if its width is ¼ inch? The density of glass is 2.5 x 10^3kg/m^3.

2. A shingle of dimensions 8"x16" has to withstand the lift force created by winds of up to 80 miles per hour.
(a) What is the total lift force (in N) created by such a wind on the shingle?
(b) The shingle is held in place by two adhesive strips, each 1" wide (see figure). What is the minimal "adhesive stress" (in N/m^2) that the strips should provide to counter the lift force? Consider the weight of the shingle as a safety factor. It does not count in the balance of the forces.

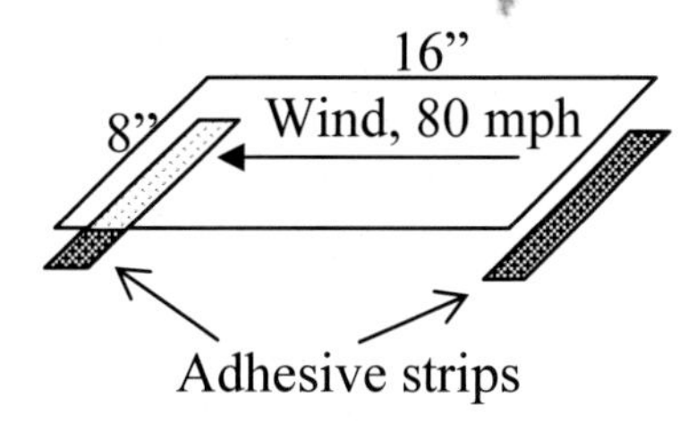

3. The surface area of an airplane wing is 35 m^2. The average velocity of the air flow above the wing is 230 mph, and that of the air flow below the wing is 190 mph. What is the lift force on the wing?

*

The Bank of China Tower in Hong Kong (Figure 5.22) is built in area prone to earthquakes and typhoons. Its architect I M Pei and its structural engineer Leslie Robertson applied an innovative approach. As for strength, the diagonal braces and the vertical columns were integrated to one frame of a three dimensional space-truss. The diagonal braces are not restricted to the exterior tube of the structure; some penetrate the inner space of that tube. Each member of the space-truss shares in supporting the vertical and lateral loads. As for aesthetics, the diagonal braces are emphasized on the skin of the building, rather than being hidden, contributing to a motif that symbolizes the client: a growing bamboo plant.

The building consists of four quadrants, each shaped as a triangular prism. The prisms that have truncated tops rise to different heights. Five columns define the edges of the prisms. Four columns rise from the corners of the square base of the building. A fifth column rises at the center of the building from the tip of the lowest prism, at the twenty fifth floor, to the tip of the

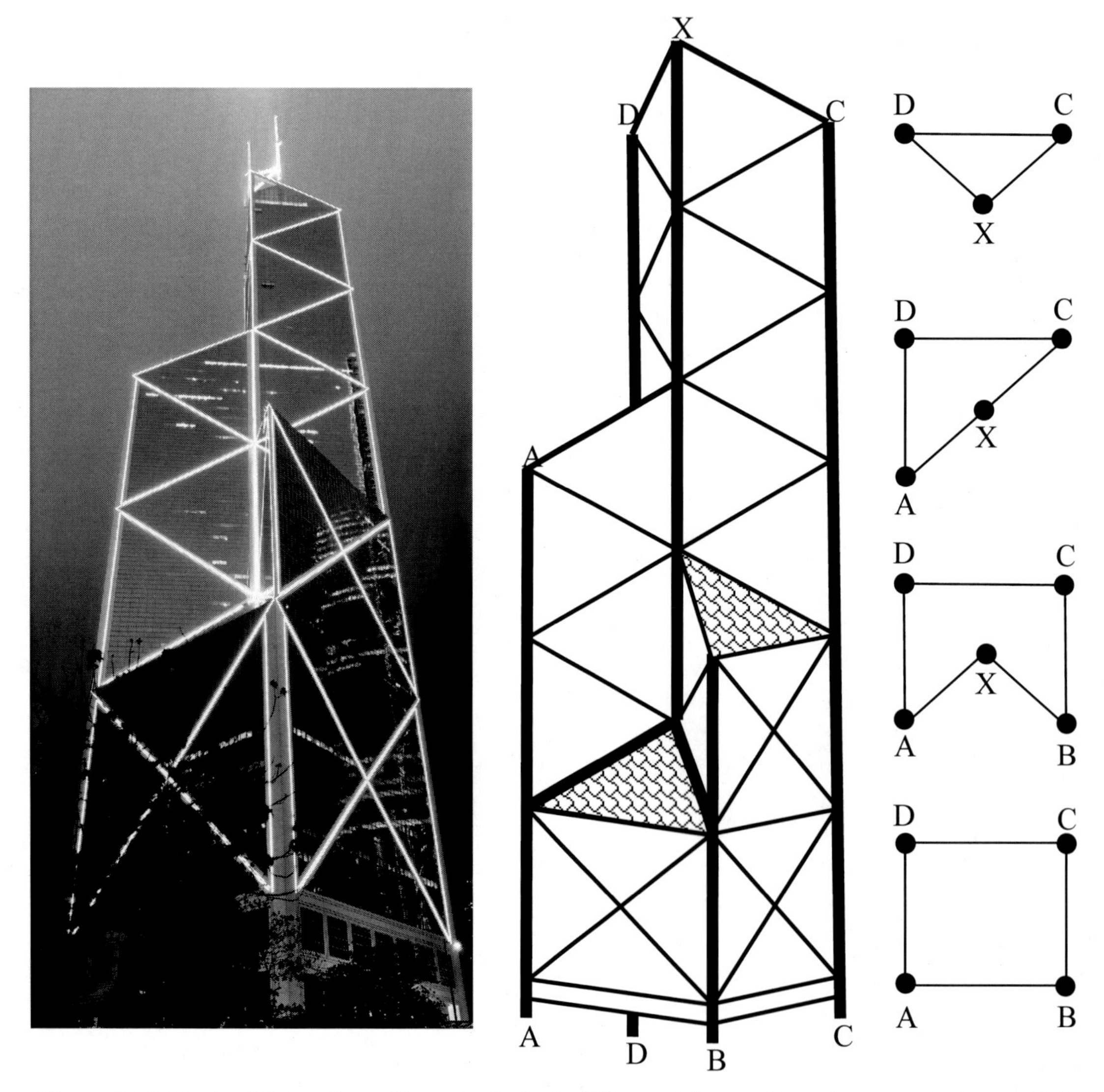

Figure 5.22: Bank of China, Hong Kong. Left; the building as it is seen from the ground . Center: Schematics of the space-truss frame of the structure, viewed from the same side as the picture. The four corner columns A, B, C, and D and the center column X are indicated by heavy lines. The two shaded triangles indicate the tops of two quadrants. Right: Floor plans at different heights. The highest suction pressures occur around the tops of the three lower quadrants.

building, at the seventy sixth floor. The central column serves as an edge to three prisms. Diagonal braces zigzag from the side columns to the center column. The rectangular bases of the prisms, which form the outside walls of the building, are braced with crossed diagonals. In addition, transverse trusses wrap around the building at selected levels.
The strongest shear forces in an earthquake-jolt are experiences by the lower part of the building, which supports the weight of the entire building. To provide the necessary shear resistance at the lower parts of the building, the columns are connected at the fourth floor to a system of interior composite walls, consisting of steel plates and concrete slabs, surrounded by slurry walls at the perimeter. The columns continue on to the foundations, to resist overturning moments.
It turned out that the cost of building this tower was significantly less than that of other high rises with comparable floor areas.

4. Windows of high-rise buildings have to withstand suction forces due to strong winds. Estimates of those forces are obtained by a combination of computational methods and model-simulations in wind tunnels. A model of a skyscraper is much smaller than the real thing, and the air speed in the wind tunnel is usually much smaller than that of a hurricane, or a typhoon as it is called in East Asia. Formulas are used to scale up the results obtained from wind tunnel models.

In wind tunnel testing of models of the Bank of China Tower in Hong-Kong it was found that suction pressure near the edges of the building are greater than those at the middle of the walls, and suction at certain corners were even higher than at the edges. Assume that for a tunnel wind-speed of 40 mph, the suction pressure at the center of an outter face of the building was 88 Pa, and at the tip of one of the quadrants it reached 160 Pa. Estimate the suction pressure at those points for wind speeds of 200 mph, assuming that but for the different speeds, the flow patterns of the wind are the same for both speeds.

Section 5.2

5. In a rigid three stories building, the weight of each story is 10,000lbs. Find the lateral force at the junction of the first and the second stories (the ceiling of the first and the floor the second) and at the junction of the second and third stories, when the earthquake's acceleration of the ground is 2m/s^2.

6. The ground-floor of a building serves as parking space. The inhabited (main) part of the building, above the parking area, rests on 9 columns that are fixed in the ground. The mass of the main part is 30,000 kg. The building is hit by an earthquake whose first jolt has an average horizontal acceleration of 0.8g. The structure was not damaged in the earthquake.
(a) What total force was exerted by the main part of the building on the tops of the columns during the first jolt?
(b) If the cross section of each column is a square of 10cm x 10 cm, what was the shear stress at the top of each column?

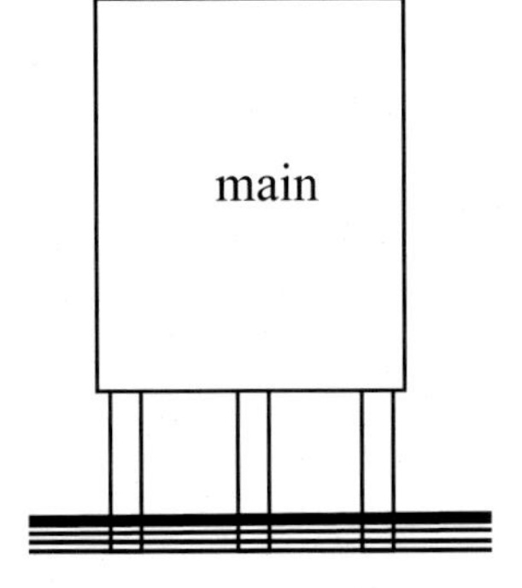

(c) The ground tremors of the aftershock of that earthquake may be approximated as a sinusoidal wave whose maximum horizontal displacement was 7 cm, and whose period was 6 seconds. The building, whose natural period is 2.5 seconds, did not enter into resonance. Assuming rigid structure, what was the maximum shear stress at the top of the columns during the aftershock? (Hint: consult section 2.6.3).

7 The building of the last problem was hit by a second set of aftershocks and entered to resonance. The amplitude of the aftershocks was 0.15 cm, and they lasted for 8 seconds. The structure can withstand 2 cm horizontal bending of the top of the columns. Did the building survive the aftershocks?

8. A cylindrical chimney stack of height 17m has a base diameter of 1.2m. The chimney is made of bricks held together by cement. Assume that the cement has zero tensile strength, and very large shear strength.
(a) What is the maximum horizontal acceleration that this chimney can tolerate in an earthquake jolt, so that cracks would not appear in its cement?
(b) What would be the answer should the shear strength of the cement be zero, and the tensile strength is large?

9. The diameter of the base of a cylindrical chimney is 6', its height is 30', and its weight is 1200lbs. What should be the smallest tie down force that will prevent its toppling in an earthquake whose ground acceleration is 2.5 m/s^2?

Section 5.3

10. Figure 5.23 shows a cross section of a retaining wall made of concrete. The thickness of the wall and its foot is 6 cm. The length of the wall is 4m. The density of the concrete is 2,300 kg/m^3. The density of the soil, which has to be taken in figuring the horizontal soil pressure is 800 kg/m^3. The density of the soil for figuring downwards pressure is 1,900 kg/m^3. Find (a) the total horizontal force that acts on the wall. (b) The total gravitational force that acts on the wall (c) The torque of the horizontal forces. (d) The torque of the gravitational vertical forces. The particular building code requires that the total gravitational torque should be twice the horizontal torque. Is the building code met?

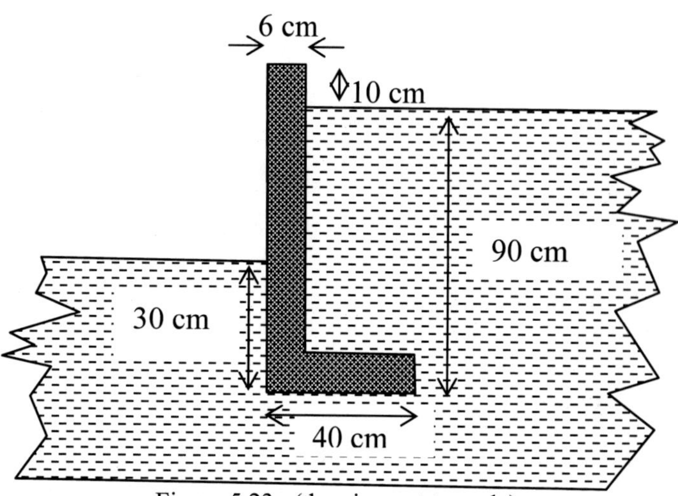

Figure 5.23: (drawing not to scale).

6. WORK, ENERGY, POWER, MOMENTUM

A: Why do we need complex concepts?
B: To simplify things….

In the previous chapters, the motion of objects was described by dealing directly with the fundamental concepts of displacement, velocity, acceleration, force, and mass. In this chapter, we will see how by defining new concepts, based on those fundamental ones, a different way of treating moving objects emerges. This new way simplifies in many cases the treatment of complex situations. The new concepts are work, energy, momentum and the laws of conservation of energy and conservation of momentum. First, we introduce the new concepts and outline the motivation behind them. Then, we show how those concepts are put to work in analyzing complex situations.

6.1 Work and kinetic energy

Consider the situation in which a block of mass m is pulled across a horizontal frictionless plane by a force **F**. The force makes an angle α with the direction of the displacement, which is the horizontal direction in this case (figure 6.1).

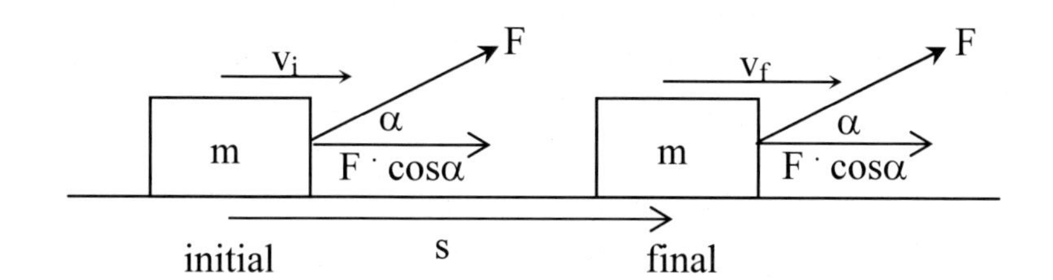

Figure 6.1: A block being pulled on a horizontal surface.

According to Newton's Second Law, the acceleration a of the object could be determined by F_S – the component of the force that acts in the direction of the motion:

$$F_S = m \cdot a \qquad [6.1a]$$

where $F_S = F \cdot \cos\alpha$. F_S is also called the **parallel force**, because it is parallel to the direction of the displacement. Since the object does not move in the vertical direction, the sum of the forces in that direction must be zero.

Equation [6.1a] enables us to find the acceleration a, and from it we can calculate other parameters of the motion. For example, if we know the initial velocity v_i and the distance traveled s, we can calculate the final velocity v_f by using equations of section 2.2.5. From equation [2.8] we can express a as:

$a = \frac{v_f^2 - v_i^2}{2 \cdot s}$ (We renamed v_f=v, and s=(x-x_i)).

By substituting this a in [6.1a], and re-arranging the terms we get:

$$F_S \cdot s = \frac{1}{2} m \cdot v_f^2 - \frac{1}{2} m \cdot v_i^2 \qquad [6.1b]$$

Equations [6.1a] and [6.1b] express the same idea (Newton's Second Law for the circumstance described in figure 6.1), but they use different variables to do it. The terms in [6.1b] were given names: F_S.s is called 'the work W done by the force F on the object' (the box in this case), or the work invested. The unit of work in the SI system is the joule: 1 joule = 1 newton x 1 meter. In the British system, the unit of work is the pound foot. 1 pound foot=1 pound x 1 foot.

$$W = F_S \cdot s \qquad [6.1c]$$

The general term $\frac{1}{2} m \cdot v^2$ is called the 'kinetic energy of the object' (KE).

$$KE = \frac{1}{2} m \cdot v^2 \qquad [6.1d]$$

Accordingly, the term $\frac{1}{2} m \cdot v_f^2$ is called the final kinetic energy of the object, and similarly, the term $\frac{1}{2} m \cdot v_i^2$ is called the initial kinetic energy of the object. In words, [6.1b] says: The work invested in the object equals to the change in its kinetic energy (final kinetic energy minus initial kinetic energy). Or, the work invested in the object was converted into kinetic energy. Again, the last two statements are rephrasing the outcomes of Newton's Second Law for the particular situation of figure 6.1.

Let a parallel force F_{S1} act on an object of mass m along a displacement s_1, and then, let another parallel force F_{S2} continue to act on the object along a displacement s_2, as shown in figure 6.2.

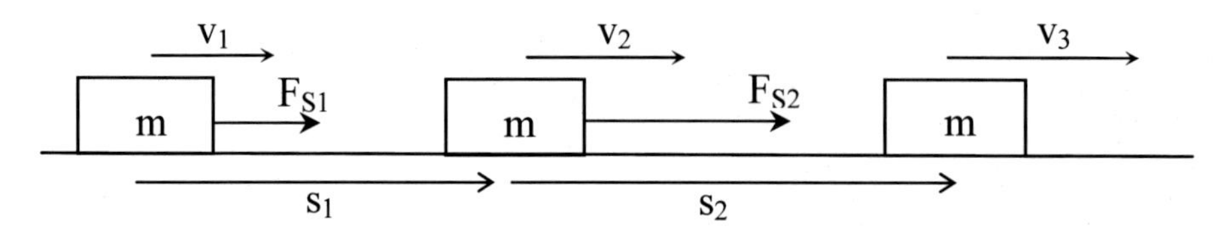

Figure 6.2: Work-energy relationship for a variable force

We want to express the relationship between work and kinetic energy for each of the forces and also for the entire process. From equation [6.1b] we get:

For F_{S1}:

$$W_1 = F_{S1} \cdot s_1 = \frac{1}{2} m \cdot v_2^2 - \frac{1}{2} m \cdot v_1^2$$

For F_{S2}:

$$W_2 = F_{S2} \cdot s_2 = \frac{1}{2} m \cdot v_3^2 - \frac{1}{2} m \cdot v_2^2$$

And by adding the two expressions:

$$W = W_1 + W_2 = F_{S1} \cdot s_1 + F_{S2} \cdot s_2 = \frac{1}{2} m \cdot v_3^2 - \frac{1}{2} m \cdot v_1^2 \qquad [6.1e]$$

In words it says: The total work invested in the entire process (W) is equal to the change in the kinetic energy of the entire process. Or, the total work invested was converted into kinetic energy. The term $\frac{1}{2} m \cdot v_2^2$, which is the kinetic energy in the middle, does not appear in the final result. The only kinetic energies that matter are the initial and the final. Equation [6.1e] is a generalization of [6.1a] for cases where the force changes as the object is moving. Although [6.1e] deals with only two consecutive forces, the same conclusion is true for any number of consecutive forces that act on an object.

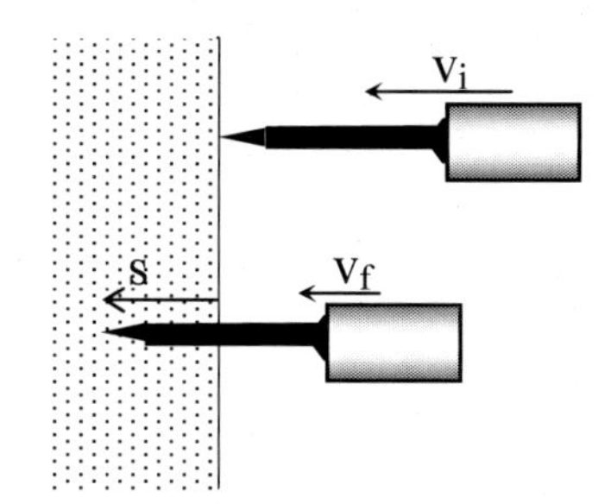

Figure 6.3: Driving a nail into a wall.

Consider now a hammer of mass m driving a nail into a wall. Figure 6.3 shows two instances in the process. In the first, the hammer has velocity v_i. Later, after the nail was driven a distance s into the wall, the hammer's velocity was v_f. At the end of the process v_f=0. The nail was driven into the wall because of the force that the hammer exerted on it. Let F_S denote the average of this force. We know that that force acted parallel to the displacement s of the nail. Therefore, a work of F_S·s was invested in the nail.

F_S is not given explicitly, but we can find it. According to Newton's Third Law, F_S has a reaction force F_R that the nail exerts on the hammer: $F_S=-F_R$. F_R is responsible for the velocity lost by the hammer, i.e. to the hammer's acceleration a. As in the previous case, $\mathbf{a} = \frac{v_f^2 - v_i^2}{2 \cdot s}$. $F_R=ma$, and

$$F_S = -F_R = -m\frac{v_f^2 - v_i^2}{2 \cdot s}$$

After rearranging we get:

$$F_S \cdot s = -(\frac{1}{2}m \cdot v_f^2 - \frac{1}{2}m \cdot v_i^2) \qquad [6.1f]$$

In word it says: The work invested in the nail is equal to the kinetic energy lost by the hammer. Equation [6.1f] rephrases the ideas of Newton's Laws for the particular circumstances.

To summarize, we have seen processes in which work is converted into kinetic energy, and processes in which kinetic energy is converted into work. In those examples, work that was invested in an object increased the kinetic energy stored in that object. When the kinetic energy of one object was decreased, it turned into work invested in the other object.

Example

A net force of 1,000N pushes a car of mass 1,200 kg horizontally along a distance of 50 m. (a) What is the change in the kinetic energy of the car? (b) What is the final velocity of the car?

Solution

(a) The work that the net force invested in the car is found from [6.1c]: F_S=1,000N and s=50 m, yielding W=50,000 J. Based on [6.1b], this work is equal to the change in the car's kinetic energy.

(b) The initial kinetic energy was zero, because the car was initially at rest [6.1d]. Therefore, the entire invested work is equal to the final kinetic energy. KE_f=50,000 J. The mass of the car is m=1,200kg, V_f=? Substituting in [6.1d] we get: $50{,}000=0.5 \cdot 1{,}200 \cdot V_f^2$, yielding V_f=9.1m/s.

Example

A hammer of mass m=1.5 kg had velocity of v_i=4 m/s right before it hit a nail. After driving the nail into the wall, the hammer came to rest. How much energy did it impart to the nail?

Solution

The hammer lost its entire kinetic energy, which was $0.5 \cdot 1.5 \cdot 4^2=12$ J. This kinetic energy was converted into work that was done on the nail.

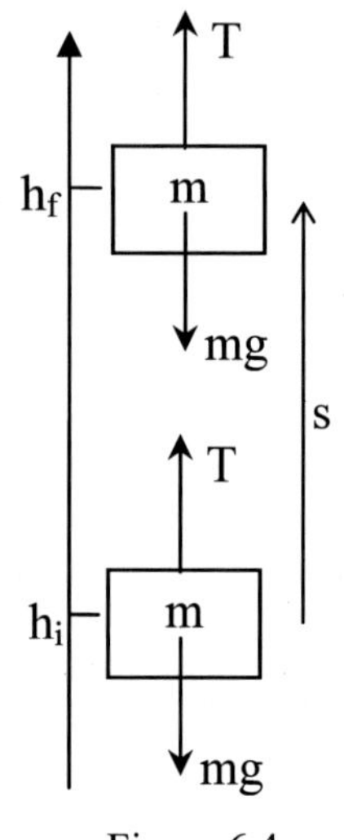

Figure 6.4

6.2 Work and gravity

Let an elevator of mass m move straight up at a constant velocity v (figure 6.4). On the vertical axis, we mark its initial position by h_i and it final position by h_f. Since it moved up at a constant velocity, the net upward force was zero. In addition to the force of gravity mg, which pulled it down, the tension of the cables T pulled it up. T was pointing in the direction of the displacement, therefore $T=F_S$, where F_S denotes the parallel force. The work invested by the cables is $W = F_S \cdot s = T \cdot (h_f - h_i)$. Since T=mg, we get: $W = mgh_f - mgh_i$.

The general term mgh is called the gravitational potential energy of the object, or in short potential energy (PE). Accordingly, mgh_f is the object's final potential energy, and mgh_i is its initial potential energy. We can write:

$$W = mgh_f - mgh_i = PE_f - PE_i$$
where [6.2.a]
$$PE = mgh$$

Equation [6.2a] says that the work invested in lifting an object without changing its velocity is equal to the change in its potential energy. Please note that the value of the potential energy of an object depends on the arbitrary reference point that was chosen as the origin of the vertical axis. However, the change in potential energy $(mgh_f - mgh_i)$ does not depend on the chosen reference point.

If the tension T in the cables is greater than the weight mg of the elevator, the elevator will be accelerated upwards at an acceleration a. According to Newton's Second Law T-mg=ma, or T=mg+ma. The tension T is still parallel to the displacement, therefore $T=F_S$. So, the work invested by T is now

$$W = F_S \cdot s = T \cdot s = (mg + ma) \cdot s$$

Let v_i and v_f denote correspondingly the initial and final velocity of the mass. The displacement s is given by $s=(h_f-h_i)$. As before, substituting for a, based on equation [2.8], and rearranging we get:

$$W = T \cdot s = (mgh_f - mgh_i) + (\frac{1}{2}m \cdot v_f^2 - \frac{1}{2}m \cdot v_i^2) \quad [6.2b]$$

That means that the work invested by the active force T was converted into two forms of energy: potential energy and kinetic energy. The sum of the changes of those energies is equal to the work invested.

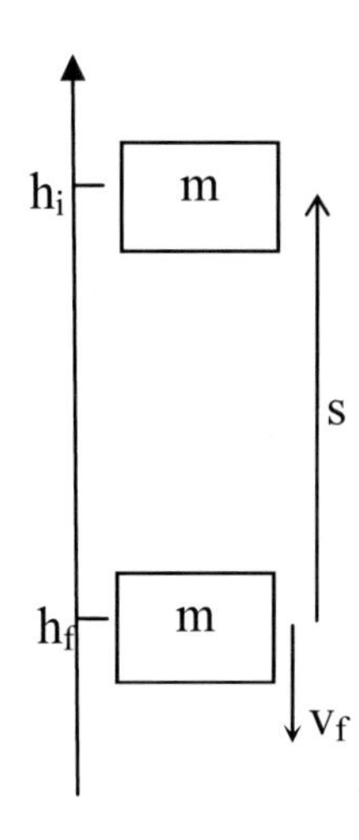

Figure 6.5

Let an object of mass m be dropped from rest at a height h_i above a reference point. When it reaches the height h_f its velocity is v_f (figure 6.5). The velocity v_f can be found by substituting the values of this case into equation [2.8]: $v_i=0$; $x_i=h_i$; $x=h_f$; and $a=-g$. After multiplying both sides by m we get:

$$mv_f^2 = 0 - 2mg(h_f - h_i)$$

and after rearranging:

$$mgh_i = mgh_f + \frac{1}{2}mv_f^2 \quad [6.2c]$$

In words it reads: The initial potential energy of the object was converted into a sum of potential energy and kinetic energy. If $h_f=0$, $mgh_f=0$, and, in this case, the final potential energy is zero. The initial potential energy was converted completely into kinetic energy.

If the object were thrown from the ground upwards, instead of being dropped, it could be shown in a similar way that its initial kinetic energy is converted to kinetic plus potential energy. At the top of its trajectory, the initial kinetic energy would be converted entirely into potential energy.

To summarize, we have seen in this section examples of processes in which work is converted into potential energy, and examples where potential energy is converted into kinetic energy. We have already seen that kinetic energy can be converted into work, so potential energy can also be converted into work.

Example
An elevator of mass 900 kg is lifted from a height of 2 m to a height of 13 m. What is the change in its potential energy?

Solution
In order to find the potential energy, we have to choose a reference plane. Let's choose the ground as a reference. Following the notations of [6.2] we get: h_i=2 m, h_f=13 m, m=900 kg. The initial potential energy was PE_i=900·9.8·2=17,640 J. The final potential energy was: PE_f=900·9.8·13=114,660 J. The change in potential energy was 97,020 J.

6.3 Dissipated energy

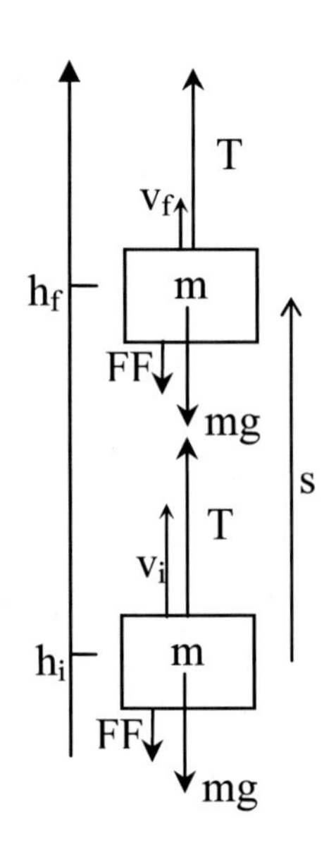

Figure 6.6

In the previous sections we saw how invested work can be transformed into kinetic and potential energy and vice versa. All the cases that we analyzed so far did not involve friction forces. Friction forces affect the motion of objects, and therefore they will affect the balance between invested work and the changes of kinetic and potential energies. As an example, consider the elevator that is pulled up by a tension force T (figure 6.4), but now a friction force FF is present and opposes the motion (figure 6.6). The new features in figure 6.6, which have been added to those of figure 6.4, are: the upward acceleration a of the elevator, the friction force FF, and the initial and final velocities v_i and v_f. Newton's Second Law for the new situation would be:

T-mg-FF=ma, or T=ma+mg+FF

Here again T=F_S is a force parallel to the displacement, and the work that it invests is given by

$$W = T \cdot s = ma \cdot s + mg \cdot s + FF \cdot s$$

By substituting in [2.8], as in the discussions above, and also substituting s=h_f-h_i , we get:

$$W = (\frac{1}{2}mv_f^2 - \frac{1}{2}mv_i^2) + (mgh_f - mgh_i) + FF \cdot s \qquad [6.3a]$$

Comparing [6.3a] with [6.2b], we notice the additional term FF·s on the right. This term is called the **dissipated work**, or **dissipated energy**. The meaning of equation [6.3a] is that the invested work is converted into kinetic energy, potential energy, and dissipated energy. The sum of these energies is equal to the invested work. Equation [6.3a] can be re-arranged and written as:

$$W_{inv} - W_{dis} = (\frac{1}{2}mv_f^2 - \frac{1}{2}mv_i^2) + (mgh_f - mgh_i)$$

or [6.3b]

$$KE_i + PE_i + W_{inv} - W_{dis} = KE_f + PE_f$$

Where $W_{inv} = F_S \cdot s$ is the work invested by the active force (other than the gravity). W_{dis} is the work that went to overcome the friction force (FF·s in this case). KE_i and KE_f are the initial and final kinetic energies of the object, respectively. PE_i and PE_f are the initial and final potential energies of the object, respectively.

Equations [6.3b] are special cases of the more general Law of Conservation of Energy. They say that the work invested in the object minus the energy dissipated is equal to the change in the object's kinetic plus potential energies (upper formula). The lower formula of [6.3b] means: The sum of the object's energies in the final state is equal to the initial sum of these energies plus work invested in the object minus work dissipated between the initial and the final times.

Example
A box slides on the floor for 6 m. The friction force that acts on the box is 11N. How much energy was dissipated during the slide?

Solution
The work done by the friction force was 11·6=66 J. This was the amount of the dissipated energy.

6.4 Conservation of energy

The equation F=ma has three types of variables: The force F describes the properties of the external factors that affect the object. The mass m describes the inertia, which is a permanent property of the affected object. The acceleration a is an acquired property of the affected object. If we know the forces that act on an object and its mass, we can find the acceleration, from which we can, in principle, find how the object moves.

Equation [6.3b] $KE_i + PE_i + W_{inv} - W_{dis} = KE_f + PE_f$ was derived from Newton's Laws that were applied to a particular group of circumstances. Each term in [6.3b] is a combination of more fundamental variables: The terms W_{inv} and W_{dis} combine in them properties of forces, and the distances along which they were affecting the objects. The KE and PE terms describe properties of the affected objects. KE combines the inertia of an object and its velocity. The potential energy combines the mass of the object and its position. It also contains a property of the force that was acting on the object (through g, which results from the force of gravity). In many cases we can derive information about the motion of the object from equation [6.3b]. Although [6.3] appear to be more complicated than F=ma, in many cases they are much simpler to handle, as will be illustrated in the next section.

Equation [6.3b] equates energies at a final state of the system with energies and work prior to that final state. It states that the sum of a certain combination of energies and works does not change, or in other words they are conserved – hence the name conservation of energy.

In words, equation [6.3b] says that work can be transformed into energy, that energy can be transformed into work, and that one kind of energy can be transformed into another. However, the final energy of the system at the end of these processes is equal to its initial energy plus the net work that was put into the system.

We derived equation [6.3a] for a particular situation: an object was moving in the vertical direction due to a driving force and against a friction force. Similar derivation, which requires more advanced mathematical techniques, can be done for a system of objects that move in three dimensions. The interesting thing is that the final expression for this general case is the same as equation [6.3b]:

$$KE_i + PE_i + W_{inv} - W_{dis} = KE_f + PE_f \qquad [6.4]$$

The only difference is that each term now has more general contents. KE is the sum of the kinetic energies of all the particles in the system. Kinetic energy can be due to translational motion of the object ($\frac{1}{2}mv^2$) and due to rotational motion, or spinning of the object. We discuss only the first type here. PE is the sum of potential energies of the objects. That includes gravitational potential energy (mgh), elastic potential energy, and electrostatic potential energy. We discuss only the first type here.

Work was defined as $W=F_S s$ [6.1c]. The work terms in [6.4], W_{inv} and W_{dis}, are due to forces that are not accounted for in the potential energy terms. W_{inv} is the sum of the work invested in all the objects. A force F_S invests work in an object if it acts in the direction of the object's motion. W_{dis} is the sum of all the energy dissipated by the objects. In general, dissipated energy is converted into heat, sound, and deformation of objects. A force F_S dissipates energy if it acts against the motion of the object. Friction and viscosity forces always dissipate energy. Other forces may contribute to the invested energy or to the dissipated energy, according to the situation. For example, if the object is a rocket moving in the field of gravity, and the force is the thrust of its engine, this force contributes to the invested energy if it acts in the direction of the motion of the rocket. In such cases it lifts and/or accelerates the rocket, thus increasing its energy. If the engine acts against the motion of the rocket, it slows it down without lifting it. In other words, it dissipates its kinetic energy.

There is an interesting analogy between energy and money. If you keep your money in your wallet and at home, (kinetic energy and potential energy), the money that you have at the beginning of the month (KE_i+PE_i) plus the money that you earn during the month (W_{inv}) minus the money that you spend during the month (W_{dis}) is equal to the total money that you'll have at the end of the month in your wallet and at home (KE_f+PF_f). This 'law of conservation of money' does not specify how the money is divided between the wallet and the home. However, you can find from it what is the most that is available for spending in a month, or what is the most that you can have in your wallet at the end of the month. The law of conservation of energy provides similar information. Based on energy considerations, it tells us what processes cannot happen, and what are the limitations on certain processes. Since it was derived from Newton's Laws, it also provides information about velocities and positions of the involved objects. We'll look into some of these applications next.

6.4.1 Applications of Conservation of Energy

In the following examples, information about velocities and positions of objects under various circumstances are found based on the principle of conservation of energy, instead of directly from Newton's Second Law $F = m \cdot a$. In some cases, using Newton's Second Law would be too complicated mathematically, so as to render that approach infeasible.

The first step in solving each problem is to declare what is the initial and what is the final state of the system. Then the appropriate energy and work terms are substituted in the equation of conservation of energy ($KE_i + PE_i + W_{inv} - W_{dis} = KE_f + PE_f$). Then, the resulting equation is solved for the unknown. In all the following examples, air resistance is neglected.

1. A rocket of total mass 500kg is launched straight up. The thrust of its engine is 1.2×10^4 N. What would be its velocity when it reaches a height of 10^4 m? The engine, whose mass was almost unchanged, was working all that time.

Solution

The chosen initial state is when the rocket is at rest on the ground. The final state is when it reaches 10^4m. The reference plane for determining potential energy is the ground. In the initial state, the kinetic energy of the rocket is zero, because it is not moving. The initial potential energy is zero, because the height is zero. On its way to the final state, the thrust acts as a force that invests energy in the system, because it is in the direction of the rocket's motion. Since air friction is neglected, there is no dissipating work (the contributions of the force of gravity are taken care of in the potential energy terms). In the final state the rocket has both kinetic and potential energy. When substituting in the equation of conservation of energy we get:

$$KE_i + PE_i + W_{inv} - W_{dis} = KE_f + PE_f$$

$$0 + 0 + 1.2x10^4 \cdot 10^4 - 0 = \frac{1}{2}500 \cdot v_f^2 + 500 \cdot 9.8 \cdot 10^4$$

which yields the final velocity v_f=533m/s

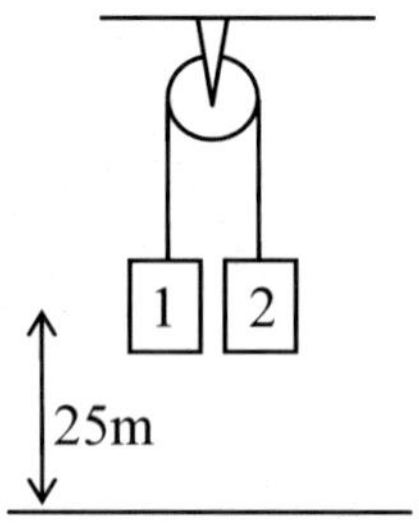

2. Two masses, m_1=40kg and m_2=50 kg are connected by a cable that is wrapped over a frictionless pulley that hangs from the ceiling (Figure 6.7). The masses are held at rest at the same height of 25m above the floor. The masses are then let go. What would be the velocity of the masses when m_2 reaches the height of 5m above the floor?

Figure 6.7

Solution
The initial state is when the two masses are at rest. The final state is when m_2 is 5 m above the floor. At that time, m_1 is 45m above the floor. The floor is chosen as the reference plane for calculating potential energy. The magnitudes of the velocities of the two masses are equal to each other at any moment, because they cover the same distance at the same time. When this information is substituted into the equation of conservation of energy we get (each of the KE and PE terms has two contributions, one from each mass):

$$KE_i + PE_i + W_{inv} - W_{dis} = KE_f + PE_f$$

$$(0+0) + (40 \cdot 9.8 \cdot 25 + 50 \cdot 9.8 \cdot 25) + 0 - 0 = (\frac{1}{2}40 \cdot v_f^2 + \frac{1}{2}50 \cdot v_f^2) + (40 \cdot 9.8 \cdot 45 + 50 \cdot 9.8 \cdot 5)$$

which yields: v_f=6.6 m/s.

3. A car of mass 2000kg is parked at the top of a hill, (inclined plane) 80m high (figure 6.8). The slope of the hill is 25 degrees. The brakes malfunction, and the car starts to roll down. (a) Neglecting friction forces, what would be its velocity at the foot of the hill? (b) What would be its velocity at the foot of the hill if a friction force of 900N were acting on it all the way down? (c) What would be its velocity if the engine was operating and pushing the car down with a force of 450 N?

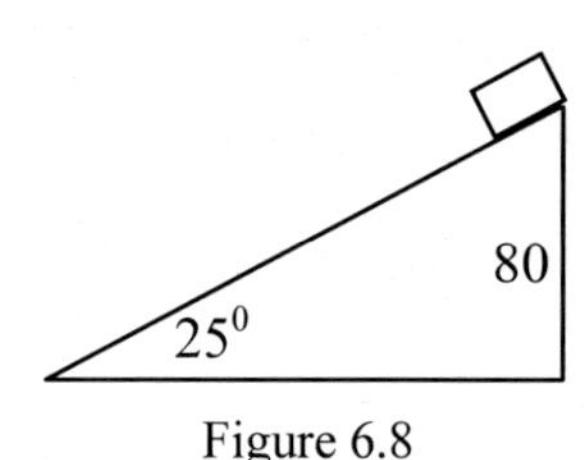

Figure 6.8

Solution
In all cases, the initial state is when the car is at the top of the hill, and the final state is when it is at the bottom. The height is determined with respect to the bottom of the hill. The distance s that the car traveled is s=80/sin(25^0)=189.3m.

(a)

$$KE_i + PE_i + W_{inv} - W_{dis} = KE_f + PE_f$$

$$0 + 2000 \cdot 9.8 \cdot 80 + 0 + 0 = \frac{1}{2}2000 \cdot v_f^2 + 0$$

which yields v_f=39.6m/s

(b)

$$KE_i + PE_i + W_{inv} - W_{dis} = KE_f + PE_f$$

$$0 + 2000 \cdot 9.8 \cdot 80 + 0 - 900 \cdot 189.3 = \frac{1}{2} 2000 \cdot v_f^2 + 0$$

which yields v_f=37.4 m/s

(c)

$$KE_i + PE_i + W_{inv} - W_{dis} = KE_f + PE_f$$

$$0 + 2000 \cdot 9.8 \cdot 80 + 450 \cdot 189.3 - 900 \cdot 189.3 = \frac{1}{2} 2000 \cdot v_f^2 + 0$$

which yields v_f =38.5 m/s

4. A pendulum (figure 6.9) starts swinging from rest 1.2 m above the floor (point a). It sweeps uninterrupted into a pit. At its lowest point (point c), the pendulum is 0.3m below ground level. (a) What was its velocity at ground level (point b)? (b) What was its velocity at the lowest point of the sweep?

Solution

For all cases, the ground is chosen as the reference for determining height. Based on that choice, the height of point a is +1.2m, the height of point b is zero, and the height of point c is -0.3m. The mass of the pendulum is m.

The initial state is at point a, and the final point is point b. Substituting into the equation of conservation of energy:

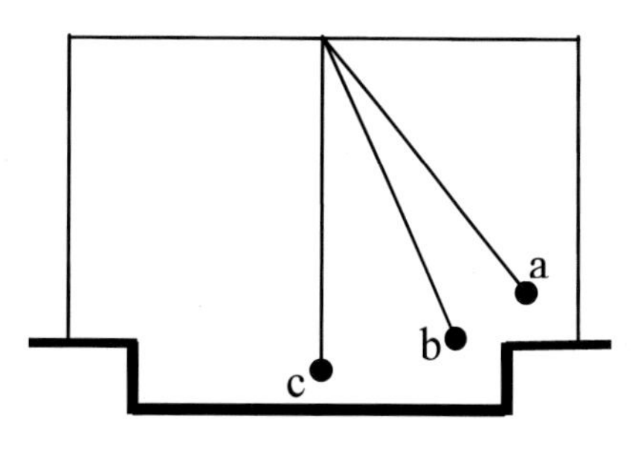

Figure 6.9

$$KE_i + PE_i + W_{inv} - W_{dis} = KE_f + PE_f$$

$$0 + m \cdot 9.8 \cdot 1.2 + 0 + 0 = \frac{1}{2} m \cdot v_f^2 + 0$$

which yields: v_f=4.85m/s

(b) The initial state is at point a, the final state is at point c.

$$KE_i + PE_i + W_{inv} - W_{dis} = KE_f + PE_f$$

$$0 + m \cdot 9.8 \cdot 1.2 + 0 + 0 = \frac{1}{2} m \cdot v_f^2 + m \cdot 9.8 \cdot (-0.3)$$

which yields v_f=5.42 m/s.

Things that cannot happen

5. Using conservation of energy, show that if a basketball is dropped to the floor of a gym, it will never bounce back to a point higher than the point from which it was dropped, no matter how flexible is the ball or the floor.

Solution
The initial state of the ball is just as it is dropped, h_i. The final state is the highest point to which it bounced back h_f. The initial and final kinetic energies are zero. There is a dissipative energy term, due to air friction, the generation of sound, and energy lost in the deformation of the ball as it hit the floor. When all these are substituted into the law of conservation of energy we get:

$$KE_i + PE_i + W_{inv} - W_{dis} = KE_f + PE_f$$

$$0 + m \cdot g \cdot h_i + 0 - W_{dis} = 0 + m \cdot g \cdot h_f$$

From this we get:

$$h_f = h_i - \frac{W_{dis}}{m \cdot g}$$

which means that the final height is smaller than the initial height. The total initial energy, which was potential energy, is divided between the final potential energy and the dissipated energy. That means that the final potential energy is smaller than the initial potential energy, and therefore the final height is smaller than the initial height.

6. Using conservation of energy, show that an object at rest on a slope of a hill can, by itself, go only downhill.

Solution
Like in the case before, the initial energy, which is potential energy, is divided between the final potential energy and dissipated energy, which in this case would be friction with the slope. Therefore, the final height must be smaller than the initial height.

6.5 Power

A racecar can accelerate from rest to 80 miles per hour in 4 seconds. It will take a resting passenger car of the same mass 10 seconds to reach that speed. The change in kinetic energy of the racecar is the same as that of the passenger car, because all the elements in the expressions $KE = \frac{1}{2}mv^2$ are identical for both cars. The amounts of work invested by each engine to generate its change in kinetic energy are the same for both cars. The only difference is that the engine of the racecar generated that energy in a shorter time. The entity that combines an amount of energy and the time that it takes to generate it (or to use it) is called power. In physics, power is defined in the following way:

$$\text{Power} = \frac{\text{Work}}{\text{time}} \qquad P = \frac{W}{t}, \qquad [6.5a]$$

where time in [6.5a] is the time period during which the work or the energy was generated or consumed. Based on this definition, we can say that the engine of the racecar has more power than that of the passenger car. The time in [6.5a] is in the denominator, and if it takes less time to generate the same amount of work, the generator of that work has more power.

In the SI system, the unit of power is watt. $\text{Watt} = \frac{\text{joule}}{\text{second}}$. In the British system the unit of power is $\frac{\text{pound} \cdot \text{foot}}{\text{second}}$, (no special name). A commonly used unit of power is the horsepower: 1horsepower=746watt. Another common unit of power, which is neither in the SI nor in the British system, is the kilowatt (kW=1000 watts).

Devices that generate energy (or consume energy) are characterized by the energy that they can generate (or consume) per unit time. For example, an engine of 1 horsepower can generate 756 joules every second of its operation. This energy can be used in many ways e.g. to lift objects, to accelerate them, to push them against friction forces, etc. An example of consumption of power is an air conditioner, say of 0.8 kW. It consumes 800 joules every second of its operation.

Equation [6.5a] can be written also as:

$$\text{Work} = \text{Power} \cdot \text{time} \qquad W = P \cdot t \qquad [6.5b]$$

We use this equation when we want to know the amount of energy generated (or consumed) when a device of a certain power P was operating during a time period of t. For example, the amount of energy consumed by a 100W TV set in 4 hours (14,400 seconds) is $W=100 \cdot 14{,}400=1.44\text{x}10^5$ joules. (Please note that W, when marked on many appliances, stands for the power in watts consumed by that appliance. In formulas, e.g. [6.5], W stands for work, which is measured in joules in the SI system.)

A commonly used unit of work–the kilowatt-hour (kWh)–is based on equation [6.5b]. A kWh is the amount of work generated (or consumed) by a device whose power is 1kW when operating for an hour.
$1\text{kWh}=1{,}000\text{watt} \cdot 3{,}600\text{seconds}=3.6\text{x}10^6$ joules.

In certain situations, a constant force F acts on an object and causes it to move at a constant velocity v along a distance d. The reason that the object is not accelerated by F is that other forces, such as friction or gravity, are also acting on the object against F and exactly balancing it. For such cases we get:

$$P = F \cdot v$$

because [6.5c]

$$P = \text{Power} = \frac{\text{Work}}{\text{time}} = \frac{\text{Force} \cdot \text{distance}}{\text{time}} = \frac{F \cdot d}{t} = F \cdot \frac{d}{t} = F \cdot v$$

Examples:

1. A racecar can accelerate from zero to 80 miles per hour in 4 seconds. The mass of the car is 2500 kg. What is the average power of its engine?

Solution
The work invested by the engine is converted into kinetic energy (neglecting friction forces). The initial kinetic energy of the car is zero. Its final velocity is 80x1610/3600=35.8m/s, and the final kinetic energy is $0.5 \times 2500 \times 35.8^2 = 1.6 \times 10^6$j. The power is $1.6 \times 10^6/4 = 4 \times 10^5$W.

2. The car of an elevator is usually connected to a cable, wrapped around a pulley system. A balancing weight is connected to the other end of the cable (figure 6.10). The weight of the car is equal to the weight of the balancing weight. A motor drives the elevator-weight system. There are three motion phases as the car moves up: First, the car is accelerated from rest to a cruising speed. Second, the car moves at a constant cruising speed. Third, the car slows till it comes to a stop. Find the average power provided by the motor in the first two phases of the following empty elevator: The mass of the car is 360 kg. A cruising speed of 2m/s was reached after three seconds, and the car continued to cruise for another 20 seconds. (Neglect friction forces).

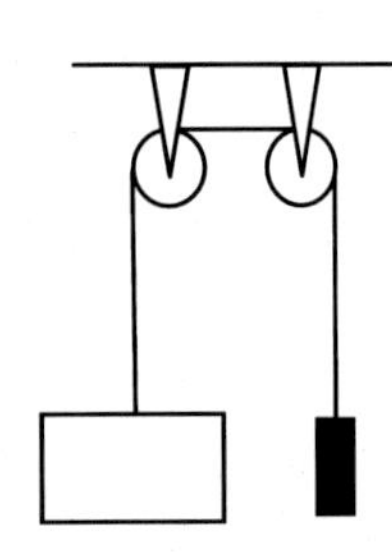

Figure 6.10

Solution
We have to find the change in energy of the system during the first phase, and divide it by the time that it took to cause this change (3sec). The distance that the car moves up from its starting point (h) is always equal to the distance that the balancing weight moves down from its starting point (-h). The total change of potential energy of the system (car and balancing weight) is always: mgh-mgh=0. (This is true only for an empty elevator). The instantaneous velocity of the car is always equal to that of the balancing weight. At the end of the first phase it was 2m/sec. The total kinetic energy after 3 seconds was: $0.5 \times 360 \times 2^2 + 0.5 \times 360 \times 2^2 = 1440$j. This is also the total energy provided by the motor. The power of the motor in this phase was: 1440/3=480W.

In the second phase (when the elevator was cruising), there was no gain in the total potential energy and in the total kinetic energy. Therefore, no work was provided by the motor. Theoretically, the motor could be idle. In reality, the motor will have to provide energy to overcome the friction forces.

Consider the same elevator, but now an average friction force of 50N was acting during the cruising phase. What was the power of the motor in this case?

Since the motor had to overcome a constant force of F=50N when the system was moving at a constant speed (v=2m/s), the power now can be found based on formula [6.5c]: $P=50\cdot 2=100W$.

6.6 Conservation of Momentum

According to Newton's Third Law, if object A exerts a force **F** of object B, object B exerts a force –**F** of the same magnitude and opposite direction on object A. This is true for any two objects and for any interacting force that exists between them. According to Newton's Second Law, if a net force is acting on an object, the object will be accelerated according to $\mathbf{F}=m\cdot\mathbf{a}$.

Consider now two interacting objects, A and B, whose masses are m_A and m_B, which are moving at the velocities $\mathbf{v}_A$ and $\mathbf{v}_B$, correspondingly. Because of the interaction force **F**, after a certain time Δt, the velocities of A and B would change to $\mathbf{u}_A$ and $\mathbf{u}_B$, correspondingly. By combining Newton's Second and Third Laws we get:

$$m_A \frac{\mathbf{u}_A - \mathbf{v}_A}{\Delta t} = -m \frac{\mathbf{u}_B - \mathbf{v}_B}{\Delta t},$$

which, after rearrangement, gives:

$$m_A \mathbf{v}_A + m_B \mathbf{v}_B = m_A \mathbf{u}_A + m_B \mathbf{u}_B \qquad [6.6a]$$

Equation [6.6a] has four terms, each of which is the mass of an object multiplied by its velocity at a certain time. The product m**v**, where m is the mass of an object and **v** is its velocity, is called the momentum of the object. By checking [6.6a] we see that the left hand side is the sum of the momentums of the objects before the interaction between them took place, and the right hand side is the sum of the momentums after the interaction took place. These two sums are equal to each other, or in other words, they are conserved (they do not change with time). Equation [6.6a] is called the Law of Conservation of Momentum. It is a way of combining Newton's second and Third Laws.

It is common to denote momentum by **P**. Momentum **P**=m**v** is a vector. The magnitude of the vector is P=mv, where v is the magnitude of the velocity of the object. The direction of the momentum is the same as the direction of **v**.

When two interacting objects move on the same line, this line can be chosen as an axis. In such cases, the direction of the momentum is encoded by the sign of P=mv. Positive P means moving in the positive direction of the axis, and negative P means moving in the negative direction of the axis, the same as we treated force, velocity, displacement, and other vectors in 1-D.

When the two interacting objects are moving in 2-D, equation [6.6a] is written separately for the x and for the y-components of the vectors:

$$m_A v_{AX} + m_B v_{BX} = m_A u_{AX} + m_B u_{BX}$$
$$m_A v_{AY} + m_B v_{BY} = m_A u_{AY} + m_B u_{BY} \qquad [6.6b]$$

The law of conservation of momentum is useful for studying interactions between objects when the forces are not known explicitly.

Example
The mass of a gun is 3.5 kg, and the mass of a bullet is 0.015 kg. When fired, the recoil velocity of the gun is 2 m/s. What is the velocity with which the bullet leaves the gun?

Solution

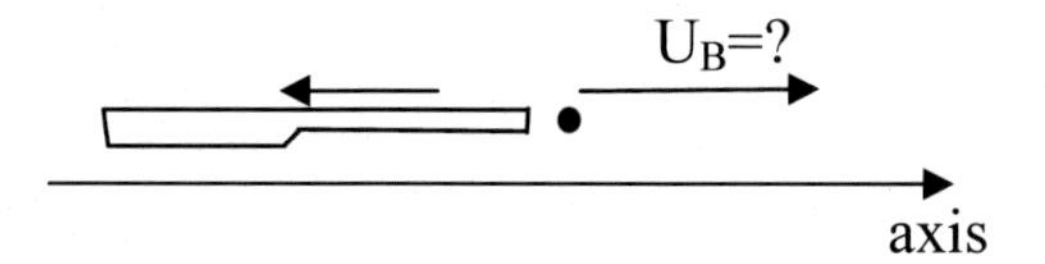

The gun recoils in the opposite direction to the motion of the bullet. Let's choose an axis, pointing in the direction of the bullet's motion. The initial state of the system is before the gun was fired. In this state, the gun and the bullet were at rest. The final state of the system is as the bullet was leaving the gun. The gun is denoted by A, the bullet by B. Following the notation of formula [6.6a] : m_A=3.5 kg, m_B=0.015 kg. V_A=0, V_B=0, U_A=-2m/s, the velocity of the bullet U_B=? Upon substituting in [6.6a] we get:
$0+0=3.5\cdot(-2)+0.015\cdot U_B$
yielding U_B=467 m/s. This method makes it possible to find the velocity of a bullet, even though it cannot be seen in its flight, and the force exerted during the explosion of the gun powder is unknown.

Example
A car of mass 2,000 kg moving at 22 m/s to the east and a car of 3,000 kg moving at 33 m/s to the north collided at an intersection. After the collision, the entangled wreck moved as one unit. What was the magnitude and direction of the velocity of the wreck right after the collision? (Neglect friction forces.)

Solution
The initial state of the two-car system was just before the collision, and the final state right after the collision. Since this is a 2-D situation, we'll define an x-y coordinate system so that the origin is at the point of collision, the x-axis points to the east, and the y-axis points to the north. The 2,000 kg car will be denoted as A, and the 3,000 kg car as B. Following the notations of [6.6b]:

m_A=2,000 kg, m_B=3,000kg, V_{AX}=22 m/s, V_{AY}=0, V_{BX}=0, V_{BY}=33 m/s. After the collision, $U_{AX}=U_{BX}=U_X$=? $U_{AY}=U_{BY}=U_Y$=? Substituting in [6.6b] we get:

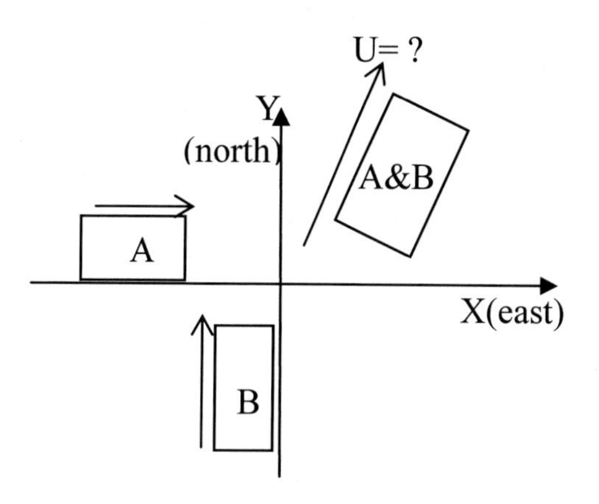

$2,000 \cdot 22+0=2,000U_X+3,000U_X$
$0+3,000 \cdot 33=2,000U_Y+3,000U_Y$

Solving yields: U_X=8.8 m/s; U_Y=19.8 m/s.
The magnitude of the velocity of the wreck is $U=\sqrt{8.8^2+19.8^2}=21.7\text{m/s}$. The direction of its motion is: $\alpha=\tan^{-1}(\frac{19.8}{8.8})=66^0$ north of east.

PROBLEMS

Sections 6.1-6.4

1. A hammer of mass 0.75kg has a velocity of 3.5 m/s just before it hits a nail. After driving the nail 2 cm into the wall, the hammer lost all its velocity.
(a) What are the initial and final kinetic energies of the hammer?
(b) What was the dissipated energy in this process?
(c) What was the average resistance force of the wall to the nail?

2. A pendulum is raised 0.70m above the floor, and then is released and swings freely. Neglecting friction forces, what would be its velocity as it swings at a point 0.25m above the floor?

3. A crate of mass 45kg was pushed from rest by a horizontal force of 120N on a horizontal plane from point A to point B. The distance between A and B was 80m. The friction coefficient between the crate and the plane was 0.2.
(a) How much work was invested by the pushing force?
(b) How much work was dissipated by the friction force?
(c) What was the change in kinetic energy of the crate as it moved from A to B?
(d) What was the velocity of the crate at point B?
(e) After the crate reached point B, the pushing force stopped, and the crate continued to move until it came to a complete stop at point C, due the friction force. How much work was dissipated by the friction force as the crate moved from A to C?

4. A basketball is dropped from a height of 2m. It lost 20% of its energy, in the bouncing from the floor and due to friction with the air, before it reaches the highest point of its first bouncing. How high did it bounce?

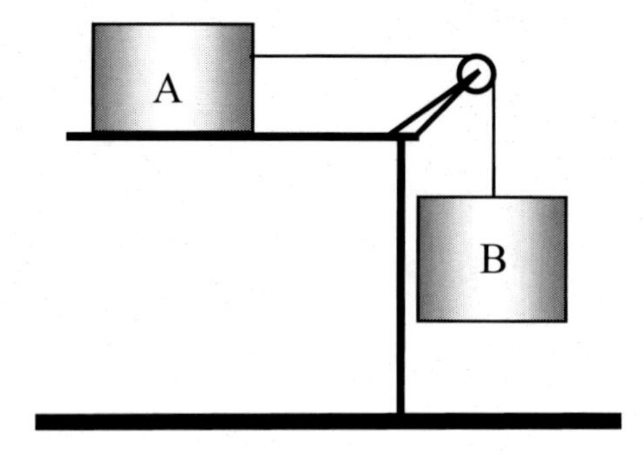

5. Mass A is of 4kg, and mass B is of 8 kg. Initially, both masses are at rest. Then, they are let go.

(a) What would be the velocity of mass A when mass B drops 0.3m? Assume that no friction forces exist in the system.

(b) What would be the velocity of mass A when mass B drops 0.3m, if a friction force of 10N acts on mass A?

Section 6.5

6. A 2,500 kg airplane can accelerate on the runway from rest to 250 mph in 15 seconds. What is the power provided by the engine? Assume that all the energy goes to the plane's kinetic energy. What other energies are neglected in this assumption?

7. A 75W light bulb was inadvertently left on in the attic. It was on for 950 hours, till it burned out. Electricity cost 9 cents per kWh. How much did the wasted electricity cost?

8. In order for an empty elevator with a balancing weight to move up at a constant velocity of 2.2 m/s, the motor has to provide 220 W.
(a) What is the magnitude of the frictional forces that opposes the motion of the elevator?
(b) Three people, whose total mass are 240 kg, entered the elevator on its way up. What was the power provided by the motor as the elevator was moving at 2.2 m/s?
(c) If the maximum power that the motor can provide during the cruising phase is 1.5 horsepower, what is the maximum load that can be lifted at a speed of 2.2 m/s?

Section 6.6

9. An asteroid of mass 4,000 kg is heading towards Earth. A rocket of mass 100 kg is sent up to smash head-on into it. It was calculated that in order to deflect the asteroid so that it won't hit Earth, its speed should be reduced from 6,000 m/s before the interception to 5,800 m/s right after the interception. What should be the velocity of the rocket just before it hits the asteroid?

10. Object A of mass 5 kg moves at 15 m/s, and object B of mass 8 kg moves at 20 m/s. What are the magnitude and direction of the total momentum of A and B if:
(a) A and B move in the x-direction?
(b) A moves in the x-direction and B moves in the y-direction?
(c) A moves in the x-direction and B moves 30^0 relative to the x?
(d) A moves 45^0 relative to the x, and B moves 120^0 relative to the x?

11. Two cars, the first of mass 1,800 kg was headed east at 26m/s, the second of mass 1,500 was headed north when they collided. Their tangled wreck moved 45^0 north of east right after the collision. What was the velocity of the second car right before the collision? (Neglect friction forces).
Hint: you may want to use the following procedure:
Choose a coordinate system such that the origin is at the point of collision, the x-axis points east and the y-axis points north.
(a) What was the total x-component of the momentum before the collision?
(b) What was the total x-component of the momentum right after the collision?
(c) What was the magnitude of the momentum of the wreck right after the collision?
(d) What was the magnitude of the y-component of the momentum of the wreck right after the collision?
(e) What was the magnitude of the total y-component of the momentum right before the collision?
(f) What was the velocity of the second car right before the collision?

7. HEAT AND THERMODYNAMICS

7.1 Temperature

The sensations of hot and cold are basic sensations of our body. The thermometer is a device that measures how hot or cold an object is, and assigns a numerical value – temperature – to that property.

A common thermometer consists of a glass bulb that a thin glass tube is coming out of it. Alcohol or mercury fills the bulb and part of the tube. As the bulb contacts a hot object, the liquid in it expands into the tube. If a contact is made with a colder object, the liquid in the tube retracts. The height of the liquid in the tube indicates the temperature of the object. A scale at the back of the tube, against which the height of the liquid in the tube is determined, provides the reading of the temperature of the object in degrees. Nowadays, there are thermometers that operate on principles other than the expansion and contraction of mercury and alcohol.

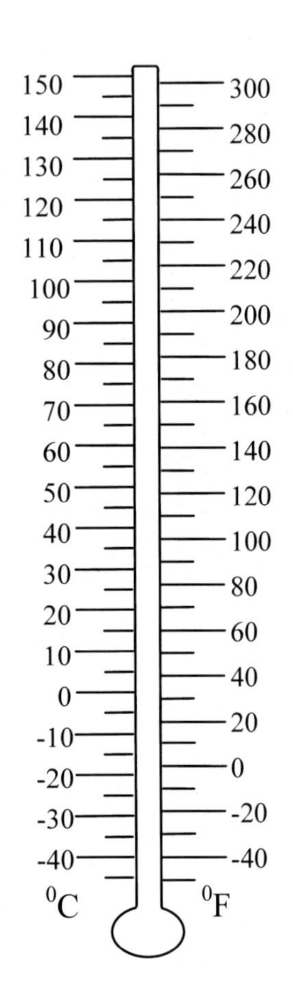

Figure 7.1

Several temperature scales are in use today, the most common are the Celsius and the Fahrenheit scales. The notation ^{0}C indicates degrees Celsius and ^{0}F indicates degrees Fahrenheit. In the Celsius scale, the boiling temperature of water is 100^{0}C, and the freezing temperature of water is 0^{0}C. The scale in between the boiling and freezing points is divided into 100 equal intervals, defining the Celsius temperatures between 0^{0}C and 100^{0}C. In the Fahrenheit scale, the boiling of water is at 212^{0}F, and the freezing of water is at 32^{0}F. In between those points, the scale is divided into 180 equal intervals, defining Fahrenheit temperatures between 212^{0}F and 32^{0}F. These scales have been extended beyond the freezing and boiling points of water (figure 7.1). Since both Celsius and Fahrenheit scales measure temperature, it is possible to convert from one scale to another. The conversion formulas are:

$$F = C \cdot 1.8 + 32$$
$$C = \frac{F - 32}{1.8} \qquad [7.1a]$$

where F is the temperature in degrees Fahrenheit, and C is the temperature in degrees Celsius.

Formulas [7.1a] give the relationships between temperatures of objects in the two scales. In order to convert change of temperature or temperature difference, we use the fact that a change of one degree Celsius corresponds to a change of 1.8 degrees Fahrenheit:

$$\Delta F = 1.8 \cdot \Delta C$$
$$\Delta C = \frac{\Delta F}{1.8} \qquad [7.1b]$$

where ΔF indicates temperature difference or temperature change in Fahrenheit, and ΔC indicates temperature difference or temperature change in Celsius.

Example
Body temperature of 40^0C indicates high fever. What is the corresponding Fahrenheit temperature?

Solution
According to [7.1a] F=40x1.8+32=104^0F

Example
The temperature inside the house is 72^0F, and outside is 18^0F. What is the temperature difference between the inside and outside in ^{0}C?

Solution
The temperature difference is 72-18=54^0F. According to [7.1b], the temperature difference in ^{0}C is:
ΔC=54/1.8=30^0C

7.2 Expansion and contraction

7.2.1 Solids and liquids

When objects warm up they expand, and when they cool down they contract. Usually, those changes are small and they are unnoticeable to the naked eye. However, in many cases those effects cannot be ignored. Bridges, railroad tracks, and some other long objects that are undergoing daily and seasonal cycles of temperature changes experience related changes in their lengths. If ignored, such changes may cause deformation or cracking of the structure. This is why spaces are left between segments of railroad tracks and between sections of bridges (figure 7.2). Those spaces allow for thermal expansion and contraction without damaging the structure.

Figure 7.2

In order to quantify thermal expansion and contraction, let's deal first with one-dimensional objects. A one-dimensional object is an object that one of its dimensions (its length) is significantly greater than its other two dimensions (breadth and width). It was found that the change in length (ΔL) of a one-dimensional object due to temperature changes depends on the material of the object, and is proportional to the initial length of the object (L_0) and to the change in the temperature of the object (ΔT). This is expressed by formulas [7.2]:

$$\Delta L = \alpha \cdot L_0 \cdot \Delta T$$
$$\text{where} \quad \Delta L = L - L_0 \quad \text{and} \quad \Delta T = T - T_0 \qquad [7.2a]$$

(L is the final length of the object and T is its final temperature).
This formula can also be written as:

$$L = L_0 \cdot (1 + \alpha \cdot (T - T_0)) \qquad [7.2b]$$

The property of the material is expressed by α–the linear expansion coefficient of that material. The linear expansion coefficient tells us by how much a rod of a unit length of that material will expand when undergoing a temperature change of one degree. The linear expansion coefficient depends on the temperature scale used (Celsius or Fahrenheit), but it does not depend on the unit of length used. The change in length (ΔL) will be in the same unit that is used to express the initial length (L_0).

Formulas [7.2] describe both expansion and contraction due to temperature changes. When $\Delta T > 0$, the final temperature is greater than the initial temperature. The formulas give $L > L_0$, or $\Delta L > 0$, meaning that the object expands. Similarly, when $\Delta T < 0$, the formulas describe contraction.

In order to figure out how two and three-dimensional objects expand and contract due to temperature changes, we can imagine that each such object is made of many straight thin wires. To a good approximation, each of those wires expands and contracts according to formula [7.2], unaffected by the presence of the other wires. Figure 7.3 illustrates how the thermal expansion of objects of any shape could be determined. The initial peripheries of two objects (rectangle and ellipse in this example) are denoted by thick lines. Their final peripheries, after thermal expansion, are denoted by thin lines. Arbitrary groups of imaginary, thin, straight wires have been chosen in each object (full lines inside objects). The wires of the rectangle form a grid, and those of the ellipse form a fan. (It should be noted that any other group of thin straight wires that are distributed throughout the object could have been used.) The thermal expansion of each wire is marked by thin dotted lines. Those expansions can be figured out according to formulas [7.2]. The ends of those thin wires trace the peripheries of the expanded object.

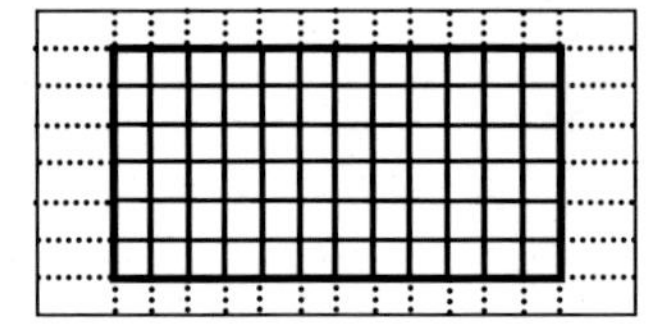

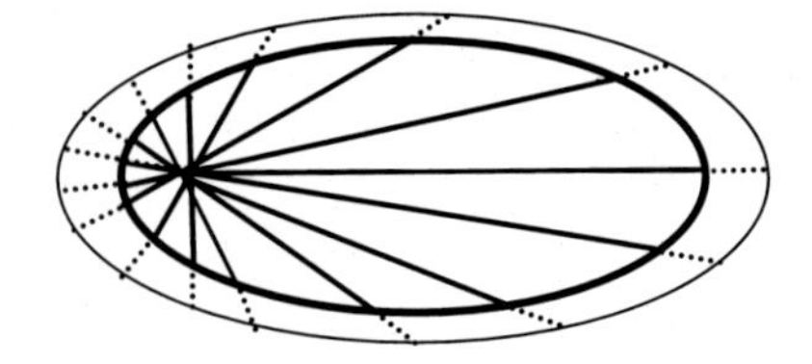

Figure 7.3: Tracing the thermal expansion of arbitrary objects.

Material	*Coefficient of linear expansion, α ($^0C^{-1}$)*	*Coefficient of volume expansion, β($^0C^{-1}$)*
Solids		
Aluminum	$25 x 10^{-6}$	$75 x 10^{-6}$
Brass	$19 x 10^{-6}$	$56 x 10^{-6}$
Iron or steel	$12 x 10^{-6}$	$35 x 10^{-6}$
Lead	$29 x 10^{-6}$	$87 x 10^{-6}$
Glass (Pyrex)	$3 x 10^{-6}$	$9 x 10^{-6}$
Glass (ordinary)	$9 x 10^{-6}$	$27 x 10^{-6}$
Concrete and brick	$12 x 10^{-6}$(approx)	$36 x 10^{-6}$(approx)
Marble	$1.4\text{-}3.5 x 10^{-6}$	$4\text{-}10 x 10^{-6}$
Liquids		
Mercury		$180 x 10^{-6}$
Water		$210 x 10^{-6}$
Gasses		
Air/ most gasses at atmospheric pressure		$3400 x 10^{-6}$

Table 7.1: Linear expansion coefficient (α) at 20^0C

Based on those thermal expansion properties of two and three-dimensional objects, formulas were derived for the change in area of two-dimensional objects, and the change in volume of three-dimensional objects. The formulas are:

$$A = A_0 \cdot (1 + 2 \cdot \alpha \cdot (T - T_0))$$
$$V = V_0 \cdot (1 + 3 \cdot \alpha \cdot (T - T_0)) \qquad [7.3a]$$

where A is the final area, A_0 the initial area, V final volume, V_0 initial volume, and the rest is as in [7.2].

Liquids are usually three-dimensional objects. It is customary to specify the expansion property of liquids by their volume expansion coefficient β. For solids, β=3α. The volume expansion of liquids and solids can also be expressed as:

$$V = V_0 \cdot (1 + \beta \cdot (T - T_0)) \qquad [7.3b]$$

Example

A steel bridge 100 m long experiences temperature changes of 20 degrees Celsius during one day. What is the corresponding change in its length?

Solution

According to the notations of [7.2], L_0=100m, ΔT=20^0C, $\alpha = 12 x 10^{-6}$ ($^0C^{-1}$) (from table 7.1), and ΔL=?. Substituting in [7.2a] yields ΔL=0.024m (approximately an inch).

Although thermal expansion and contraction are relatively small, they may create large forces in structures. Those forces are called thermal loads. In general, if the thermal expansion or contraction of a structural element are restricted by other elements of the structure, that element will exert thermal loads on the restricting elements. For example, a glass pane that is tightly fitted in an aluminum frame will experience a thermal load when the temperature drops. Let the thermal expansion of an unrestricted element be denoted by ΔL. When restricted, the thermal load of that element would be equal to the elastic force associated with the same expansion ΔL. For example, if an unrestricted steel beam expands by 1mm due to a temperature change, the thermal load that it would exert when restricted would be equal to the force needed to compress it back by 1mm against its elasticity, without changing its temperature.

It happens that the linear expansion coefficients of steel and concrete are very close to each other (see table 7.1). Because of that, it is possible to reinforce concrete with rods of steel. Concrete is not reinforced with other metals because their different expansions and contractions would eventually weaken the structure.

Unwanted effects of thermal expansions and the resulting thermal loads can be accommodated in several ways. The first is to leave appropriate spaces between structural elements, so that they could expand and contract without affecting each other. This is done not only in railroad tracks and in bridges, but also in windows, where spaces are left between the frame and the glass pane, and in floors, where spaces are left between the floor-tiles and the walls.

When leaving spaces is not feasible, for example in heating and air-conditioning pipes that carry liquids, flexible elements are introduced in the otherwise solid structure. A pipe shaped as a U can bend around the center of the U. When a pipe is curved as a U in between points at which it is securely anchored, the natural flexibility of the U can accommodate the thermal expansion and contraction of the straight parts of the pipe.

The Water Anomaly

Water contracts as it cools. However, when it is cooled from 4^0C to 0^0C water expands, and as it freezes it expands even more. The volume of ice is greater than the volume of the water from which it was made. (The volume of ice at 0^0C is 9% greater than its volume as water at 4^0C, 8% greater that its volume as water at 50^0C, and 4.5% greater than its volume as water at 100 ^{0}C.) If water freezes in a confined space, where there is no room to accommodate the expanding ice, pressure builds against the walls of the container. If the walls are not strong enough, they would crack. This is what happens sometimes when water freezes in water pipes and water mains. Unlike water, most other liquids contract as they cool and continue to contract as they freeze. That different behavior of water is called the water anomaly.

Because of its expansion at freezing, the density of ice is smaller than the density of water. (Density is inversely proportional to the volume). In everyday language, we could say that "ice is lighter than water". Because of that, ice floats on water.

7.2.2 Gasses

Gasses are kept in containers, except for atmospheric air and gasses that are mixed in it. A gas fills the entire volume of its container. If a container expands, the gas in it expands by itself, and fills the entire volume.

When contained gas is heated, it tends to expand. Whether it would expand or not depends on the nature of the container. If the container is totally rigid, the heated gas will not be able to expand. In such cases the pressure that the gas exerts on the container will increase. If the container is flexible (e.g. a balloon), or adjustable, (e.g. a cylinder with a piston), the heated gas could expand. The amount of the expansion in such cases will depend on the resistance of the container. In general, both the pressure and the volume of a heated gas may change as its temperature changes. This is different from solids and liquids, in which only the volume is affected.

Experiments were done to find relationships between the temperature of a gas and its volume and pressure. Two sets of experiments tested those relationships in simplified situations. First, when the container was rigid and its volume could not change. Those experiments provided information on how the pressure of the gas changes with temperature, when the volume is kept constant. In a second set of experiments, the volume of the container was allowed to change while the pressure of the gas was kept constant. Figures 7.4 illustrate the results of such experiments.

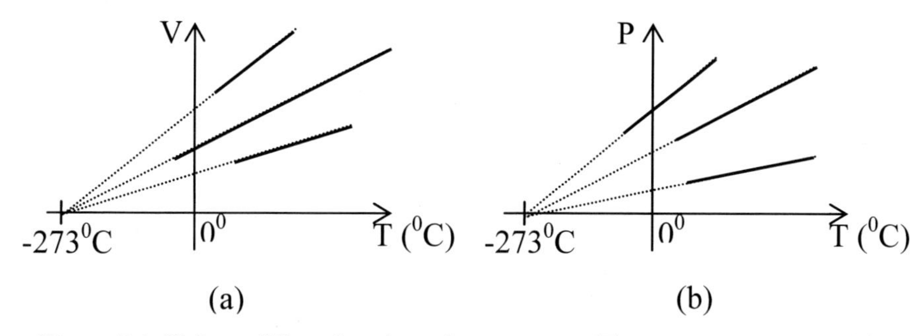

Figure 7.4: Volume (V) as function of temperature (T) at constant pressure (a), and pressure (P) as function of temperature at constant volume (b) for ideal gasses. Solid lines: actual experiments. Dotted lines: extrapolations.

It can be seen in figure 7.4(a) that under constant pressure the volume of a gas changes linearly with its temperature. From figure 7.4(b) we get that under constant volume the pressure changes linearly with the temperature. When the lines that describe those observations are extrapolated, they all intersect on the temperature axis (x-axis) at the temperature of -273^0C. This means that if the reaction of the gasses to cooling would not change with temperature, then when they reach -273^0C they lose all their volume and all their pressure. Once they get to that state, there is nothing more to cool. Therefore, -273^0C is called the absolute zero. That observation prompted the introduction of two additional temperature scales–the absolute scales. The zero of the absolute scales is the absolute zero: -273^0C or -460^0F. In the Kelvin (absolute) scale, a change of one degree is the same as in the Celsius scale. So the freezing of water is at 273K. In the Rankine (absolute) scale, a change of one degree is the same as in the Fahrenheit scale. So, the freezing of water is at 460^0R. In general, the conversion to the absolute scales is given by:

$$K = {}^0C + 273 \qquad \text{and} \qquad {}^0R = {}^0F + 460 \qquad [7.4]$$

(Note: The exact value of the absolute zero was determined to be -273.15 ^{0}C. Therefore, to a higher level of accuracy, equations [7.4] are: K=^{0}C+273.15 and ^{0}R=^{0}F+459.67.)

It was found that the relationship between the temperature (T) of an ideal gas and its pressure (P) and volume (V) could be expressed as:

$$\frac{P \cdot V}{T} = \text{constant}$$

which is the same as [7.5]

$$\frac{P_1 \cdot V_1}{T_1} = \frac{P_2 \cdot V_2}{T_2}$$

where the indices 1 and 2 indicate two states of the same confined gas. The temperature T must be expressed in an absolute scale (Kelvin or Rankine). The constant depends on the amount of the gas in the container and on its kind.

Equations [7.5] include in them as special cases the expansion of gas at constant pressure and the change of pressure at constant volume when the temperature changes, as illustrated in figures 7.4. They also describe the relationships between the volume and pressure of a gas when its temperature T is not changing. When P is constant, the volume V is proportional to the temperature T (this is Charles and Gay-Lussac's Law). When T is constant, the product of the pressure and the volume is constant, or P is inversely proportional to V (this is Boyle's Law).

Example

The pressure of the gas in a container is 2 atmospheres, when the temperature is 20^0C. A fire broke out in the room, and the temperature of the gas rose to 120^0C. Assuming that the volume of the container did not change (which is a reasonable approximation), what would be the pressure of the gas?

Solution

The situation is described by formulas [7.5]. Let the index 1 denote the state of the gas before the fire, and index 2 denote it after the fire. Accordingly, P_1=2 atmospheres, P_2=? atmospheres, $T_1=20^0C$, $T_2=120^0C$, $V_1=V_2$. Since P_1 and P_2 are both expressed in atmosphere, they can be substituted in [7.5] without any conversion. However, the temperatures in [7.5] must be expressed in Kelvin. Therefore, they will be converted according to [7.4]: T_1=293K, T_2=393K. When all that is substituted into [7.5] we get: $\frac{2 \cdot V_1}{293} = \frac{P_2 \cdot V_2}{393}$. Since V_1=V_2, they can be eliminated, leaving us with one equation with one unknown (P_2). Solving yields P_2=2.68 atmospheres.

7.3 Heat

7.3.1 What is heat?

"He who got scalded by milk is wary of yogurt."

In many cases, there is no way of telling how hot an object is by just looking at it. This is because the temperature of an object depends on the motion of atoms, molecules, nuclei, or electrons within it. Those tiny particles are in constant random motion. The kinetic energy of the random motion of those particles determines the temperature of the object. The higher that random kinetic energy, the higher is the temperature. However, if the object as a whole is moving, the kinetic energy of that motion is not a part of the random kinetic energy that expresses itself as the object's temperature.

Molecules of gasses move around with very small interactions among themselves. The empty spaces between the molecules are large compared to the size of the molecules. The entire situation may be visualized as a few ping-pong balls moving freely within the volume of a room, colliding occasionally with each other and with the walls. Molecules in liquids may be visualized as ping-pong balls in a loose bag. They are densely packed, but they still can slide on each other. If transferred to another container, they will rearrange themselves to fit the new container. Molecules in solids can be visualized as ping-pong balls that have been glued together. They are packed tight, and they can just slightly vibrate around their positions. In some solids, electrons lose their strong connections to the nuclei. In such cases the nuclei are held in place, and those electrons are free to move inside the solid. When the temperature of a gas or a liquid increases, the average random velocity of its molecules increases. When the temperature of solids is increased, the molecules or the nuclei vibrate more vigorously around their equilibrium points. The random velocity of free electrons, if they are present, increases with temperature. In all those cases, the random kinetic energy of the involved elements of the substance increases when the temperature rises, and decreases when the temperature drops.

When the temperature of many solids increases beyond a critical point, the vibrations of their molecules become so strong that the bonds between them break, and the solid becomes a liquid. When the temperature of liquids increases beyond a critical point, the molecules break their loose bonds with their neighbors and move freely in space, thus becoming a gas. The transitions from solid to liquid, from liquid to gas, and their inverses are called phase transitions.

Heat is that energy that when provided to an object would cause its temperature to rise, or would change its phase from solid to liquid or from liquid to gas. When heat energy is taken away from an object, the opposite processes occur. Thus, heat is a form of energy that manifests itself in the random motion of atomic-size elements that make up the object. The difference between heat energy given to an object and kinetic energy given to the object is that heat energy affects the random motion of particles at the atomic scale that make up the object. It affects the temperature of the object, but not its motion. On the other hand, the kinetic energy of an object relates to the motion of the object as a whole, and is not manifested in its temperature.

7.3.2 The First and Second Laws of Thermodynamics

As a form of energy, heat can be generated by the conversion of other forms of energy. For example, when we rub our hands, the mechanical energy of the motion of our hands is converted to heat energy, which causes the temperature of our hands to rise. When we boil water on a gas stove, the chemical energy of the gas is converted into the heat energy of the flame, which is converted into heat energy of the water. "The First Law of Thermodynamics" recognizes heat as a form of energy, which obeys the law of conservation of energy like any other energy form.

Heat energy can be transferred from a hot object to a colder object or from a hot part of an object to a colder part of the same object. The random kinetic energy of the particles of a hot object is greater than the random kinetic energy of the particles of a cold object. When the two objects come into contact, particles at the interface collide with each other. As a result of those collisions, some particles lose random kinetic energy and others gain it. Overall, the particles of the hot object loose more random kinetic energy than they gain. That manifests itself in the cooling of the hot object. Overall, the particles of the cold object gain more random kinetic energy than they loose. That manifests itself in the heating of the colder object. When objects of equal temperatures touch each other, their exchange of thermal energy is at equilibrium. Each object looses and gains the same amount of random kinetic energy. Their temperatures do not change.

When a hot object contacts a cold object, the hot object will never become hotter while the cold object becomes colder. For example, when it is hot outside, heat energy will not leave the cool room (thus cooling it even more) and be transferred to the hot outside (thus making it even hotter). Such a process would not contradict the First Law of Thermodynamics, because heat that would have been lost by the colder room would have been gained by the warmer outside. Another example of a similar situation is a glass of water. The water will never separate by itself into two parts: a freezing part and a boiling part. These are two examples of a law of nature, called "The Second Law of Thermodynamics". According to this law, heat can flow on its own from a hot object to a colder object. Heat cannot flow on its own from a cold object to a hot object. It is possible, though, to cool a room when it is hot outside, and to deposit the heat extracted from the room in the outside. However, that process cannot happen on its own. It requires the participation of some external force, such as an air conditioner. Another aspect of the Second Law of Thermodynamics deals with the efficiency of converting heat energy to mechanical energy. It is impossible to convert the entire random kinetic energy of the atoms in heat sources into useful work. So, in practice, machines will always exhaust heat-waste.

7.3.3 Units of Heat

A unit of heat-energy has to be defined in order to be able to measure heat and to treat it in scientific and engineering applications. Since water is one of the most common substances, both the SI and the British system rely on it to define associated units of heat. In the British system, the unit of heat is the British Thermal Unit (BTU). A BTU is the amount of heat that is needed to raise the temperature of one pound of water by one degree Fahrenheit. Kilocalorie is a common unit of heat, defined in conjunction with the SI system. A kilocalorie is the amount of heat needed to raise the temperature of one kg of water by one degree Celsius. It should be noted that under standard conditions, the volume of one kg of water is one liter. So, it may also be said that one kilocalorie is the amount of heat needed to raise the temperature of one liter of water by one degree Celsius. The relationship between a BTU and a kilocalorie is:

1 BTU=0.252 kilocalories   1 kilocalorie=3.97 BTU

There are 1000 calories in a kilocalorie. A kilocalorie is sometimes also referred to as a Calorie, written with an uppercase C. Since Calorie and BTU measure heat, and heat is a form of energy, it is possible to convert them to the conventional energy units:

1 kilocalorie=4,190 J and   1 BTU=778 ft·lb=1056 J

(Note: The unit of heat in the SI system is the joule. Calorie is associated with kg, similarly to the BTU, which is associated with pound. In the following, we may say loosely that Calorie is an SI unit and BTU is a British system unit.)

The cooling and heating power of devices such as air conditioners, furnaces, ovens, etc. is measured in the amount of heat that they can exchange per unit time, e.g. BTU/hr (BTUh), Calorie/second, etc. Since heat is a form of energy, these quantities can be expressed also by units of power (which is energy per time):

1 BTU/hour = 0.293 watt = 0.216 ft·lb/s

Please note that the second, and not the hour, is the unit of time in the British system.

Example

What is the relationship between BTU/hour and horsepower?

Solution

Horsepower is a unit of power. One horsepower is equal to 745 watt, which are 745 joules per second. BTU/hour is also a unit of power. The conversion between the two is:

$$\frac{1\text{BTU}}{1\text{hr}} = \frac{1056\text{J}}{3600\sec} = 0.293\text{Watt} = \frac{0.293\text{Watt} \cdot \text{horsepower}}{745\text{watt}} =$$
$$= 3.9\text{x}10^{-4}\,\text{horsepower.}$$

Example
The writing on an air-conditioning unit is 12,000 BTUh, 1hp (horsepower). What does it mean?

Solution
The air conditioner can extract 12,000 BTU of heat from the room every hour, and deposit it in the outside. To be able to do so, the air-conditioner's motor consumes electrical energy at the rate of 1 horsepower. One horsepower is 2,564 BTU/hr. So, energy of 2,564 BTU has to be provided to the motor for every 12,000 BTU of heat that it moves from the room to the outside.

7.3.4 Heat, Temperature, and Matter

When a Calorie is provided to a kilogram of a substance other than water, the temperature change of that substance would be usually different from 1^0C. In the SI system, the specific heat of a substance (c) is defined as the heat in Calories needed to raise the temperature of one kilogram of that substance by 1^0C. In the British system, the specific heat (c) would be the number of BTU's needed to raise one pound of the substance by 1^0F. The numerical value of the specific heat (c) of any substance is the same in the SI and in the British system. For example, the specific heat of water is 1 Calorie/(kg ^{0}C) in the SI system, and 1 BTU/(pound ^{0}F) in the British system.

The amount of heat ΔQ needed to raise the temperature of a certain amount m of a substance is proportional to m, and to the desired temperature change ΔT. The proportionality constant, which is a property of the substance, is c–the specific heat of that substance. The relationship between those variables is expressed by formula [7.6]:

$$\Delta Q = c \cdot m \cdot \Delta T \qquad [7.6]$$

where $\Delta T = T_{final} - T_{initial}$ of the substance.

In the British system, ΔQ is in BTU, m is in pounds, and ΔT in ^{0}F. In the SI system ΔQ is in kilocalories (or Calorie), m is in kg, and ΔT is in ^{0}C. The numerical values of the c are the same in both unit-systems. The sign of ΔQ is the same as the sign of ΔT. When heat is provided to the substance, its final temperature is greater than it initial temperature. Both ΔQ and ΔT are positive in those cases. When heat is taken away from a substance, it cools off, and both ΔQ and ΔT are negative.

Consider the process of heating a bucket of ice whose temperature is below freezing. The ice turns into water, and then the water turns into steam. When analyzed closely, a number of stages could be identified in that process. First, the temperature of the ice rises till it reaches 0^0C. Then, the ice melts into water. During the entire melting stage, the temperature of the ice-water mixture stays at 0^0C (provided that the mixture is stirred thoroughly). After all the ice has completely melted, the temperature of the water rises till it reaches boiling at 100^0C. Then, as more and more water turns into steam, the temperature of the boiling water stays at 100^0C, until all the water has turned into steam. The entire process of turning one kg of ice at -20^0C into steam at 120^0C is illustrated in

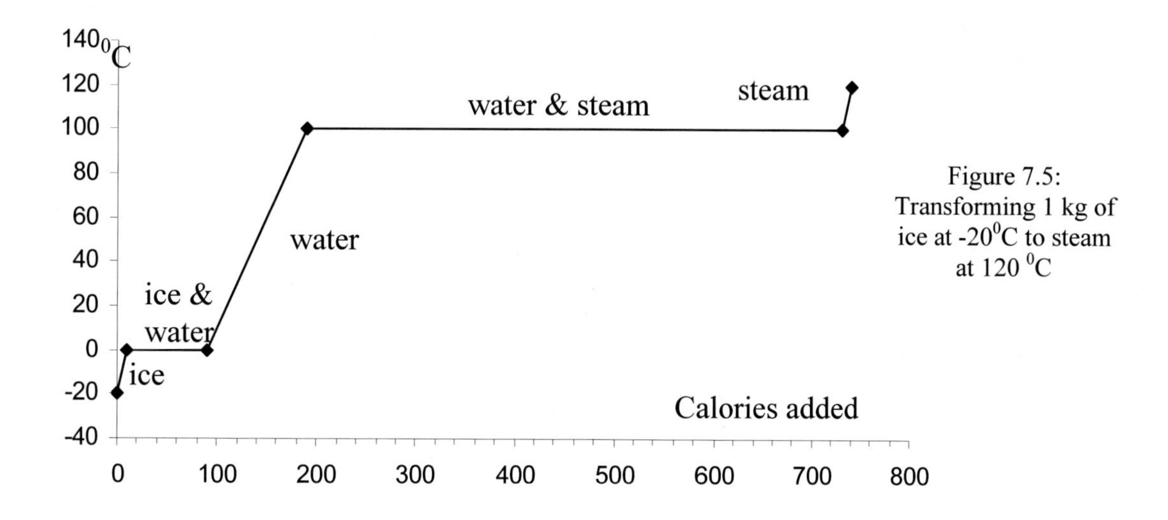

Figure 7.5: Transforming 1 kg of ice at -20^0C to steam at 120 ^{0}C

figure 7.5. The amount of heat needed for each stage is indicated too.

Formula [7.6] describes what happens when heat is provided to a solid or liquid substance, and the substance does not change its phase (solid remains solid and liquid remains liquid). When the provided heat causes a phase transition (solid to liquid or liquid to gas), formula [7.6] does not describe what happens, because heat is provided, and the temperature does not change. Formula [7.6] is also not applicable for phase transitions in cooling (gas to liquid, and liquid to solid), where heat is extracted form the substance. The relationship between ΔQ, the heat exchange in a phase transition and the amount of substance m that changes its phase is given by [7.7]

$$\Delta Q = L \cdot m \qquad [7.7]$$

L is called the latent heat of the substance. Substances have two kinds of latent heat: latent heat of fusion and latent heat of vaporization. In the SI system, the latent heat of fusion provides the number of Calories required to melt one kg of the substance that is already at its melting temperature. In the British system, the latent heat of fusion provides the number of BTU's required to melt one pound of the substance at its melting temperature. Similarly, the latent heat of vaporization provides the amount of heat required to vaporize one unit of the substance, when the substance is already at its boiling point. In cooling processes, those latent heats provide the amount of heat released (or extracted) in the corresponding phase transitions.

	Melting point (^{0}C)	Boiling point (^{0}C)	Specific heat		Heat of fusion		Heat of vaporization	
			kcal/kg ^{0}C Or BTU/lb ^{0}F	J/kg^{0}C	kcal/kg	J/kg	kcal/kg	J/kg
Aluminum	660	2057	0.22	920	76.8	3.21×10^5		
Brass	840		0.092	390				
Copper	1083	2330	0.092	390	49.0	2.05×10^5		
Glass			0.21	880				
Ice	0		0.51	2100	80	3.35×10^5		
Iron, steel	1540	3000	0.115	481	7.89	3.30×10^5		
Lead	327	1620	0.031	130	5.86	2.45×10^5		
Mercury	-38.9	357	0.033	140	2.82	1.18×10^5	65	2.72×10^5
Silver	961	1950	0.056	230	26.0	1.09×10^5		
Steam			0.48	2000				
Water	0	100	1.00	4190	80	3.35×10^5	540	2.26×10^6

Table 7.2: Specific and latent heats

Phase transitions may also take place not at freezing and boiling temperatures. For example, wet clothes dry on the cloth line in temperatures much lower than 100 ^{0}C. Evaporation is a phase transition that happens not necessarily at the boiling temperature. Formula [7.7] applies also to evaporation processes. The value of L will depend on the temperature at which the evaporation occurs. For water, the latent heat of vaporization at 100^{0}C is 539 Calorie/kg, and at 20^{0}C is 585 Calorie/kg. The difference between evaporation and boiling is that evaporation occurs at the surface of the liquid, while in boiling the gas emerges from the entire volume of the liquid.

The boiling temperature and the latent heat depend on the external atmospheric pressure. The values given above are for standard pressure of 1 atmosphere. If the external pressure is lower, such as in mountain areas, the values of L and the boiling point decrease with the external pressure.

Heat associated with processes that do not involve a phase change (equation [7.6]) is called sensible heat. Heat associated with a phase change (equation [7.7]) is called latent heat.

7.3.5 Exchange of heat in closed systems

Since heat is a form of energy, which is stored by objects in the form of random kinetic energy, the Law of Conservation of Energy applies to the transfer of heat between objects. A special case of conservation of energy deals with situations where a group of objects exchange heat with each other, but they do not exchange heat with any other object. Also, no other form of energy is exchanged or transformed by those objects. For example, when a thermos contains ice and water, heat is exchanged between the ice and the water, but, to a good approximation, heat is not exchanged between those ice and water and anything else. If the objects in such a system are denoted by numbers 1, 2, 3,..., and the amounts of heat exchanged by the objects are denoted by ΔQ_1, ΔQ_2, ΔQ_3,... respectively, the Law of Conservation of Energy could be expressed as:

$$\Delta Q_1+\Delta Q_2+\Delta Q_3+\ldots=0 \qquad [7.8]$$

Formula [7.8] states that the total heat exchanged in a closed system is zero. Any heat lost by an object is gained by other object/s, and any gained heat comes from heat lost by other object/s. Heat gained is expressed by positive ΔQ, and heat lost is expressed by a negative ΔQ. Heat is always transferred from a hot object to a colder object: Heat cannot move by itself from a cold object to a hot object or between objects at the same temperature. Therefore, if objects at a closed system have different temperatures, they will exchange heat until they reach equilibrium at the same temperature. That equilibrium temperature would be somewhere between the initial highest and initial lowest temperatures. The exact value of the final temperature depends on the details of the system.

Example

Ten liters of water at 80^0C are poured into a tub containing 25 liter of water at 15^0C, and mixed thoroughly and quickly till the entire mixture reaches a final uniform temperature. Assume that the mixing was done so quickly that no significant heat exchange with the tub or the air could take place. What was that final temperature of the mixture?

Solution

The heat lost by the poured water was gained by the water initially in the tub. Let an index 1 denote the poured water, and an index 2 denote the water initially in the tub. According to [7.8] $\Delta Q_1+\Delta Q_2=0$. In order to find ΔQ_1 and ΔQ_2 we use equation [7.6]. According to that notation, and the definition of ΔT in [7.2a] we get: $c_1=1$ Calorie/(kg·^{0}C), $m_1=10$kg, $T_{1,initial}=80^0$C, $T_{1,final}=T_{final}=?$,
$\Delta T_1=(T_{final}-80)$; $c_2=1$ Calorie/(kg·^{0}C), $m_2=25$kg, $T_{2,initial}=15^0$C, $T_{2,final}=T_{final}=?$, $\Delta T_2=(T_{final}-15)$. Substituting these data in [7.8] gives:

$$1 \cdot 10 \cdot (T_{final} - 80) + 1 \cdot 25 \cdot (T_{final} - 15) = 0$$

Solving for T_{final} yields: T_{final}=33.6^{0}C.

7.3.6 Heating and cooling of air

Air is a mixture of gasses. Around ground level air consists of 78% nitrogen, 21% oxygen, 0.1% argon, 0.03% CO_2 and the rest are other gasses. The concentration of water vapors in the air varies with time and location. It may reach 4% in warm tropical locations and only a small fraction of a percent in polar areas. The **absolute humidity** is a measure of the amount of water vapor in the air. The absolute humidity is the ratio between the mass of water vapor and the mass of all the other gasses.

The gasses that make up dry air (all the gasses except water vapor) remain in their gas phase in the entire range of air temperatures that occur naturally on earth. Water vapors are different. They may turn into water or ice under conditions that happen naturally. Another property of water vapor is that it has a saturation point. At any given temperature, only a certain amount of water vapor could be contained in the air. Any excess water vapor would condensate into water. The higher the temperature of the air, the more water vapor it can contain. The top curved line in figure 7.6 is the saturation curve. It gives the highest possible absolute humidity (saturation) as a function of temperature.

The relative humidity is a measure that combines the actual amount of water vapor in the air and its amount at saturation at the same temperature. The relative humidity is the ratio between the actual mass of water vapor in the air to the mass of water vapor at saturation. For example, if at 80^{0}F 1 lb of air contains 0.010 lb of water vapor, the absolute humidity is 0.010/1=1%. From figure 7.6 we get that at saturation the absolute humidity of that air is 0.022 or 2.2%. The relative humidity of that air is therefore 0.010/0.022=45%.

As unsaturated air is cooled, it can hold less and less water vapor. Eventually, it may reach the temperature at which it becomes saturated. Beyond that temperature, excess water vapors liquefy into water droplets. That temperature is called the **dew point** of that air. In a hot and humid summer day, water droplets condensate on the outside of a glass of iced water. The water vapor in the room-air is unsaturated. The air that comes in contact with the cold glass cools to its dew point (or below it), and cannot contain all its water vapor. The excess liquefies and condensates on the glass. The process is similar to the formation of dew outdoors on plants and on the ground that are colder than the surrounding air. If the ground's temperature is below freezing, excess water vapors may freeze directly into small particles of ice, forming frost.

Water that condensate regularly on the cooling elements of air-conditioners and refrigerators have to be discarded. Water may condensate also on water pipelines whose temperatures are below the dew point of the surrounding room-air. Proper insulation of the pipelines may resolve that problem. The insulation keeps the temperature of the air film that comes in contact with the pipeline above the dew point. Similar situation may occur in basements, whose walls are colder than the air in the basement, or on sinks and bowls that the water in them lowers their temperatures below the temperature of the (humid) air around them. If a wall is insulated by a porous insulator, such as fiber glass, it is common to cover that insulator by a sheet of plastic or thick paper that block humid air from seeping through the insulator to the cold wall and condensate there.

Water on wet surfaces evaporates into the surrounding unsaturated air. This process would continue until the surfaces dry out. However, if the surrounding air were saturated, that evaporation could not take place, because saturated air cannot absorb any more water vapors. In such cases, the surfaces would remain wet. In humid days, or in marsh areas where the air is almost saturated, sweat released from the sweat glands remain on the skin. Its evaporation into the humid air is too slow or completely impossible. The evaporation rate depends on the relative humidity of the surrounding air. The higher the relative humidity, the slower is the evaporation.

In order to cool room-air by 1^0C, approximately 0.2 kilocalorie of heat has to be extracted from each **kilogram** of air. (Under standard conditions, 1 kilogram of air occupies about 0.77 m^3). This amount of sensible heat does not depend very strongly on the humidity or the temperature of the air. However, if the dew-point temperature is crossed in the cooling process, significantly more heat has to be extracted. Heat has to be extracted to cool the air and also to liquefy the water vapors. Each **gram** of water vapor that condensates require the additional extraction of approximately 0.5 kilocalories of latent heat. So, an air conditioning unit has to spend most of its energy to condensate water vapors, without cooling the air, as the dew point is crossed. Once the cooled air becomes unsaturated again, the energy spent by the air conditioning unit goes again only into cooling the air.

It is possible to cool hot dry air without using an air conditioner. If water droplets are sprayed into hot dry air, they evaporate. The evaporation process requires approximately 0.5 kilocalories for each 1-gram of evaporated water. That heat is extracted from the hot dry air, thus cooling it. One gram of evaporating water can lower the temperature of 1 kg (approximately 0.77 m^3) of air by approximately 2.5 ^{0}C. The amount of evaporation, and the cooling that comes with it, depend on the relative humidity of the air. They decrease with increased humidity, and stop when the air is saturated. So, it is impossible to cool saturated air by spraying water droplets into it.

7.3.7 Sensible heat, latent heat, and enthalpy

Heat is a form of energy stored by an object. The heat of humid air consists of two kinds of stored energies: sensible heat and latent heat.

Sensible heat of air is the energy due to the random velocities of air and water-vapor molecules. The higher is the temperature of the air, the faster is that random motion. Therefore, increasing the temperature increases the sensible heat of air, and decreasing the temperature decreases it.

Latent heat of humid air is the energy stored by the water molecules, due to them being in the state of vapor, rather than in the state of liquid. It takes energy to convert water to water vapor. That energy is stored by the vapor. If the vapor condensates to water, it releases back that energy. Latent energy has to be taken from the vapor in order to condensate it. Therefore, latent heat does not depend on the temperature of the air.

When the temperature of the air changes without changing the amount of the water vapor in it, the sensible heat of the air changes whereas its latent heat is unchanged. When the temperature of the air does not change, but the amount of vapor in it changes (decreases due to condensation or increases due to added vapors) the latent heat changes whereas the sensible heat does not change.

Enthalpy of air is the sum of its sensible and latent heats. Specific enthalpy is the enthalpy per unit mass (or per unit weight). Common enthalpy units are BTU per pound of dry air, or kilocalorie per kg of dry air. If we know the enthalpies of a certain mass of air at two different temperature states, we can find out how much heat is involved in the transition from one state to the other.

Example

The specific enthalpy of air at 90 ^{0}F and 80% relative humidity is 48.5 BTU/pound of dry air. The specific enthalpy of air at 72 ^{0}F and relative humidity of 50% is 26 BTU/pound of dry air. How many BTU's have to be extracted from 2,000 pounds of air at 90 ^{0}F and 80% relative humidity in order to cool it to 72 ^{0}F and de-humidify it to 50% relative humidity?

Solution

In order to cool and de-humidify one pound of air between those two states, 48.5-26=22.5 BTU of enthalpy will have to be extracted from it. To do the same to 2,000 pounds of air, 22.5x2,000 =45,000 BTU will have to be extracted. This amount of heat has to be extracted to cool the air and the vapor in it from 90 ^{0}F to 72 ^{0}F, and to condensate some of the vapor so that the relative humidity drops from 80% to 50%.

(Note. The specific weight of dry air at 90 ^{0}F is approximately 2% less than when its relative humidity is 80%. We have approximated the weight of the humid air by the weight of dry air.)

The definition of enthalpy given above and the way that it is used are the ones common in heating and air conditioning applications. A more general definition, for other uses and circumstances, has not been discussed here.

7.3.8 Heat Comfort (Psychrometry)

Organisms always generate heat as a byproduct of their activities. They are also exposed to their surroundings, and exchange heat with it. Hot surroundings transfer heat to the organism, and cold surroundings extract heat from it. In order to function properly, all organisms have to maintain their body temperature within strict margins, and to balance between their internal heat generation and the exchange of heat with the surroundings.

The human body uses several mechanisms to regulate its temperature. Heat is carried from internal parts of the body to its surface by the blood flow. At the skin, heat is exchanged with the surrounding air using a variety of mechanisms. Heat is also exchanged by radiation with objects that do not have a direct physical contact with the body, such as the sun or other heat-radiation sources. An important cooling mechanism is by evaporation. Sweat that evaporates into the air extracts some of the required heat from the skin of the body, thus cooling it off. Heat exchange occurs also in internal surfaces of the body, such as the stomach and the lungs. Cold drinks and foods are warmed in the stomach to body temperature, thus extracting heat from the body. The temperature of exhaled air is usually higher than that of the inhaled air. Water vapor, evaporated at the lung's linings, extract heat from the body. All those heat losses cool the body.

Since all those mechanisms are involved with regulating the body's temperature, it is not surprising that our sense of heat-comfort depends both on the temperature of the surrounding air and on its humidity. If the air is too hot or too cold, our body would feel a corresponding discomfort. If the air is too dry, our skin and the linings of the mouth, throat, and lungs won't be able to maintain their normal moisture, resulting in discomfort. If the air is too humid, heat loss through evaporation won't be effective, resulting in discomfort.

Standards are available that define the ranges of air temperature and humidity that are considered comfortable by the majority of people. Roughly, temperatures between 68-78^{0}F and relative humidity between 20-80% are considered comfortable. More detailed characterization of comfort zones take into consideration the season (that affects clothing), air circulation, and available radiant heat. Figure 7.6 is a psychrometric chart, which shows the comfort zones. The shown comfort zones are for average values of air circulation and radiated heat. The thin film of air around our body is warmed by our skin and absorbs evaporated sweat. Circulating air replaces that film of air by cooler, less humid film, thus assisting the body in cooling itself. Radiant heat, from manmade sources or due to direct sun radiation, warms the body in addition to the direct heat exchange with the air. The comfort zones of figure 7.6 have to be adjusted if those conditions are different from the standard ones.

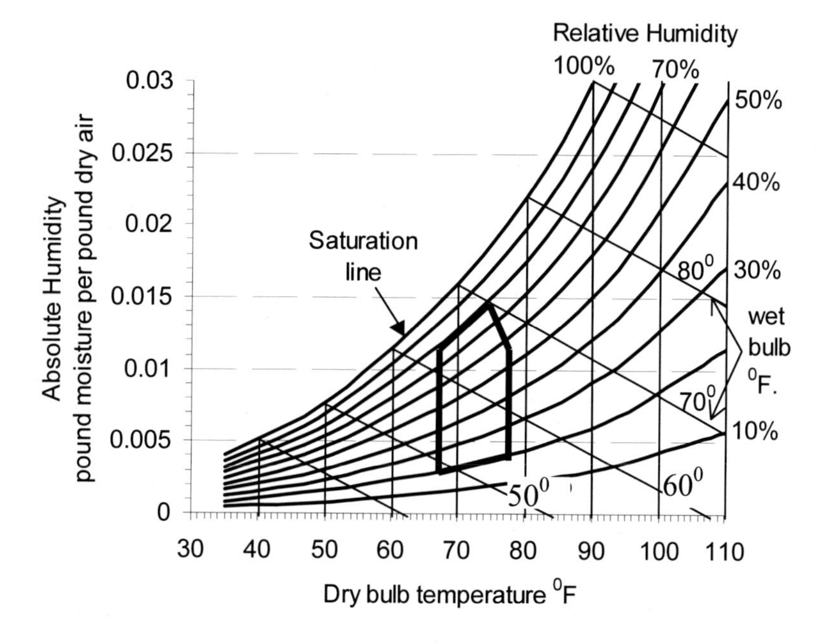

Figure 7.6: A psychrometric chart and thermal comfort zone (thick lines). The x-axis is the temperature of the air, and the y-axis is the absolute humidity of the air. Each point in the x-y plane is thus characterized by a pair of values: air's temperature and air's absolute

humidity. In addition, there are line families super imposed on those points. Each such line provides additional information about the points that it covers. In this chart, there are lines for relative humidity (the curved lines that fan out from the left end), and wet bulb temperature (the diagonal lines). For example, when the air's temperature is 90 ^{0}F, and its absolute humidity is 0.015, its relative humidity is 50% and its wet bulb temperature is 75 ^{0}F. (The latter value is interpolated based on the 70 ^{0}F and the 80 ^{0}F wet bulb temperature lines.) Wet bulb temperature is measured by a thermometer whose bulb is wrapped with a wet sleeve. It is simple to measure wet bulb temperature, and find the air's humidity from it, using the psychrometric chart. Common psychrometric charts include additional families of lines, such as enthalpy lines, which provide additional information. (Adapted from Lechner).

Example: The limits of evaporative cooling of air

Consider 1 m^3 of dry air, 0% humidity, at 90^{0}F. When droplets of water are sprayed into it, they evaporate by absorbing heat from the air, thus cooling it while raising its humidity. The evaporation, and with it the cooling of the air, will stop when the air reaches saturation. At what temperature will that happen, and how much water would have been evaporated by then?

Solution:

Let m be the mass of the evaporated water, T the final temperature of the cooled air, and T_0=90^{0}F the initial temperature of the air. The latent heat of evaporation of water is approximately 500 kilocalories per kg. The mass of 1 m^3 of air is 1.29 kg. The specific heat of dry air is approximately 0.14 kilocalories /(kg$\cdot^0$F).

The heat lost by the air is equal to the heat gained by the evaporating water. Based on [7.7] and [7.7] we get:

$$500\cdot m=0.14\cdot 1.29\cdot(90-T)$$

The mass m (kg) of water in 1 kg of saturated air at the temperature T degrees Fahrenheit is given by:

$$m=0.0004\cdot T-0.012$$

(This is an approximation to the saturation line in figure 7.6, for the temperature range $55\ ^0F \leq T \leq 75\ ^0F$.)

We have here two equations with two unknowns, m and T. Solving yields:

T=59^{0}F, and m=0.012 kg

7.4 Heat transfer

There are three mechanisms by which heat can be transferred from one object to another: conduction, convection, and radiation.

When we hold one end of an iron bar and put the other end in a fire, heat will reach our end after a short time. This propagation of heat did not involve any visible motion of the bar. It was accomplished completely due to small motions of atomic-scale particles that make up the bar. This is an example of heat transfer by conduction. Some materials, such as metals, are good heat conductors. Others, such as plastic or wood, do not conduct heat as well. We can hold one end of a stick of wood when the other end is burning, and not feel the heat. Such materials are called heat insulators.

A different mechanism of transferring heat is the one used in forced-air heating. Air is heated by a central furnace, and then blown to the rooms by a fan. That is an example of convection. In convection, heat is carried from one place to another by flows of matter.

The third mechanism is radiation. Rays from the sun warm the Earth. Those rays can move through empty space, and carry the heat-energy with them. This is an example of heat transfer by radiation. It can be accomplished even if there is no matter between the source of the heat and its recipient.

In real situations, heat may be transferred by any combination of those basic mechanisms. Here are some examples.

Heat is transferred from a hot room to the cold outside through walls, windows, roofs, and floors. If the walls don't have any cavities, the heat transfer through them is by pure conduction.

If the windows are made of a single glass pane, the heat transfer through them is by conduction and by radiation. A glass pane conducts heat like regular walls. In addition, glass allows heat radiation, such as that coming from the sun, to pass through it.

A double-pane glass window has a layer of air between two glass panes. Heat is transferred through such windows by conduction, convection, and radiation. Heat is conducted and radiated through the first pane, then through the air layer, and then through the other pane. In addition, small air currents develop in the air layer between the panes. Those currents transfer heat by convection.

The amount of heat transferred across the envelope of a structure is an important design factor, and it can be calculated or estimated according to the properties of the envelope and the heat transfer mechanisms.

7.4.1 Heat conduction

R-number, U-value, and thermal resistance

Most often, in order to feel comfortable, the air inside the house is maintained at a different temperature than the air outside. The air inside is in contact with the walls, which are in contact with the air outside. Because of the temperature differences between the inside and the outside, heat will flow from the high to the low temperature.

Assume that the inside is warmer than the outside. What would be the temperature of the wall? As heat flows through the wall from the inside out, some of the heat may be absorbed by the wall, and change the wall's temperature. After a while, a steady state is reached, in which a stable temperature gradient is formed across the wall. At the room side, the wall's temperature is the same as that of the room. At the outside, it is the same as the outside's temperature. Inside the wall, the temperature changes gradually between those two values. If the wall is large, solid (no cavities), and homogeneous, the temperature changes linearly across its thickness.

After reaching the steady state, the heat that flows through the wall does not change the temperature of the wall. However, as the indoor or outdoor temperatures change with time, a new steady state has to be reached. In order to reach that steady state, the wall may absorb some of the heat that flows through it, or it may release some heat to the surroundings. In hot climates, walls that get warm during the day may release at night the heat stored in them, thus reducing the effect of the night's cooler air.

Common estimates of heat flows through walls are done under the assumption that the heat flows through the wall at a steady rate. No net heat is absorbed or released by the wall itself. It was found that for flat, large, solid, and homogeneous walls, the amount of heat (ΔQ), that flows in the time interval (Δt) through a wall is proportional to the area of the wall (A) and to the temperature difference (ΔT) between the two sides of the wall. The flow is inversely proportional to the thickness of the wall Δx. All those relationships are contained in formula [7.9a]:

$$\frac{\Delta Q}{\Delta t} = \frac{\sigma \cdot A \cdot \Delta T}{\Delta x} \qquad [7.9a]$$

where σ is a constant, called the specific thermal conductivity, which depends on the material of the wall (see Table 7.3).

It is common to combine the thickness of the wall (Δx) and the specific thermal conductivity σ to one constant:

$$R = \frac{\Delta x}{\sigma}.$$

When R is substituted in [7.9a] we get

$$\frac{\Delta Q}{\Delta t} = \frac{A \cdot \Delta T}{R} \qquad [7.9b]$$

R is called the R-value. The R-value is proportional to the thickness (Δx) of the material. When [7.9b] is solved for R we get: $R = \frac{\Delta t \cdot A \cdot \Delta T}{\Delta Q}$. The common units in which the R-value is given are neither in the British nor in the SI system. They are in $hr \cdot ft^2 \cdot F/BTU$. So, when those units are used for the R-value in any formula, time has to be expressed in hours, length in foot, area in $foot^2$, temperature in degrees Fahrenheit, and heat in BTU. R is also called sometimes the thermal resistance. However, we will use the term "thermal resistance", denoted by r, to describe a different property, to be discussed soon.

Example

What is the R-value of a panel of glass, 0.5 inch thick?

Solution
The thermal conductivity of glass (σ) is 0.5 BTU/(ft ^{0}F h) (Table 7.3). The thickness of the glass is 0.5", which is 0.5/12=0.042 ft. The R-value is R=0.042/0.5=0.083 hr·ft^2·F/BTU.
Note: In order to find the R-value of glass windows, the effects of the surrounding air has to be considered. This is discussed in the following.

	BTU/(ft ^{0}F hr)	*cal/(cm ^{0}C hr)*
Air	0.011	0.16
Aluminum	140	2000
Brass	62	930
Brick	0.42	6.2
Cellulose fiber (loose fill)	0.023	0.34
Concrete blocks	0.35	5.1
Copper	220	3300
Corkboard	0.022	0.32
Glass	0.50	7.4
Gypsum board (sheetrock)	0.092	1.4
Mineral wool	0.026	0.39
Plaster	0.083	1.2
Polystyrene (expanded)	0.020	0.30
Polyurethane (expanded)	0.014	0.21
Steel/ iron	26	390

Table 7.3: Thermal conductivities (σ)

Formula [7.9b] can also be written as

$$\frac{\Delta Q}{\Delta t} = U \cdot A \cdot \Delta T \qquad [7.9c]$$

U is called the thermal conductivity or the U-factor of the wall.

In practice, it is quite common to use tables of R and/or U values, which already incorporate the width of the medium, instead of using tables of thermal conductivities (σ). In cases of solid uniform mediums, such as wood, it is justified to use $U = \frac{\sigma}{\Delta x} = \frac{1}{R}$ in order to calculate R and U values for boards of other thickness. Tables of R and U values are also available for non-uniform mediums, for air films that are in contacts with surfaces, and for air cavities. In those cases R is not proportional to the thickness, and such extrapolations to other widths are not justified. However, any given R and U values may be used in order to find the heat conduction through those elements according to formulas [7.9].

Material	*R-value*	*Material*	*R-value*
Plywood 0.25"	0.31	Plaster board 0.5"	0.45
Plywood 0.5"	0.62	Glass 1"	0.20
Hardwood 1"	0.91	Soft wood 1"	1.25
Common brick 4"	0.80	Asphalt shingles	0.21
Fiberglass board 1"	4.0	Acoustic tile 0.5"	1.19
2" by 4" wood stud, nominal	4.35	Sheathing 0.5"	1.32

Table 7.4a: R-values of common building materials (© ASHRAE)

	Position of Surface	*Direction of Heat Flow*	*R*	*U*
Still air				
	Horizontal	Upward	0.61	1.63
	Sloping 45^0	Upward	0.62	1.60
	Vertical	Horizontal	0.68	1.46
	Sloping 45^0	Downward	0.76	1.32
	Horizontal	Downward	0.92	1.08
Moving air				
15mph (winter)	Any	Any	0.17	6.00
7.5mph (summer)	Any	any	0.25	4.00

Table 7.4b: R (hr-ft^2-^{0}F/BTU) and U (BTU/hr-ft^2-^{0}F) values for air films at non-reflective surfaces. (© ASHRAE)

Position of air space	*Direction of heat flow*	*Thickness of airspace, inches*			
		0.5	0.75	1.5	3.5
Horizontal	Up	0.91	0.93	0.97	1.03
45^0 slope	Up	1.02	1.00	1.04	1.06
Vertical	Horizontal	1.13	1.18	1.12	1.14
45^0 slope	Down	1.15	1.26	1.27	1.27
Horizontal	Down	1.15	1.30	1.49	1.62

Table 7.4c: Thermal resistance (R-values), hr-ft^2-^{0}F/BTU, of air spaces. Values are for 0^0 mean temperature and 20^0 temperature difference, and effective space remittance of 0.82. (© ASHRAE)

Equations [7.9] can be rearranged in yet another way: The rate of heat flow through a conductor is affected by two major factors: the resistance of the conductor to heat flow and the temperature difference that drives that flow. If we denote the rate of the heat flow by I ($I = \frac{\Delta Q}{\Delta t}$), the thermal resistance of the conductor by r ($r = \frac{\Delta x}{\sigma \cdot A}$), and the temperature difference across the conductor by V (V=ΔT= the temperature difference), [7.9a] becomes:

$$V = I \cdot r, \quad \text{or} \quad I = \frac{V}{r} \qquad [7.9d]$$

where

$$V = \Delta T \qquad I = \frac{\Delta Q}{\Delta t} \qquad r = \frac{\Delta x}{\sigma \cdot A} = \frac{R}{A}$$

Equations [7.9d] look like equations that describe another phenomena – the flow of electricity through electric conductors. In that context (see equations [8.5]) they are called Ohm's Law. The similarity between heat flow and electricity flow does not stop here. It holds also for heat flow through assemblies of heat conductors.

Example

(a) What is the R-value of a 5/8" gypsum board? (b) What is the thermal resistance of a 5/8"x4'x8'gypsum board? (c) What is the heat flow rate through such a board when the temperature difference across it is 30^0F?

Solution

(a) From table 7.3 we get that the thermal conductivity of gypsum board is σ=0.092 BTU/(ft ^{0}F hr). The thickness of the board is Δx=5/8"=0.052'. Since $R = \frac{\Delta x}{\sigma}$ we get R=0.57 hr-ft^2-^{0}F/BTU.

(b) The thermal resistance of the board is

$$r = \frac{\Delta x}{\sigma \cdot A} = \frac{R}{A} = \frac{0.57}{4' x 8'} = 0.018\ {}^0F \cdot hr \cdot BTU^{-1}$$

(c) The heat flow rate is $I = \frac{V}{r} = \frac{30^0 F}{0.018} = 1{,}667\ BTU/hr$.

A short summary: Thermal properties of objects are described by three related constants: σ, R, and r.

σ, the thermal conductivity, is a property of the material from which the object is made. It does not depend on the dimensions of the object.

R, the R-value, is a property of flat objects. It incorporates the properties of the material/s and the thickness of the object. It does not take into account the surface area of the object.

r, the thermal resistance of an object, is also a property of flat objects. It incorporates the properties of the material/s from which the object is made and its dimensions.

Equations [7.9d] give the relationships between σ, R, and r for flat objects that are made of one material. Next, we will see how to calculate thermal properties of objects that are combinations of several flat thermal conductors.

Thermal resistors in series

Quite often, a wall is made of a number of "sandwiched" layers that have different R-values. All those layers have the same area A (figure 7.7). In such cases we say that the thermal conductors are connected in series. In stationary situations, all the heat that enters a layer from its warmer side leaves it through its colder side. No heat stays in any layer, or in the wall as a whole, because that would change the layer's and/or the wall's temperature, thus violating the stationary condition.

The equivalent R-value of layers connected in series is the sum of the R-values of all the layers. Say that a wall is made of three layers whose R-values are R_1, R_2, and R_3 (figure 7.7). The equivalent R-value of the entire wall R would be:

$$R=R_1+R_2+R_3 \quad [7.10a]$$

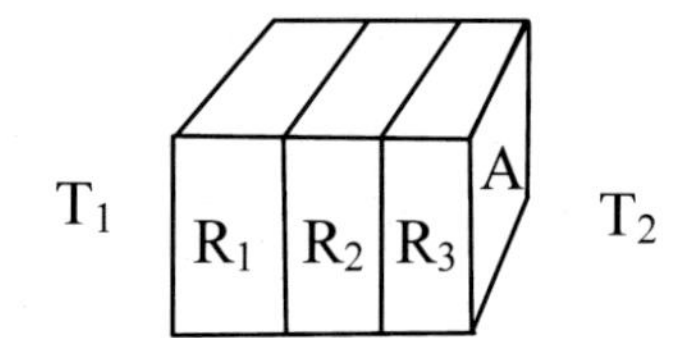

Figure 7.7: Three layers of thermal conductors connected in series. All have cross sectional area of A. The temperature at one side is T_1, and at the other is T_2. The equivalent thermal resistance R is $R=R_1+R_2+R_3$

That equivalent R-value can be used in formulas [7.9] to calculate the heat flow through the entire wall.

Since the area of each conductor is A, we get for the equivalent thermal resistance r of all the serial layers

$$r=r_1+r_2+r_3 \quad [7.10b]$$

The same amount of heat ΔQ passes through each layer, one layer after the other. In the notations of [7.9d], we can say that

$$I=I_1=I_2=I_3 \quad [7.10c]$$

The temperature difference across the combined conductors is equal to the sum of the temperature differences across the individual conductors, or in the notations of [7.9d]:

$$V=V_1+V_2+V_3 \quad [7.10d]$$

The temperature difference ΔT_i across the i'th thermal conductor can be found by combining equations [7.10] and [7.9]:

$$V_i = \Delta T_i = \frac{R_i}{R} \cdot \Delta T = \frac{r_i}{r} \cdot \Delta T \qquad [7.10e]$$

where ΔT is the temperature difference across the entire wall.

Still air is quite a good thermal insulator. The tiny air pockets that are trapped in wool fibers and in down feathers make those materials good thermal insulators. In buildings, a film of air attaches itself to surfaces of the structure such as walls and windows. Those air films act as insulators. Air cavities inside walls also act as thermal insulators. The R-values of those air films and air cavities add to the R-value of the solid wall according to formula [7.10a]. Unlike R-values of solid walls, R-values of air films and air cavities depend on the orientation of the wall, on the direction of heat-flow, and, in the case of outside walls, on the speed of the wind. The reason for this is that air both conducts and convects heat, and the motion of air, which results in convection, depends on the orientation of the wall along which the air in moving. Table 7.4b contains the equivalent R-value for air films and table 7.4c contains equivalent R-values for air cavities for a variety of situations.

Example
Find the R value of a thermal window, consisting of two panes of 0.5" glass (R-value 0.08 $hr \cdot ft^2 \cdot F/BTU$ each), separated by a layer of air 0.5" thick.

Solution
Looking from inside the room out, we have here five layers of insulation, "sandwiched" in series: 1) a thin film of air, 2) a glass pane, 3) an air cavity between the panes, 4) a glass pane, and 5) a thin film of air. The R-value of the window as a whole is the sum of the five R-values, according to [7.10a]. The air films and the air cavity are positioned vertically, and the heat flow through them is in the horizontal direction. Based on Tables 7.4 we get: R_1=0.68, R_2=0.08, R_3=1.13, R_4=0.08, R_5=0.17 $hr \cdot ft^2 \cdot F/BTU$. R_1 is the value for sill air, because it is inside the room. R_5 is the value for air in motion, when the wind velocity is 15 mph. The R-value of the window is the sum of those R's: R=2.14 $hr \cdot ft^2 \cdot F/BTU$. This is much higher than the R-value of the two glass panes combined (0.16 $hr \cdot ft^2 \cdot F/BTU$). It is in the same order of magnitude as plywood, but much smaller than R-19 fiberglass insulation, that is commonly used inside external walls in the northern parts of the U.S.

Example
An attic in the middle of a row house has a flat R-12 roof and R-8 floor, which is also the ceiling of the room underneath. The temperature of the room is kept at 72^0F and the outside temperature is 20^0F. Assume that heat flows only in the vertical direction from the room through the attic, through the roof, to the outside. Assume that the R-

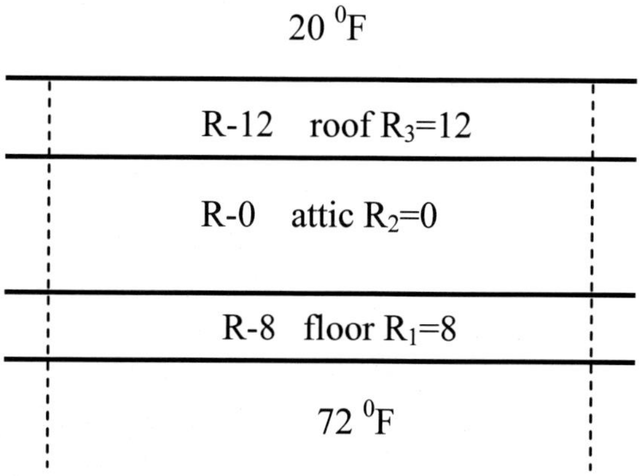

$20\ ^0$F

R-12 roof R_3=12

R-0 attic R_2=0

R-8 floor R_1=8

$72\ ^0$F

value of the air in the attic is zero. (Ignoring the R-values of the air films attached to the attic's surfaces, R=0 is a good assumption. Convection of heat by the air equalizes the temperature in the attic, which is the same as having an R-value of zero). What is the temperature in the attic?

Solution

The three layers of the attic are illustrated here:

According to [7.10e] we get:

$$\Delta T_1 = (72-20)\frac{8}{8+0+12} = 20.8^0\,\mathrm{F}$$

$$\Delta T_2 = (72-20)\frac{0}{8+0+12} = 0^0\,\mathrm{F}$$

$$\Delta T_3 = (72-20)\frac{12}{8+0+12} = 31.2^0\,\mathrm{F}$$

Therefore, the temperature inside the attic is 51.2^0F. (Starting from the room 51.2=72-20.8, and also, starting from the outside 51.2=20+31.2).

Thermal resistors in parallel

In order to find the heat flow through all the external surfaces of a house that are exposed to air (walls, roof, doors, windows), the heat flow through each such surface separately can be calculated according to equations [7.9], and then all those contributions are added up. Since the temperature difference is the same for all those surfaces, it can be factored out in equations [7.9] to give:

$$\frac{\Delta Q_{total}}{\Delta t} = \frac{\Delta Q_1}{\Delta t} + \frac{\Delta Q_2}{\Delta t} + \frac{\Delta Q_3}{\Delta t} + \ldots = \left(\frac{A_1}{R_1} + \frac{A_2}{R_2} + \frac{A_3}{R_3} + \ldots\right) \cdot \Delta T$$

This is called adding heat conductors in parallel. The same temperature difference drives the heat flow through each individual conductor. Figure 7.8 illustrates four conductors connected in parallel.

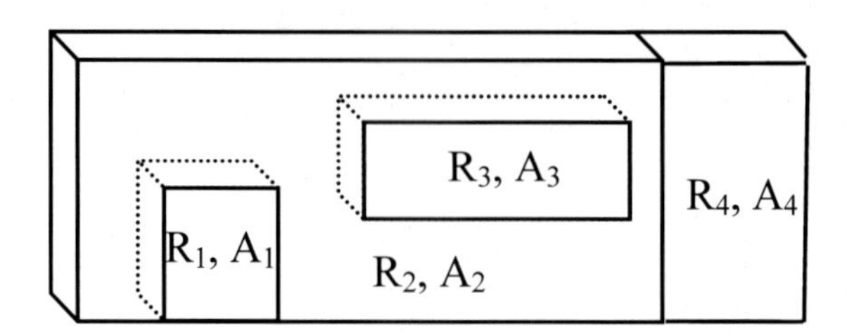

Figure 7.8: Four conductors connected in parallel. All their fronts are at temperature T_1, and all their backs are at T_2. Thus, the same temperature difference $\Delta T = T_2 - T_1$ drives heat through each conductor. The R's indicate individual R-factor, and the A's individual surface area (formula [7.10b])

In the notations of [7.9d] this can be written as:

$$I = I_1 + I_2 + I_3 + ... = (\frac{1}{r_1} + \frac{1}{r_2} + \frac{1}{r_3} + ...) \cdot V \qquad [7.10f]$$

which means that the combined thermal resistance r of all the resistances that are connected in parallel is related to the individual resistances through:

$$\frac{1}{r} = \frac{1}{r_1} + \frac{1}{r_2} + \frac{1}{r_3} + ... \qquad [7.10g]$$

Thermal resistors in series and parallel
Many walls are not uniform. For example, the inner part of a frame wall consists of air cavities separated by wood studs. In order to calculate the equivalent thermal resistance of such walls they are treated as a network of thermal conductors connected in series and in parallel, according to the situation. Consider a simple frame wall, consisting of a plasterboard plate on the inside, 2x4 wood studs in the middle, and an external sheathing on the outside. This wall is a combination of two kinds of "sandwiches". One kind (a in the figure 7.9) is made of plasterboard, wood stud, and external sheathing. The second kind (b in

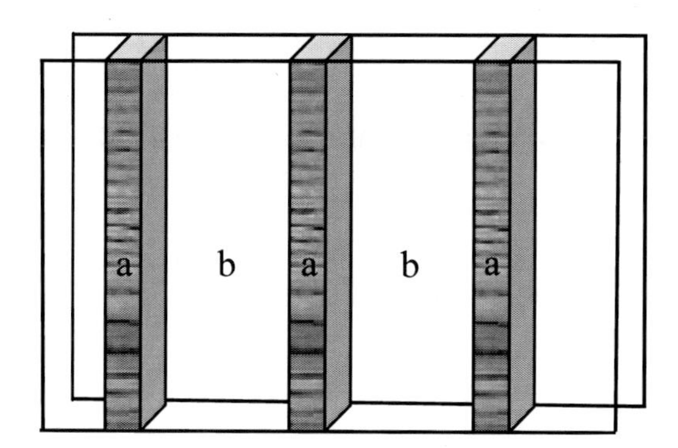

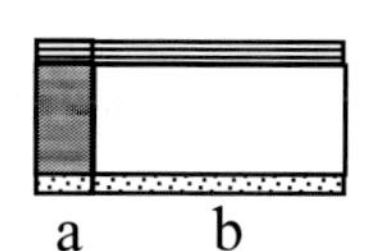

Figure 7.9: The thermal resistance of a frame wall

figure 7.9) is made of plasterboard, air cavity, and external sheathing. The R-value and the thermal resistance r of each "sandwich" is calculated according to [7.10a], because its layers are connected in series (figure 7.7). The r of the entire wall is calculated by adding all those "sandwiches" according to [7.10f] and [7.10g], because they are connected in parallel, as in figure 7.8.

Example (r and R of a frame wall).

In a typical framing (figure 7.9), the studs occupy 15% of the space between the boards, and the rest 85% are either air spaces or are filled with a thermal insulator. A 11'x9' frame wall is made of a 0.5" plaster board on the inside of the room, nominal 2x4 studs, and a 0.5" sheathing on the outside. What is the resistance (r) of the board and what is its equivalent R-value?

Solution

The wall can be divided into two kinds of "sandwiches": The first, ((a) in figure 7.9) consists of (1) an inside air layer, (2) a plaster board, (3) a stud, (4) an outside sheathing, and (5) an outside air layer. The second kind of sandwiches is (b) in figure 7.9: An inside air layer (1), a plaster board (2), a 3.5" air cavity between the boards(6) , an outside sheathing (4), and an outside air layer (5).
The corresponding R-values of these layers are (based on tables 7.4): R_1=0.68, R_2=0.45, R_3=4.35, R_4=1.32, R_5= 0.17, R_6=1.14.
Each sandwich consists of layered conductors connected in series. Therefore:

$R_a=R_1+R_2+R_3+R_4+R_5=0.68+0.45+4.35+1.32+0.17=6.97$

$R_b=R_1+R_2+R_6+R_4+R_5=0.68+0.45+1.14+1.32+0.17=3.76$

The surface area of the wall is 11x9=99ft^2. The total surface area of the stud sandwiches is: A_a=0.15x99=14.85ft^2. The total surface area of the air cavity sandwiches is A_b =0.85x99=84.15ft^2.

The total thermal resistance of the stud sandwiches is $(r=\frac{R}{A})$:

$$r_a=\frac{6.97}{14.85}=0.47\frac{hr\ ^0F}{BTU}.$$

The total thermal resistance of the air cavity sandwiches is $(r=\frac{R}{A})$: $r_b=\frac{3.76}{84.15}=0.045\frac{hr\ ^0F}{BTU}$.

The stud sandwiches are connected in parallel to the air cavity sandwiches (see figure 7.8). Therefore, the total thermal resistance of the wall is $\frac{1}{r}=\frac{1}{r_a}+\frac{1}{r_b}$, which gives after substitution: $\frac{1}{r}=\frac{1}{0.47}+\frac{1}{0.045}=24.3$,

yielding $r=0.041\frac{hr\ ^0F}{BTU}$.

The equivalent R-value of the entire wall is ($R=r\cdot A$):
R=0.041x99=4.06 hr-ft^2-^{0}F/BTU.

Basement walls and floor slabs

In order to find the heat flow through underground basement walls and floors, the ground temperature, and not the outside air temperature, has to be considered. Equivalent R-values are available for various basement wall materials and typical ground temperatures. A reasonable value for concrete basement floors is U=0.1 and for walls below ground is U=0.2 BTU/hr-ft^2-F. A common average deep ground temperature is 50 ^{0}F.

Formulas [7.9] are **not** applicable to floor slabs that lay on the ground in small buildings without basements. In such cases, most of the heat flow passes through the edges of the slab, and is affected by the temperature of the outside air, rather than by the temperature of the ground underneath the slab. The edges of the slabs have to be insulated to reduce those heat losses (figure 7.10). Formula [7.11] describes heat flow through such slabs.

$$\frac{\Delta Q}{\Delta t} = E \cdot L \cdot \Delta T \qquad [7.11]$$

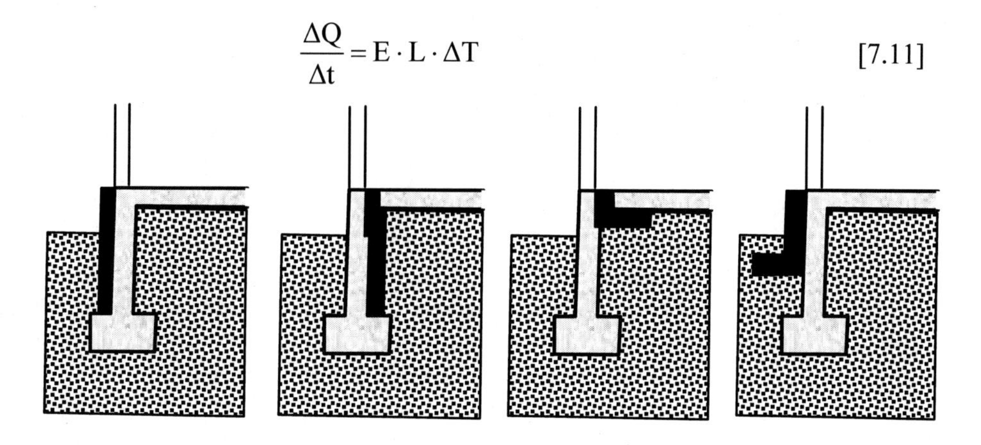

Figure 7.10: Insulating the edges of slabs. (Based on Lechner).

where E is the edge factor, and L is the length of the exposed edge of the floor. The edge factor E, which depends on the properties of the slab and on the edge insulation, can be found in tables.

Example

A 24'x36' concrete floor in a shed has R-5 edge insulation and an edge factor E=0.55 BTU/(hr ft F). The inside temperature is maintained at 65 ^{0}F. What is the heat loss through the floor when the outside temperature is 25 ^{0}F?

Solution

The circumference of the floor is (24'+36')x2=120'. According to formula [7.11], the rate of heat loss is 0.55x120x(65-25)=2640 BTU/hr.

7.4.2 Heat convection.

Hot air is lighter than cold air. Like cork that is lighter than water, hot air tends to float on colder air. If some parts of a room are warmer than others, the warm air moves upward and the cold air moves downwards. That creates air circulation that convects heat in the room. In rooms that have high vaulted ceiling, hot air that reaches the ceiling is trapped in the vault, and does not participate in the circulation. For that reason, high vaulted ceilings are desirable in hot climates. They keep the hot air up, and the cool air stays at the floor level, where it is appreciated. In colder climates, high vaulted ceiling increase the need to heat the livable lower parts of the room.

7.4.3 Heat radiation

In addition to heat radiated by the sun, radiation of manmade devices is used to direct heat to where it is needed. Heated filaments of wire are common sources of radiated heat. They are found in toasters and space heaters, and they radiate their energy without making any physical contact with the heated object.

Part of the heat radiation that hits an object is reflected and the rest is refracted into the object. The total energy of the incident radiation is divided between the reflected and refracted parts. So, if more radiated heat is reflected, less is left to be refracted into the object. The amount of reflected radiation depends on the material of the object, on properties of the reflecting surface, and on the angle of incidence. Glass partially coated with metal reflects better than plain glass. Light colored surfaces reflect more than dark colored surfaces. Polished surfaces reflect more than rough surfaces. Oblique rays are reflected more effectively than rays that hit the reflecting surface perpendicularly.

As the refracted radiation traverses through the object, it is gradually absorbed by it. If the object is opaque, it will absorb the entire radiation at the surface. If the object is transparent, some of the radiation will be absorbed as it traverses the object, and some will pass through. Absorbed radiation is converted into thermal energy, expressed by the random kinetic energy of the atomic-size particles of the object. Some of that absorbed energy is eventually re-radiated out.

A typical glazing reflects some of the radiation, absorbs some of it, and lets the rest pass through. A practical design problem is to figure out the amount of radiation that reaches the inside **(admitted radiation)** for various glazing configurations and external conditions.

Everything else being equal, the admitted radiation is proportional to the area (A) of the glazing. The larger the area of the glazing, the more the admitted radiation is.

All the other variables that affect the admitted radiation are expressed by two factors: the Solar Heat Gain Factor (SHGF) and the Shading Coefficient (SC).

The SHGF merges all the external variables: it provides the power density of admitted radiation through a standard clear glass pane 1/8" thick. The SHGF takes into consideration the reflected radiation, the inflows of direct radiation coming from the sun, the indirect diffused radiation from the sky, and the reflected radiation from the ground. Those inflows depend on the incident angle of the radiation, and on the path of the radiation through the atmosphere. In SHGF tables, that information is organized according to the orientation of the window, the geographical location, the date, and the time of day. The units of SHGF are BTU's per hour per square foot.

The shading coefficient (SC) takes into consideration the properties of the glazing itself: its absorption characteristics, thickness, shading devices that might be placed on its inside (e.g. Venetian blinds), and the reflective properties of its surface. All those reduce the amount of admitted radiation compared to the amount of radiation that would have passed through a standard clear 1/8" glass pane under the same external conditions. The SC gives the ratio of that reduction. Thus, a SC value of 0.6 indicates that the particular glazing admits 60% of the radiation that would have been admitted under the exact same conditions by the standard glass pane.

date	Solar time, A.M./ P.M.	Sun position		Direct normal $BTUh/ft^2$	Solar heat gain factor $BTU/(hr\ ft^2)$								
		Alti-tude	Azi-muth		N	NE	E	SE	S	SW	W	NW	horizontal
2/21	7/5	4.3	72.1	55	1	22	50	47	13	1	1	1	3
	8/4	14.8	61.6	219	10	50	183	199	94	10	10	10	43
	9/3	24.3	49.7	271	36	22	186	245	157	16	16	16	98
	10/2	32.1	35.4	293	20	21	142	247	203	20	20	20	143
Half day totals					81	144	634	1035	813	81	81	81	546

Table 7.5a: Solar position and intensity and solar heat gain factors (SHGF) for 40^0 north latitude. (© ASHRAE)

Formula [7.12] combines all those factors. It gives the amount of admitted radiation (ΔQ) during the time interval (Δt) for a window of area (A), shading coefficient (SC), and solar heat gain factor (SHGF).

$$\frac{\Delta Q}{\Delta t} = A \cdot SC \cdot SHGF \qquad [7.12]$$

It should be noted that the shading coefficient (SC) incorporates in it only effects of the glazing and the shading that are placed on its inside. If a window is in the shade of an outside overhang, the window (or the shaded part of it) does not receive direct sun radiation. However, it receives indirect radiation from the sky and reflected radiation from the ground. When using [7.12], the SHGF of a window in the shade (or of the shaded part of a window) may be approximated by the SHGF of an unobstructed window, facing north.

Type of glass	*Normal thickness*	*Shading coefficient**
Single glass		
Regular sheet	1/8	1.00
Regular plate/float	1/4	0.95
	3/8	0.91
	1/2	0.88
Grey sheet	1/8	0.78
	1/4	0.86
Heat absorbing plate/float	3/16	0.72
	1/4	0.70
Insulating glass		
Regular sheet out, regular sheet in	1/8	0.90
Regular plate/float out, regular plate/float in	1/4	0.83
Heat absorbing plate/float out, regular plate/float in	1/4	0.06

Table 7.5b: Shading coefficient (SC) for single glass and insulating glass. (© ASHRAE)
*Wind velocity 7.5 mph

Example
Find the admitted radiation between 9:45AM and 10:15AM through a 1/4" thick, 2ftx4ft gray sheet glass window, facing South East, on February 21, at an altitude of 40^0 North.

Solution
We'll use equation [7.12]. The admitted radiation is ΔQ=?. The area of the window is $A=2x4=8ft^2$ (units consistent with the length unit used in Table 7.5a). The shading coefficient of that glass is SC=0.86 (from Table 7.5b). The time interval for the admitted radiation is 30 minutes. To be consistent with the time unit used in Table 7.5a (hour) we have Δt=0.5hr. The value of the SHGF for the given time of day, date, and window orientation is taken from Table 7.5a: SHGF=247 BTU/(hr · ft^2). Substituting in [7.12] yields ΔQ=1,274 BTU

When heat-radiation falls on an opaque object, such as a roof or an outside wall, the refracted radiation is absorbed. The absorbed radiation is converted into heat inside the opaque object. At the beginning, that heat is concentrated near the outside surface of the object, and heats it up. The surface of the object may become hotter than both its inside and the surrounding air. Consequently, heat will be conducted deeper into the object, and also to the surrounding air. This is a dynamic process that may reach a stationary equilibrium, in which the temperature gradient in the object does not change with time. Most often, since both the air temperature and the incident radiation changes during the day, equilibrium is not reached. Rather, there is a lag between the heat released by the wall or the roof and the changing air temperature. This lag and the amount of released heat will depend on the material of the wall and on its thickness. Estimates can be done based on experimental data on the amount of heat that reaches the indoor due to absorbed radiation in roofs and external walls. In practice, equations [7.9] are used for that purpose, but instead ΔT–the difference between the inside and outside air temperatures– TETD is used. TETD stands for: Total Equivalent Temperature Difference. It defines an equivalent temperature difference that would generate a conduction heat flow of the same amount as that released by the wall or the roof. Typical values of TETD can be found in tables. The tables list the external conditions for which those tables are applicable, including the orientation of the wall or the roof, its outside color, and its consistency. Table 7.5c list the TETD of some roofs.

Example
A dark colored, flat, 5000 ft^2 roof, made of 2" insulation and steel siding, has a U value of 0.125 BTU/(hr-ft^2-^{0}F).
(a) How much absorbed radiation does it transmit to the space underneath it in 24 hours?
(b) By how much would that amount of heat be reduced if the roof were painted in white?

Solution
We will use equation [7.9c] and data from Table 7.5c. The second row of the table corresponds to the roof in question. Under the columns "Sun Time", the

Roof description	lb/ft^2	U BTUh/ft^2 ^{0}F	A.M.						P.M.											
			8		10		12		2		4		6		8		10		12	
			D	L	D	L	D	L	D	L	D	L	D	L	D	L	D	L	D	L
1"insulation + steel siding	7.4	0.21	28	11	65	31	90	48	95	53	78	45	43	27	8	6	1	1	-3	-3
2"insulation + steel siding	7.8	0.12	24	8	61	29	88	46	96	53	81	46	48	30	10	8	2	2	-3	-3
1" insulation + 1" wood	8.4	0.20	12	2	47	21	77	39	92	50	86	48	61	36	25	16	7	5	0	-1
2" insulation + 1" wood	8.5	0.12	8	0	41	18	72	36	90	48	88	40	65	38	30	19	9	7	1	0
1" insulation + 2.5" wood	12	0.19	2	-2	23	8	48	23	70	36	79	42	71	40	30	29	29	17	15	9
1" insulation + 2.5" wood	13	0.11	1	-2	19	6	43	20	65	33	76	41	72	40	53	31	33	20	18	11

Table 7.5c: Total equivalent temperature differentials (TETD) for flat roof, exposed to the sun. (D=dark, L=light). These values may be used for normal air conditioning estimates in the latitude 0^0 to 50^0 north or south. Corrections may be needed if the inside temperature is not 75^{0}F and the maximum outdoor temperature is not 95^{0}F with a daily range of 21^{0}F, as well as a variety of wind conditions. (based on ASHRAE, Lechner)

table provides the equivalent temperature differences ΔT that have to be substituted in [7.9c]. Those temperature differences are given for different hours of the day, for dark roofs (D) and for lightly colored roofs (L). In order to find the transmitted heat during the day, we will have to add the transmitted amounts of heat in the two hours time intervals during the day (from 8AM to 12PM). The resolution of the time interval of the table is 2 hours. Therefore, the time interval for formula [7.9c] is Δt=2hr. Since U=125 BTU/(hr-ft^2-^{0}F), the roof's area A=5000ft^2, and Δt=2hr are the same for all the time intervals, we can factor them out and get:

For the dark roof:

$\Delta Q=0.125 \cdot 5000 \cdot (24+61+88+96+81+48+10+2-3) \cdot 2=508{,}750$ BTU

For the lightly colored roof:

$\Delta Q=0.125 \cdot 5000 \cdot (8+29+46+53+46+30+8+2-3) \cdot 2=273{,}750$ BTU

And the difference between dark and lightly colored roof would be: 508,750-273,750=235,000 BTU.

7.4.4 The greenhouse effect

Life on Earth is made possible by the sunbeams that reach the Earth, carrying with them electromagnetic energy generated by the sun. This electromagnetic radiation consists of waves of different wavelengths such as visible light, infrared (IR) radiation and ultraviolet (UV) radiation. As a matter of fact, all objects, including Earth itself, radiate energy in the form of electromagnetic waves. The amount of energy that an object radiates increases with the temperature of the object and its surface area. That radiated energy is proportional to T^4, the fourth power of the object's temperature, expressed in degrees Kelvin. The distribution of the radiated energy among the different wavelengths (the spectrum of the radiated energy) also depends on the temperature of the object. The sun radiates most of its radiation as visible light. Objects on the surface of the Earth radiate most of the energy as infrared radiation.

The Earth and objects on it absorb electromagnetic radiation emitted by the sun. Most of that absorbed radiation is converted by the absorbing objects into heat. In addition to that absorption, the Earth and the objects on it radiate to the atmosphere infrared radiation. So, the Earth absorbs sun radiation of all wavelengths, and radiates back mostly infrared waves. There is a balance between the radiation energy that the Earth absorbs and the energy radiated by it as infrared radiation. Because of that balance, the Earth retains its average temperature in the way it does.

Some of the gasses in the atmosphere absorb certain wavelengths of the electromagnetic radiation more than other wavelengths. Carbon dioxide (CO_2), methane, nitrous oxide, and water vapor absorb infrared radiation, while other wavelengths can pass through them. These gasses then emit back the absorbed infrared radiation. As a result, they allow the visible light that comes from the sun to reach the Earth and warm it, and at the same time send back to Earth some of the infrared radiation that the Earth has emitted. This is similar to what the glass panes do in a greenhouse. They allow all light in, and reflect back into the greenhouse some of the infrared radiation emitted by the plants and the objects inside. That warms the inside of the greenhouse. Carbon dioxide keeps the Earth warmer the same as the glass keeps the greenhouse warmer. This is called the greenhouse effect. Without those gasses, the Earth would have been much colder, and its temperature would have been around -18^0C (0^0F).

The amount of CO_2 in the atmosphere has increased significantly during the last century, due to burning of fossil fuels, such as oil, gas, and coal. Higher concentrations of CO_2 and other "greenhouse gasses" in the atmosphere may cause retention of more infrared radiation, which would raise the Earth's temperature. This process is referred to as global warming. If the concentration of CO_2 in air would continue to increase at the current rate, it is estimated that the average temperature of the atmosphere would rise by 2^0C by the year 2100. This may have serious consequences for life on Earth, including melting of polar ice caps, higher sea levels resulting in flooding of many coastal cities, and climate changes that will adversely affect food production.

The greenhouse effect may be used beneficially in the design of buildings. Sunrooms and greenhouses use the sun's energy to heat their own space. Many designs of sun collectors also use the greenhouse effect. A collecting element, such as water, a wall, or rock pebbles, is heated during the day by the sun. The heat collected is then used when needed. If the collecting element is placed in a greenhouse-like enclosure, some of the heat that it loses by radiation is recaptured and returned to the collector by the infrared reflection off the glass panes.

7.4.5 Heating and cooling of buildings

In order to feel comfortable inside a house, the temperature of its air has to be within the comfortable range. If the outside air temperature is within that range, it can be circulated into the house. In climates where this is the case, the effects of the sun's radiation are an important design consideration. The roof and the walls have to allow air circulation, so that the heat that they absorb and re-emit does not accumulate inside the house. Any inside hot air could be promptly replaced by the breeze, or through chimneys and other ventilation means.

In temperate climates, the outside air is in the comfortable range only during parts of the day. During that time, inside and outside air can be exchanged. During the rest of the day, air is more or less sealed inside the house. The envelope of the house has to provide enough thermal insulation so that the inside temperature does not rise, due to conduction, to the uncomfortable range. The sealed air would be replaced by fresh air when the outside is again in the comfortable range. In climates where the days are hot and the nights are cold, the envelope of the house could be designed such that, in addition to thermal insulation, it stores heat during the day and releases it during the night. In those geographical regions, generations of trial and error have resulted in building wisdom that combines available materials, thickness of walls, and shapes of structures to make life pleasant throughout the day.

In most areas of the industrialized world the passive measures mentioned above are not sufficient to maintain the inside of the house at a comfortable temperature around the clock and throughout the year. Those passive measures have to be supplemented by active heating and cooling, done by energy consuming devices. In order to implement those devices it is necessary to know how much heat they have to provide to the house or to extract from it in a given time period. For clarity, in the following paragraphs we'll discuss only heating calculations; cooling situations are very similar. Instead of dealing with providing heat, they deal with heat that has to be extracted.

There are two general types of heating situations: transient and stationary. In transient heating, the house and its contents are initially colder than the comfortable temperature range, and they have to be brought up to that level. For that to happen, the air is heated and it warms all other objects inside the house. The amount of heat that has to be supplied depends on the specific heats of all the inside objects, on their mass, and on the difference between their initial and final temperatures. Those processes are handled by formulas [7.6], and sometimes by [7.7]. Once the inside temperature of the air reaches the comfortable level and stabilizes there, it has to be maintained at that level. Heat losses to the outside have to be replenished by the heating system. The heat loss processes are described by formulas [7.8] through [7.12]. Usually, the transient situations are more demanding on the heating system than the stationary ones. In some cases this has to be considered when deciding on the heating system. For example, if a concert hall is utilized one evening a week, it might require a more powerful heating system to raise its temperature from, say 40^0F to 72^0F, than to just keep its temperature at 72^0F. A heating system that was designed just to replenish heat loses at 72^0F may be too weak to raise its temperature from 40^0F to 72^0F in the available time. In such cases, the weak heating system will have to heat the hall when it is not in use, so that its temperature does not drop too low.

In order to figure out the specifics of a heating system for stationary situations, heat losses through all relevant structure elements are calculated for the prevailing outside temperatures and the comfortable inside temperature, according to formulas [7.8] through [7.12]. All those amounts are added up. To those are added heat losses due to ventilation and due to infiltration through cracks and other openings (these effects are not discussed here). The contribution of heat sources in the structure such as lamps, computers, appliances, and heat generated by the bodies of the inhabitants may be subtracted from the mentioned heat losses. The rate at which heat has to be provided to a house in order to maintain its comfortable temperature is called the heating load of the house. The rate at which heat has to be extracted in order to maintain the comfort temperature is called the cooling load. A common unit of heating and cooling loads is BTU/hr. The heating and cooling loads of a house are used to determine the specs of the appropriate heating and air-conditioning system.

PROBLEMS

Section 7.1

1. The cooking instructions call for a 275^0 (Fahrenheit) oven. How much is it in Celsius?

2. On a winter day, the temperature dropped to -5^0 Celsius. How much is it in Fahrenheit?

3. The temperature difference between the inside and the outside of a room is 36^0 Fahrenheit. How much is it in Celsius?

Section 7.2

4. A straight copper pipeline of length 35m carries hot water in a building. The temperature of the water (and the pipeline) varies between 50^0F and 110^0F. What are the corresponding changes in the length of the pipeline?

5. In the morning, when its temperature is 65^0F, the length of a steel bridge is 320 feet. What would be its length at noon, when its temperature is 120^0F?

6. A window frame is made of aluminum. The size of the window is 2’ x 4’. A space of 1/32” is left around the frame, between the aluminum and the glass

pane, to allow for thermal expansion and contraction. Is it enough? Justify your answer.

7. A steel bucket is full to the rim with water at 70^0F. The bucket and the water are then cooled to 45^0F. Did some water spill out or did all of it stayed in the bucket? Would the answer be the same if instead of cooling the water and the bucket they would be heated to 115^0F? Justify your answers by relying on the data in table 7.1 and on appropriate formulas.

8. Normal body temperature is 36.5^0C. Express this temperature in Fahrenheit and in Kelvin scales.

9. When a car was empty, the pressure in its tires was 35 psi (pound per square inch). After the car was loaded, the volume of the air in its tires decreased to 95% of its original value. What was the air pressure then? (Assume that the temperature of the air did not change.)

10. A room was hermetically sealed when its air temperature was 4^0C and its pressure was 1 atmosphere. The air was then heated to 22^0C. What would be the pressure of the heated air?

11. A balloon of volume 3 m^3 is filled with helium at ground level, where the temperature is 20^0C and the pressure is 1 atmosphere. It then rises to a high altitude where the pressure is 0.6 atmospheres and the temperature is -30^0C. What would be its volume there? Neglect the elastic forces of the balloon's envelope.

Section 7.3

12. (A heat pump is "an air-conditioner in reverse". An air-conditioner takes heat from a room and deposits it in the outside, thus cooling the room and heating the outside. A heat pump takes heat from the outside and deposits it in the room, thus heating the room and cooling the outside.)
The writing on a heat pump says 0.5hp, 1,500 BTU-h. What does it mean? What happens to the energy provided to the motor of the heat pump?

13. How much heat does it take to raise the temperature of an empty copper pot of mass 0.25 kg from 20^0C to 100^0C?

14. How much heat is released by a 1.4 kg iron rod as it cools from 1200^0C to 20 ^{0}C?

15. How much heat has to be extracted from 2 pounds of water at 200^0F to get ice at 15^0F?

16. How many calories does it take to melt 3kg of copper at 800^0C?

17. A piece of iron of mass 4 kg was heated and dropped into 2 kg of water at 20 degrees. The water was mixed quickly and thoroughly till it reached its final temperature of 21^0C (assume that there were no heat exchanges with the surroundings). What was the initial temperature of the iron?

18. A piece of an alloy of mass 0.5 kg at 100^0C was dropped into 4 kg of water at 22^0C. The water was mixed quickly and thoroughly till it reached its final temperature of 23.9^0C (assume that there were no heat exchanges with the surroundings). What was the specific heat of that alloy?

Using figure 7.6, answer questions 19-22:

19. What is the absolute humidity of air whose temperature is 65^0F and relative humidity is 20%?

20. What is the relative humidity of air whose temperature is 85^0F, and whose wet bulb temperature is 68^0F?

21. The size of a room in the desert is 3mX4mX2m. The air temperature inside the room is 90^0F, and its humidity is zero.
How much water vapors (in grams) have to be added to the air so that its relative humidity becomes 80%?

22. Consider 1 m^3 of air at 90^0F and relative humidity of 25%. Droplets of water are sprayed into this air until it is saturated. Calculate at what temperature it will saturate, and how much water will it absorb.

23. The specific enthalpy h of air (BTU/(pound of dry air)) is given by $h=0.24 \cdot T+A \cdot (1061+0.45 \cdot T)$, where T is the temperature in ^{0}F and A is the absolute humidity in (pound of water vapor)/(pound of dry air).
(a) How much heat is needed to raise the temperature of one pound of air from 50 ^{0}F and relative humidity of 20% to 70 ^{0}F and relative humidity of 60%? Water droplets are sprayed into the air in order to raise its humidity.
(b) How much water droplets (in pounds) turned into vapor in part (a)?

Section 7.4

24. What is the R-value of a cork board, one inch thick?

25. Which is a better insulator: a 4" thick wall of common brick, or a ¾" wall of plywood ?

26. Find the R value of a thermal window, consisting of two panes of 0.25" glass (R-value 0.08 $hr \cdot ft^2 \cdot F/BTU$ each), separated by a layer of air 0.5" thick.
How much heat flows through such a window, whose area is 6 ft^2, in an hour, when the temperature inside the house is 72^0F, and outside is 40^0F?

27. The same window as in the previous problem is now used as a horizontal window in the ceiling (a sky-light) of an air conditioned room. Find its R value in the summer, when the outside is warmer than the inside, and in the winter, when the inside is warmer than the outside.

28. In an old house, a fireplace in the first floor keeps its temperature at a cozy 74^0F, while there is no heating in the two floors above. Assume that the heat flows vertically. What would be the temperature in the second and in the third floors if the R value of each floor surface is 8, and the R value of the roof is 15? The outside temperature is 28^0F.

29. (a) Find the thermal resistance r and the equivalent R-value of a typical frame 14'x8.5' wall consisting of a 0.5" plaster board on the inside of the room, nominal 2x4 studs, and a 0.5" sheathing on the outside. The spaces between the studs are filled with an R-16 insulating foam.
(b) What is the heat loss through this wall in three hours if the inside temperature is 72^0F, and the outside temperature is 32 ^{0}F?

30. Find the heat loss through a circular concrete floor of radius 7m of a storage tank that lies on the ground. Its edge factor is 0.40 BTU/(hr-ft-^{0}F); the temperature inside the tank is 65^0F; outside air temperature is 42^0F

31. Find the admitted radiation between 2:30PM and 3:30PM through a 1/8" thick, 1.5ftx3ft gray sheet glass window, facing South West, on January 21, at an altitude of 40^0 North.

32. A bright colored, flat, 3,500 ft^2 roof, made of 1" insulation and 1" wood, has a U value of 0.206 BTU/(hr-ft^2-^{0}F).
(a) How much absorbed radiation does it transmit to the space underneath it in 24 hours?
(b) By how much that amount of heat would change if the roof were painted in black?

33. Assume that an apartment looses heat to the outside only by conduction through walls, windows, and the roof and that all the lost heat is replenished by an oil furnace. In January, the average outside temperature was 28^0F and the inside was 68^0F. The cost of the heating oil was $100.00. Estimate the cost of heating that apartment if the inside temperature were 72^0F.

34. An external 14'x9' wall is made of 8" concrete blocks. It has a 3'x4' glass window, 1/4" thick.
(a) Find the R-values of the concrete and the glass (use Table 7.3).
(b) How much heat is conducted through that wall in 24 hours, if the average outside temperature is 45^0F and the inside temperature is 68^0F?

35. Figure 7.11 shows elevations and section of a small house, located at an altitude of 40^0N. The rooms (not the basement) have to be maintained at 72^0F

when the outside temperature is 20^0F in the winter and 85^0F in the summer. Assume that the ground temperature is 40^0F year round. The R-values (in hr-ft^2 ^{0}F/BTU) of the various elements are as follows: external walls R=4.5; roof R=2.5; floor R=2.0; basement walls and floor R=1.2; widows (insulating glass, ¼" thick) R=0.3; doors R=1.5. Those values do not include the resistance of air films. Take prevailing wind speed as 7.5 miles/hr.
Find the air temperature in the basement.
Find the heating load of the house. Ignore radiation effects.
Find the cooling load of the house. Include radiation effects through the windows
Find the heating load if the R-values of the walls and roof are doubled.
.

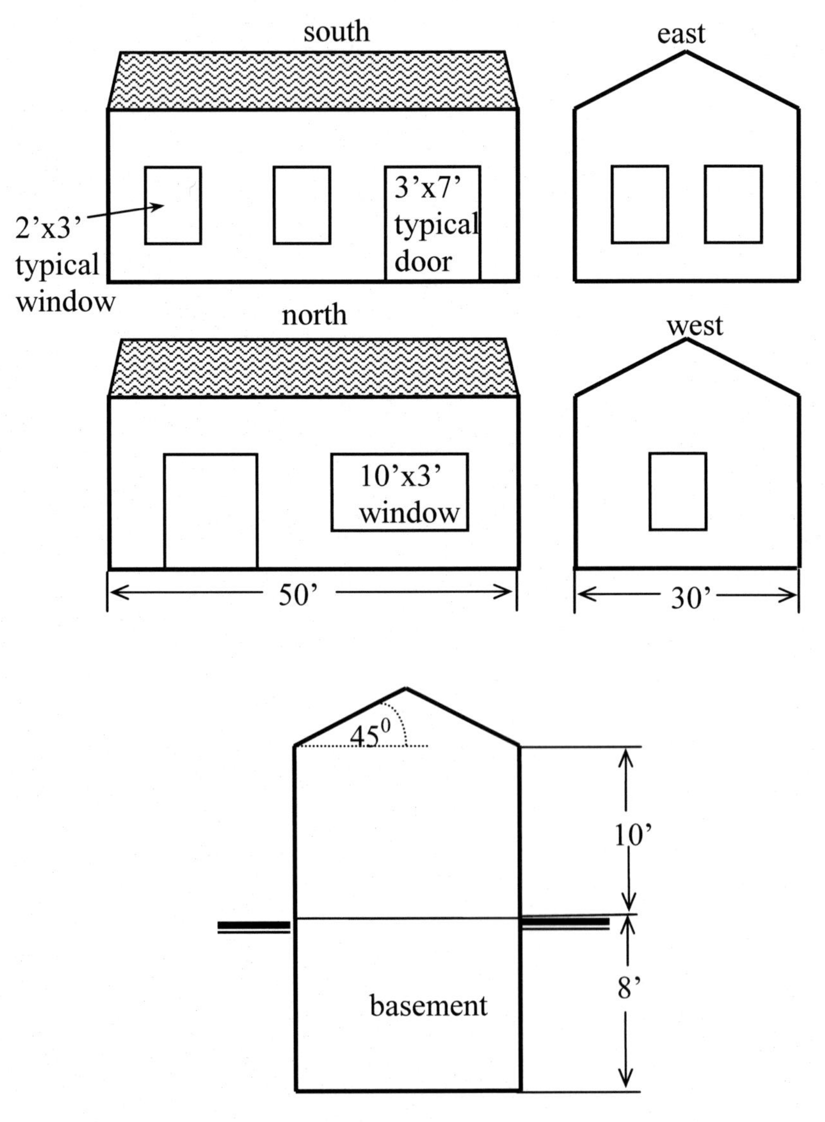

Figure 7.11: problem 33. (not to scale)

8. ELECTRICITY

8.1 Electric charges and electrostatic force

Atoms are the basic building blocks of molecules, which make up the matter in and around us. Atoms themselves are made up of three elementary particles: protons, neutrons, and electrons. Protons and neutrons consist of quarks and gluons, while electrons have no sub-particles. The mass of a proton is approximately the same as the mass of a neutron. The mass of an electron, which is 9.1095×10^{-31}kg, is approximately 1840 times smaller than that of the proton (1.6726×10^{-27}kg) or the neutron (1.6750×10^{-27}kg). Protons and neutrons bond together and form the nuclei of atoms. The electrons of an atom orbit around its nucleus. The typical distance between the orbits and the nucleus is 10,000 times greater than the size of the nucleus. (So, most of the volume of all matter, including us, is void). The number of electrons in an atom is equal to the number of protons in its nucleus. Atoms of different chemical elements have different numbers of protons in their nuclei.

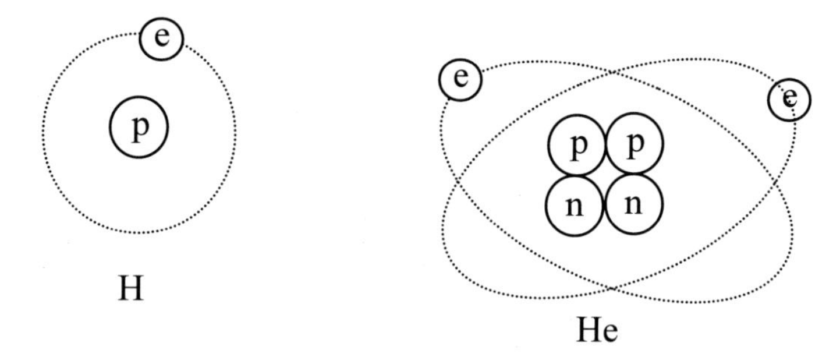

Figure 8.1: Schematics (not to scale) of two atoms. The hydrogen atom (H) has one proton in its nucleus and one electron orbiting around it. The helium atom (He) has two protons and two neutrons in its nucleus, and two orbiting electrons.

The protons and the neutrons are held together in the nucleus by the nuclear force, which is strong and have short-range effects. Its range is only a few femtometer (10^{-15} meter), which is within the radius of the nucleus. The nuclear force does not affect any particle outside the nucleus.

The electrons, on the other hand, are kept in their orbits by the electrostatic force, which is weaker than the nuclear forces, but has a long range effect. The protons in the nucleus are the source of the electrostatic force, which attracts the electrons. The neutrons do not exert any electrostatic force on the electrons or on the protons. Since electrons and protons can experience electrostatic force while neutrons cannot, it is said that electrons and protons have an electric charge, while neutrons do not have an electric charge. It was found that both electrons and protons repel their own kind by the electrostatic force.

It was concluded that there are two kinds of electric charges. Particles of the same kind of electric charge repel each other, while particles of different kinds of electric charges attract each other. Opposites attract each other–when we deal with electric charges. It was agreed to denote the charge of the proton as positive, and that of the electron as negative.

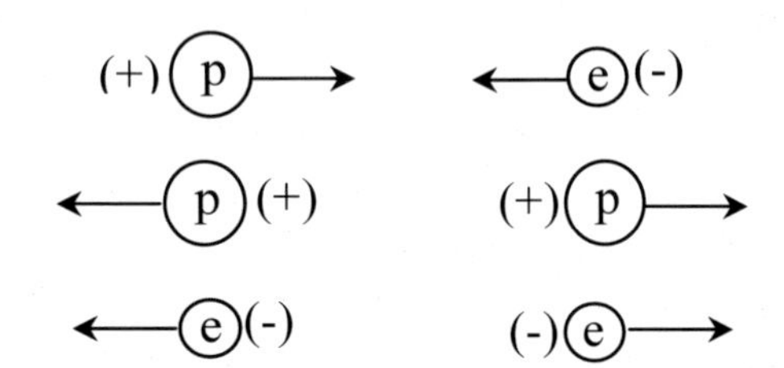

Figure 8.2: Electrostatic forces between elementary particles. If the separations within the pairs are the same, the magnitudes of the forces are the same.

In order to determine the amounts of charges carried by electrons and protons, the magnitudes of the forces that they exert on each other were compared. First, consider three pairs of particles that the distances between their members are identical: two electrons, two protons, and an electron and a proton (figure 8.2). It was found that the magnitudes of all the electrostatic forces between the members of any pair are the same for all those pairs, as long as the separations between the members of the pairs are the same. Therefore it was concluded that the magnitudes of the electric charges of the electron and of the proton are equal to each other. The SI unit of electric charge is the coulomb (C). The charge of the electron is approximately -1.6×10^{-19} coulombs and that of the proton is $+1.6 \times 10^{-19}$ coulombs.

When a number of charged elementary particles (e.g. electrons) are concentrated in a small region (a point), they may be treated as if they are one point charge. The charge of that point charge is equal to the sum of the elementary charges that make it. The magnitude of the electrostatic force between two point charges is inversely proportional to the square of the distance between them. All those facts are summarized in Coulomb's Law ([8.1]):

$$F = k\frac{Q_1 \cdot Q_2}{r^2} \qquad [8.1]$$

where F is the magnitude of the electrostatic force between two small charged particles, whose charges are Q_1 and Q_2 respectively, separated by the distance r. In the SI system, the charges are expressed in coulombs, the distance in meter, the force in Newton, and Coulomb's constant $k=8.98 \times 10^9\ N.m^2/C^2$. Equation [8.1] gives the magnitude of the electrostatic force that acts on each of the two particles. The directions of those forces depend on the signs of the charges, and they are determined as in figure 8.2. Equation [8.1] may be considered also as a way of defining the coulomb: Two identical charges of 1 coulomb each will exert on each other a force of $8.89x10^9$ N when separated by 1 meter.

When a group of charged particles exert forces on another charged particle, each group member contributes a force according to [8.1], as if the other group members did not exist. The total force is then found by adding up all those contributions. Since force is a vector, all those contributions have to be added as vectors. Formula [8.1] provides the magnitude of each contribution. The direction of each contribution can be found according to figure 8.2.

Example
Find the force that a hydrogen nucleus exerts on an electron located at a distance of 1.5 femtometer.

Solution
Following the notations of [8.1]: The hydrogen's nucleus consists of one proton, whose charge is $Q_1=1.6x10^{-19}$ coulombs. The charge of the electron is $Q_2=-1.6x10^{-19}$coulombs. The distance between the two is $r=1.5x10^{-15}$m. The force exerted is

$$F = 8.89x10^9 \frac{1.6x10^{-19} \cdot 1.6x10^{-19}}{(1.5x10^{-15})^2} = 101.1N.$$

We ignored the minus sign of the charge of the electron because [8.1] deals with the magnitude of the force, which is always positive. Since the electron and proton have opposite charges, they will attract each other. The directions of the forces that they will exert on each other are given in figure 8.2.

Example
Compare the electrostatic attraction force between the electron and the proton of example (a) above with the force of gravity that acts on the proton.

Solution
The force of gravity (weight) is given by: F=mg. In this case, the mass of the proton is $m=1.67x10^{-27}$kg. The proton's weight is $F=1.64x10^{-26}$N. The force of gravity that acts on the proton is extremely small compared to the typical electrostatic attraction between a proton and an electron in the atom.

Although an atom contains both positive and negative charges that attract each other (protons and electrons), the electrostatic forces that an atom exerts on charged particles further away are practically zero. This is because the effects of the positive and negative charges of an atom neutralize each other, when it comes to affecting charged particles on the outside. An atom as a whole is electrically neutral. The same is true for molecules as a whole. Usually, atoms and molecules do not exert electrostatic force on outside charges and are not affected by them. However, in many cases, molecules that are overall electrically neutral, but whose charges are not distributed evenly, do exert electrostatic forces on outside charged particles.

If an atom or a part of a molecule attaches or loses an electron, they are not electrically neutral any more. They are called positive ions or negative ions, according to their excess charge. An ion can exert electrostatic force on other charged particles, and experience such forces.

Example
Find the electrostatic force exerted by an ion of Ca^{++} on an ion of Cl^- found at a distance of 2 kilometers.

Solution
Ca^{++} is positively charged because it has a deficit of two electrons. Cl^- is negatively charged because it has acquired an additional electron. The distance between the two ions is r= 2,000m. When these data are substituted in [8.1] we get:

$$F = 8.89x10^9 \frac{(2x1.6x10^{-19})x(1.6x10^{-19})}{2{,}000^2} = 1.14x10^{-34}N$$

(This force is much smaller than the weight of both ions).

Two major groups of objects that are overall electrically neutral contain charged particles that are free to move within the confines of the object. The first are metals, and the second are electrolytic solutions. Some of the electrons in a metal become detached from their nuclei and move around inside the bulk of the metal. The positively charged nuclei form a fixed matrix through which those free electrons move. In an electrolytic solution, neutral molecules (such as NaCl) are dissociated into positive and negative ions (Na^+ and Cl^-), which move around among the molecules of the liquid (H_2O). A material that contains free charges that can move inside it is called an **electrical conductor** or an **electrical resistor.**

Although all those charged particles can move inside their conductor, they normally distribute themselves so that each small region is electrically neutral (has the same number of positive and negative charges).

8.2 Electric currents

8.2.1 Electric currents and their utilization

When water flows in a hose, the intensity of the water current is defined as the amount of water that passes through an imaginary cross section of the hose per unit time. For example, gallons-per-minute is a unit of the intensity of water current. Similarly, when free charges in a conductor move in the same direction, we say that an electric current is present in that conductor. The intensity of the electric current (I) is the amount of electric charge (Q) that passes through an imaginary cross section of the conductor per unit time (t).

$$I = \frac{Q}{t} \qquad [8.2]$$

In the SI system the unit of electric current is the ampere (A). One ampere is one coulomb per second.

Example

An electric current of 2 amperes flows through a metal conductor. How many electrons pass through a cross section of this conductor each second?

Solution

According to the notations of equation [8.2]: I=2A, t=1sec, Q=?. Substituting we get Q=2 coulombs. The charge of each electron is 1.6×10^{-19}C. The number of electrons that make that charge is $2/1.6 \times 10^{-19} = 1.25 \times 10^{19}$ electrons (every second).

Free electrons move at random velocities within their metal. That motion does not create an electric current. Only when electrons move as a group in one direction an electric current is present. The coordinated velocity of such electrons is called their drift velocity, and it is much smaller than their random velocities.

Most applications of electric currents involve free electrons that flow in conducting metal wires, such as copper or aluminum. Those electrons collide repeatedly with the matrix of the positive ions that make up the metal. In the collisions, energy is transferred from the electrons to the atoms, resulting in the heating of the conductor. This is how electric energy is converted into heat energy, which is used in a variety of domestic and industrial applications. For example, the filament of a light bulb glows because it gets very hot due to the electric current that flows through it.

When a conducting wire that does not contain iron is placed next to a magnet, no force exists between the two. However, if an electric current is flowing in the wire, the magnet and the wire will exert forces on each other. These are called electromagnetic forces. Electromagnetic forces also arise between two wires that are carrying electric currents. The magnitude and direction of such electromagnetic forces will depend on the relative configuration of the elements (magnets and/or wires), on the intensity of the currents, and on the strength of the magnets that are present. Those electromagnetic forces can cause the wires that carry the currents to move. They are the basis of all electric motors that are so common around us. In those applications, the electromagnetic forces help to convert the energy of the electric current into the mechanical energy and work of moving objects, such as the rotors of electric motors.

When charges oscillate, electromagnetic waves are produced. Those waves propagate away from the oscillating charges. Electromagnetic waves carry radio and television broadcasts, and other kinds of wireless communication, such as cellular phone conversations. In those cases, the energy of the electric oscillating currents is converted into the energy of electromagnetic waves.

So, electric currents are used to generate heat, to provide the forces that run electric motors, and to generate electromagnetic waves.

8.2.2 Generating electric currents

How are electric currents generated in conductors?

First, let's discuss a mechanism that might look as if it could be used to make currents, but in reality it is not practical. The positive and negative charges in a metal are distributed evenly, so that the conductor as a whole and every small region inside it are electrically neutral (figure 8.3a). Consider a positive charge placed to the right of a metal conductor (figure 8.3b). The free electrons in the metal will be attracted to the positive charge and move to the right. This coordinated motion of the free electrons constitutes an electric current. However, that current would last for a very short time. As the free electrons move to the right, the even distribution of charges in the metal is disturbed. Electrons that reach the right end of the conductor accumulate there and form a layer of negative charges. At the same time, ions at the left end of the metal loose their neutralizing electrons, which have moved to the right. As a result, a layer of positive charges is formed at the left end of the conductor (figure 8.3b). The positive layer attracts the rest of the free electron to the left. The negative layer repels those free electrons also to the left. Overall, the electrostatic forces of the two induced layers (F_{layers}) oppose the electrostatic forces that the external charge exerts on the free electrons inside the conductor ($F_{external}$). As more free electrons reach the right end of the conductor, the forces of the two layers on the inner free electrons increase. Eventually, those forces will balance the attraction forces of the external charge, and bring the current to a halt. This entire process occurs almost instantaneously. Overall, the induced layers shield the inner free electrons from the electrostatic force of the external charge. In more figurative language it could be said that the electrons that reach the end of the conductor and cannot continue in their motion crowd there and create a jam that halts the motion of all the other electrons in the conductor.

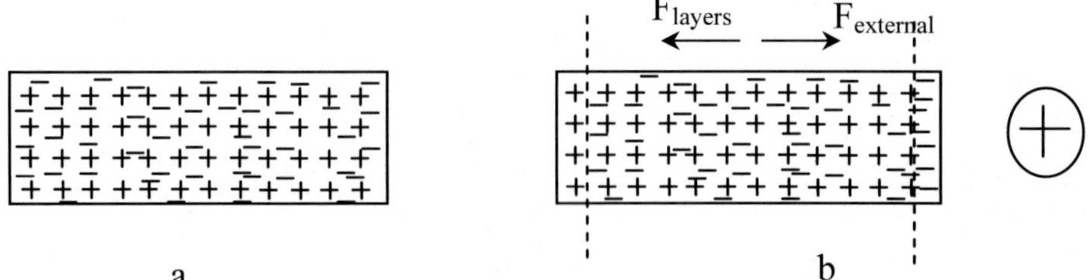

Figure 8.3: The effects of an external charge on the distribution of free electrons in a conductor. (a) Free electrons are distributed evenly across the conductor, rendering it electrically neutral. (b) An external positive charge attracts free electrons. A layer of negative electrons accumulates on the right end of the conductor, and a layer of positive charges is exposed at the left end. The forces that those layers exert on the free electrons act to the left. The forces of the external charge on the free electrons act to the right. Eventually, the forces of the layers balance the forces of the external charge, and no more current is generated.

In order to create a sustained electric current in a conductor, two conditions have to be met. First, the electrical energy of the current that is converted into heat or other forms of work and energy, as mentioned earlier, has to be provided by an outside energy source. Second, the excess of electrons that tend to accumulate at one end has to be removed, and the deficiency of electrons at the other end has to be replenished, so that every small region of the conductor remains electrically neutral throughout the entire process.

The first condition is satisfied by devices that are called **electric power sources** or **power sources** in short. A power source provides the energy needed to keep the electric current flowing. Without that supply of energy, the current of the electrons will loose its energy and come to a halt. According to the particular circumstance, that energy would be converted to heating the conductor; to mechanical energy and work if the conductor is affected by electromagnetic forces (as is the case in electric motors); or to electromagnetic waves. In general, the object that converts the supplied electrical energy to another form of energy is called **load**.

The second condition is met by making the conductor part of a closed conducting loop. Each power source has two ends called terminals or electrodes. Each electrode is connected to an end of a conductor, thus forming a closed loop through which the current flows. A loop consists of the conductor, the power source, and connections between the two (figure 8.4). Current will flow in the conductor as long as the loop is closed. Once an electrode is disconnected from the conductor, the current will stop. Charges (the circles in figure 8.4) circulate around the loop (dotted arrows). The power source provides the energy for the circulation, and it is also part of the path around which the charges circulate. The same amount of charges that leave the conductor at the right end is replenished by charges that enter through its left end. That prevents the formation of charged layers at the ends of the conductor.

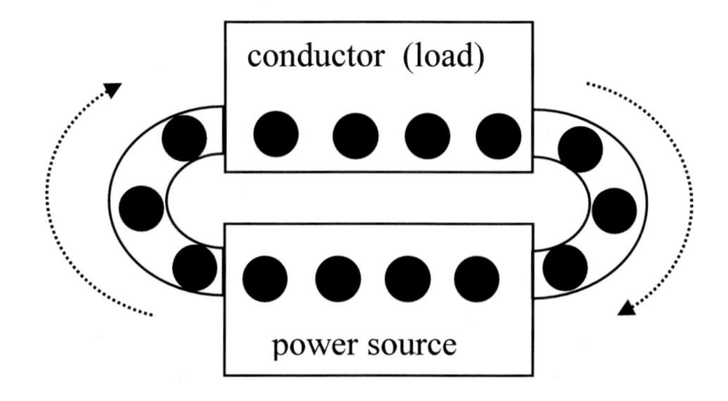

Figure 8.4: Schematics of a closed electric loop. The power source circulates charges (circles) around the loop. The amount of charges that leave the conductor at its right end is replenished at the left end.

The whole situation is similar to filtering water in an aquarium. A pump (the power source) circulates the water (the charges) through the filter (the conductor) and the pipes and the rest of the aquarium (the connections). Any water that leaves the filter to the aquarium is replenished by water from the aquarium that enters the filter from its other side. The closed loop of aquarium-pump-filter guarantees that the filter will always have water in it. The pump provides the energy needed to push the water through the filter, and to overcome the friction and viscosity forces in the pipes and aquarium (the connectors). This is similar to the heat and mechanical work generated by the electric current flowing in a conductor.

8.2.3 Electric energy and power

The intensity of the water current that the pump circulates through the filter depends on the strength of the pump and on the resistance of the filter and the aquarium to the water flow. The same pump will generate different water currents when connected to different filters. Similarly, the same power source will generate different electric currents when connected to different conductors. Because of that, it is impossible to characterize a power source only by the current that it generates. In order to define a unit with which intensities of power sources could be described, an ideal situation is assumed. It is assumed that the two electrodes of the power source are surrounded by vacuum, and that a test charge of one coulomb is placed next to the electrode that repels it. This charge will gain kinetic energy until it reaches the other electrode. The kinetic energy of that unit test charge as it reaches the other electrode indicates the strength of the power source. The larger is the final kinetic energy, the stronger is the power source. The intensity of the power source is called **voltage**, or the **potential difference** between the electrodes. By definition, if the final energy of that one coulomb test charge is one joule, the voltage of the power source is one **volt**. The letter that symbolizes volts is V. The voltage in volts is equal to the final energy in joules of that one coulomb test charge.

In general, if a charge of Q coulombs is let go in vacuum from one electrode of a power source whose voltage is V volts to the second electrode, that charge would gain kinetic energy of W joules, where [8.3]:

$$W = Q \cdot V \quad \text{or} \quad V = \frac{W}{Q} \qquad [8.3]$$

Example

An object whose charge is 2 coulombs is released from rest next to one electrode of a power source. As it reaches the other electrode, it has gained 8 joules of energy. What is the voltage (potential difference) between the two electrodes?

Solution

According to the notations of [8.3] Q=2C, W=8 joule, and V=? Substituting gives V=4 volts.

Equation [8.3] applies also to situations where the charge Q moves between any two points (not necessarily electrodes) that are affected by other charges. The voltage V (or potential difference) between any such points is given by [8.3], where W is the change in energy of the charge Q, as it moved from one point to another. A positive charge is pushed from a point of a high potential towards points of lower potential.

Equation [8.3] defines the intensity of the power source based on the energy gained by a free charged particle that moves from one electrode to the other, without the interference of any other factor. This charge does not collide with other particles, nor is it affected by other electromagnetic forces. If the charge is a free charge in a conductor, and not in free space, it will interact with the particles around it, as it moves from one electrode to the other. It turns out that that charge will give almost all the energy imparted to it by the electrodes to the particles around it. That will heat up the conductor. The charge itself will drift very slowly from one electrode to the other. All the free charges in the conductor will drift like that charge, thus creating an electric current in the conductor. If a charge Q is passing through the electrode, and hence through any cross section of the conductor, in a time t, the power provided by the power source is given by the relation

$$P = \frac{W}{t} = \frac{Q \cdot V}{t} \quad \text{or} \quad P = I \cdot V \qquad [8.4]$$

{Proof of [8.4]
If a charge Q is passing through every imaginary cross section i in a time interval t, the energy imparted to it (based on [8.3]) would be $Q\Delta V_i$, where ΔV_i is the potential difference that the Q passed in the time interval. Overall, the total energy imparted to all the charges between the two electrodes would be $W = \sum Q\Delta V_i = Q\sum \Delta V_i = QV$. The power (P) provided by that power source, whose voltage is (V), is equal to the work (W) provided by the power source divided by the time (t) that the power source was pushing the current (I). Based on [8.3] and [8.2] we get [8.4].}

In the SI system, the unit of current is ampere, the unit of voltage is volt, and the unit of power is watt. Sometimes, based on [8.4], the unit ampere-volt, or volt-ampere, is used to express electric power. That unit is a synonym of watt. Similarly, kilovolt-ampere (kVA) is a synonym of kilowatt (kW).

Example
The voltage of a car battery is 12V. What current flows through a 25W light bulb in that car ?

Solution
That light bulb consumes 25 watts. Following the notations of [8.4], P=25W, V=12 volts, I=?. Substitution yields I=2.1 ampere.

The current (I) drawn from a power source by a conductor will depend on the voltage (V) of the power source, and on the properties of the conductor. The ratio between V and I is called the resistance of that conductor. The SI unit of resistance is the ohm, whose symbol is Ω.

$$R = \frac{V}{I} \quad \text{or} \quad V = I \cdot R \qquad [8.5]$$

Formula [8.5] is called Ohm's Law. It was found that for a large variety of conductors the resistance R does not depend on the voltage. Those conductors are called ohmic conductors or ohmic resistors. All the metals fall in this category.

The electrical energy of the current that flows in a conductor is converted into heat energy. The rate at which that conversion occurs (the power) can be obtained by combining [8.4], which holds for any process that involves current, and [8.5], which deals only with currents through resistors. The result is:

$$P = I \cdot V \quad \text{or} \quad P = I^2 \cdot R \quad \text{or} \quad P = \frac{V^2}{R} \qquad [8.6]$$

Example
What is the power that a resistor of 2,500Ω consumes when connected to a power source of 120V?

Solution
Using the notations of [8.6] we have: R=2,500 ohms, V=120 volts, and P=? Substitution yields: P=5.76 watts.

Example
What is the current drawn by a microwave oven rated at 1.5kW and 120V?

Solution
That oven consumes power at the rate of P=1,500 watts, when connected to a power source of V=120 volts. Substituting in [8.6] yields I=12.5 amperes.

8.2.4 Resistance and resistivity

The value of the resistance R depends on the material of the conductor, on its geometry, and slightly on its temperature. In general, the resistance increases with temperature. Often, cylindrical resistors, such as metal wires, are compared to water pipelines. The resistance of a pipeline to water flow through it increases with the length of the pipeline and decreases with its cross sectional area. Similarly, the resistance R of a cylindrical conductor to the flow of electric current through it is found to be proportional to its length, and inversely proportional to its cross sectional area. This is expressed by formula [8.7]

$$R = \rho \frac{L}{A} \qquad [8.7]$$

where R is the resistance of the conductor, L its length, A its cross sectional area, and ρ is the resistivity–a coefficient that depends on the material of the wire. Table 8.1 shows the resistivity of some materials.

Material	*Resistivity ρ (Ω m)*
Silver	1.59×10^{-8}
Copper	1.68×10^{-8}
Aluminum	2.65×10^{-8}
Glass	10^{9}-10^{12}
Rubber	10^{13}-10^{15}

Table 8.1: Resistivity at 20^0C

Example
Find the resistance of a copper wire 1 m long and of diameter of 1 mm.

Solution
We will use equation [8.7]. The resistivity is $\rho=1.68 \times 10^{-8}\ \Omega$ m (table 8.1), the length l=1m, and the cross sectional area
$A=3.14 \times (0.5 \times 10^{-3})^2\ m^2$. Substituting in [8.7] yields $R=0.021\Omega$.

8.2.5 Electric power sources

There are two general kinds of power sources: direct current (DC) and alternating current (AC) sources. A DC source generates a current that flows in one direction in the connected conductor. An AC source generates a current that changes its direction and intensity in a periodic (sinusoidal) way. A common source of DC current is the battery. A common source of AC current is electrical power stations that deliver their electricity to wall outlets in houses.

DC Current

In DC sources one of the electrodes is called positive (+) and the other negative (-). If the current in the load consists of positive ions, they flow from the positive electrode to the negative electrode. If the current consists of negative ions or electrons, they flow from the negative to the positive electrode. All the equations pertaining to current that were introduced above are the same for both kinds of moving charges. Therefore, when those equations are used, it is customary and convenient to say that the electric current in a load consists of positive charges that flow from the positive to the negative electrode.

The electrical battery was invented in 1800 by Alessandro Volta. Volta found that if a piece of cloth soaked with acid is sandwiched between two plates of different metals, such as zinc and copper, a voltage is formed between the plates.

If the plates are then connected by a metal wire, a sustained DC current flows through it. The same thing happens if the two plates are immersed in a bath of acid. That arrangement is called a voltaic cell. The voltaic cell is a power source. Volta found that if a number of such cells are arranged as a battery (hence the name) and connected by metal wires as shown in figure 8.5, the voltage between the end plates of the battery would be the sum of the voltages of the individual cells. That kind of connection between DC power sources is called a series connection. In a series connection of batteries, the positive electrode of one cell (or battery) is connected to the negative electrode of another cell (or battery), and the entire arrangement has a free positive and a free negative electrode that serve as the electrodes of the battery.

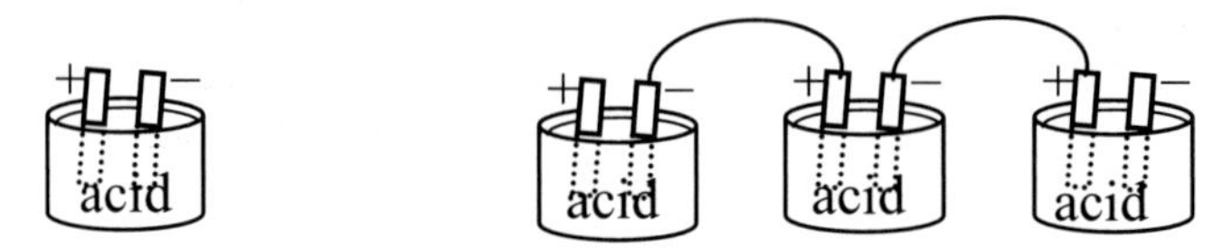

Figure 8.5: A voltaic cell (left), and a battery consisting of three cells connected in series.

Nowadays, batteries and voltaic cells are based on the same old principles, but they use improved materials and technologies. The voltage of a standard cell is 1.5 volts. Commercially available batteries are made of standard cells connected in series. Most car batteries are 12 volts. They are made of eight cells connected in series. Many household electronic devices, such as portable radios and remote control units, require DC sources other than 1.5 volts. They have built-in battery holders that accept 1.5 volt batteries, and provide the series connections to produce to required voltage. Thus, a remote control that accepts four 1.5V batteries connects them in series to provide 6 volts.

A battery is a power source. According to formulas [8.6], the power P and the current I that a load whose resistance is R draws from a battery of a given voltage V depends on the value of R. The smaller the R, the more current and power it will draw. If R is extremely small, the current and the power would be extremely large. In principle, formulas [8.6] imply that the amount of power that a battery can provide is unlimited. This, of course, does not make sense. One cannot replace the 12V battery of a car by eight small 1.5V flashlight batteries connected in series, and expect to start the car with them. When a battery delivers current to a load, the same current has to pass also through the battery. This is similar to the water pump that circulates water through the filter and the aquarium. The water has to pass also through the pump itself. Part of the energy of the pump goes into pushing the water through the pump. Similarly, part of the energy of the battery goes into maintaining the current through the battery itself. Equations [8.5] and [8.6] ignore that fact. As long as the battery does not deliver too much current, those equations are a good approximation. If the current is too large, they have to be modified. In such cases the effective voltage that acts on the load is less than the nominal voltage V of the battery. The less current a load draws, the more accurate are the equations. The technical specification of batteries list the maximum current for which equations [8.5] and [8.6] still hold with sufficient accuracy.

A battery converts chemical energy into electrical energy. Each battery has a certain amount of energy that it stores, and consequently can deliver. Energy (W) is power (P) multiplied by time (t). Based on equations [8.6], the total energy that a battery can deliver can be expressed as:

$$W = P \cdot t = V \cdot I \cdot t \qquad [8.8]$$

where W is the total energy that the battery can deliver. Since the voltage of the battery V is its primary specification number, specifying a value of I.t implies the value of W. The common unit of I.t is ampere hour.

To summarize, three parameters are used to describe the properties of a real battery: its nominal voltage V; the maximum current that it can deliver so that the actual voltage across the load is not much smaller than the nominal voltage; and something related to its stored energy: its ampere hour.

Example

A radio transistor requires a power source of 6 volts. When connected to the appropriate 6 volts battery, it drew a current of 0.2 milli-ampere for 20 hours, before the battery "died". How much energy was consumed by that radio? Was this all the energy stored in the battery?

Solution

Following the notations of [8.7], V=6 Volts, $I=0.2x10^{-3}$ampere, t=20hours=72,000sec, and W=?. Substituting in [8.7] yields W=86.4 joules. The battery stored more energy than that, because additional energy was needed to circulate the current through the battery itself.

AC Sources

Alternating currents (AC) are generated by AC generators, also called alternators. Household AC is generated in power plants, and delivered to the users by electric power lines. An alternator consists of fixed magnets and a coiled metal-wire with open ends that rotates between the poles of the magnets. Because of that rotation, a voltage is formed between the open ends of the coil, making the alternator a power source. The intensity of the generated voltage and its orientation change in a sinusoidal way as the coil rotates. In the U.S., the maximum of that voltage V_{max} (the amplitude, peak voltage) at the standard home wall outlet is around 160V.

The current that flows through a conductor connected to an AC generator obeys Ohm's Law (equation [8.5]). Since the voltage V is alternating, the current I is also alternating with it. The maximum value of the current I_{max} is related to V_{max} through Ohm's Law: $V_{max}=I_{max} \cdot R$. The power consumed by that conductor will alternate together with the current and the voltage. Its instantaneous value is given by equations [8.6]. For practical applications, the instantaneous value of the power is not so important. More important is the average value of the power during a cycle. The average value of the power ($P_{average}$) is given by:

$$P_{average} = \frac{I_{max} \cdot V_{max}}{2} = \frac{I_{max}}{\sqrt{2}} \cdot \frac{V_{max}}{\sqrt{2}} = I_{rms} \cdot V_{rms}$$

and consequently [8.9]

$$P_{average} = I_{rms}^2 \cdot R = \frac{V_{rms}^2}{R}$$

The value of V_{rms} of a standard home wall outlet in the U.S. is around 110-120V, and the frequency is 60 Hz. In most European countries the V_{rms} is 240V, and the frequency is 50 Hz.

Electrical appliances list the voltage required to operate them and the average power that they consume. Some devices also list the current that they draw. Devices operating on DC list the voltage of the required DC source. The current that they draw and the power that they consume are given by equation [8.6]. Devices operating on AC list the required V_{rms} and the average power that they consume ($P_{average}$). If an AC device lists a current, it lists the I_{rms}. The relationships between those parameters are given by [8.9], which are similar to [8.6]. In the following, when we discuss issues of power consumption, we will use the notations of [8.6] for both DC and AC circuits. In AC circuits I and V would mean I_{rms} and V_{rms}, respectively.

Example

The readings of a light bulb are: 120V, 75W. (a) What is its resistance when operating normally? (b) What is the I_{rms} current that it draws?

Solution

According to the notations of [8.9] V_{rms}=120V, $P_{average}$=75watt, R=?, and I_{rms}=?

Substituting yields: (a) R=192Ω, and (b) I_{rms}=0.625ampere.

8.2.6 Generation and manipulation of AC

Generators in power plants convert thermal, hydraulic, or nuclear energy into electrical energy, in the form of AC. The energy that the generators receive is used to rotate coils of conducting wires between the poles of fixed magnets. That motion generates the AC that is delivered to the loads. The configuration of the coils determines the kind of the generated AC voltage. When one coil is used, a single phase AC voltage is generated. Figure 8.6a shows the generator and figure 8.6b shows the voltage as function of time for a single phase AC voltage (a typical sine function).

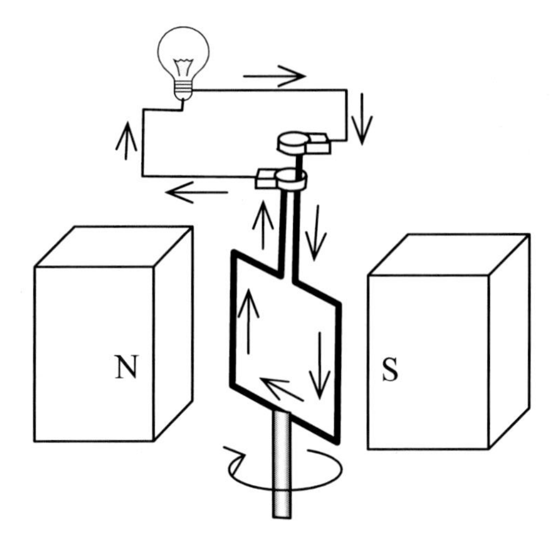

Figure 8.6a: Generating one-phase AC. A conducting loop that rotates between the poles of a magnet (N and S) generates an AC current. The current is collected by brushes that slide on rings connected to the terminals of the loop. From the brushes the current flows through the load (light bulb here). The straight arrows indicate the current at a certain time. The current changes direction and magnitude according to figure 8.6b, as the loop rotates.

It is quite common to use generators that have three coils, separated by 120^0 from each other, rotating between the poles of the same fixed magnet. Each coil generates its own single phase voltage. All together, such a generator generates a three-phase voltage. Figure 8.6c shows the relationships between the three phases generated by a three-phase generator.

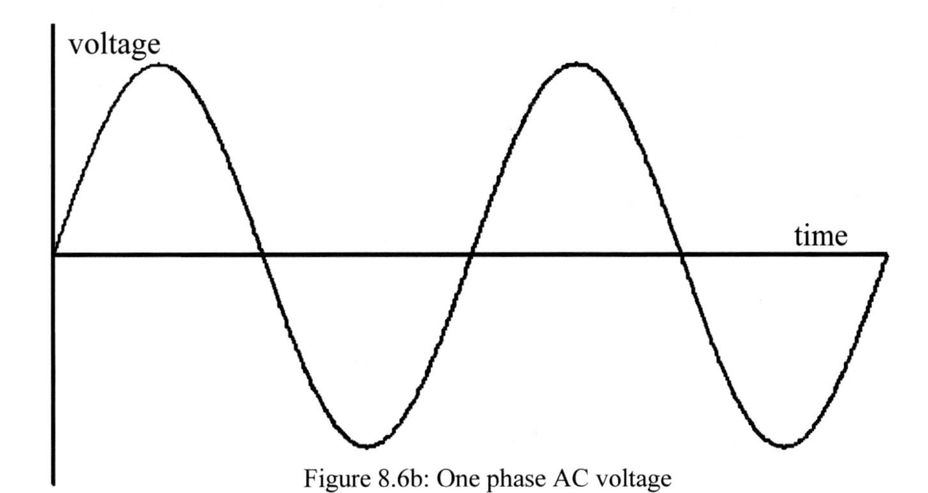

Figure 8.6b: One phase AC voltage

A generator that generates one phase voltage has two terminals. A load is connected by conducting wires to two terminals, thus forming a closed electric loop. A three phase generator has four terminals. Each phase has one terminal of its own, and the other terminal is common to all phases. A load may be connected to any of the three phases. Certain devices are designed to be connected to all three phases together.

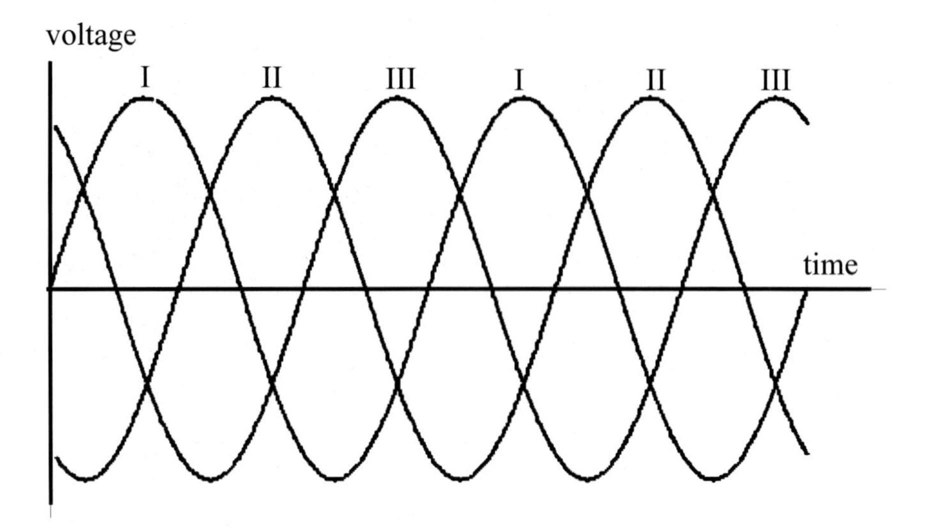

Figure 8.6c: Three-phase AC voltages. Three coils, separated by 120^0, rotate on the same axis between the poles of the same magnet. Each phase (I, II, and III) is delivered by a separate wire. All phases share a common second terminal.

A typical household in the US receives one or two inputs of one-phase AC voltage of 120 volts. If a household receives two one-phase sources, they would usually be in opposite phases to each other (figure 8.6d). Each phase would have one terminal of its own and one common terminal. This makes it possible to use the two phases together as one source of 240 volts, for heavy duty appliances such as electric driers and larger air conditioners.

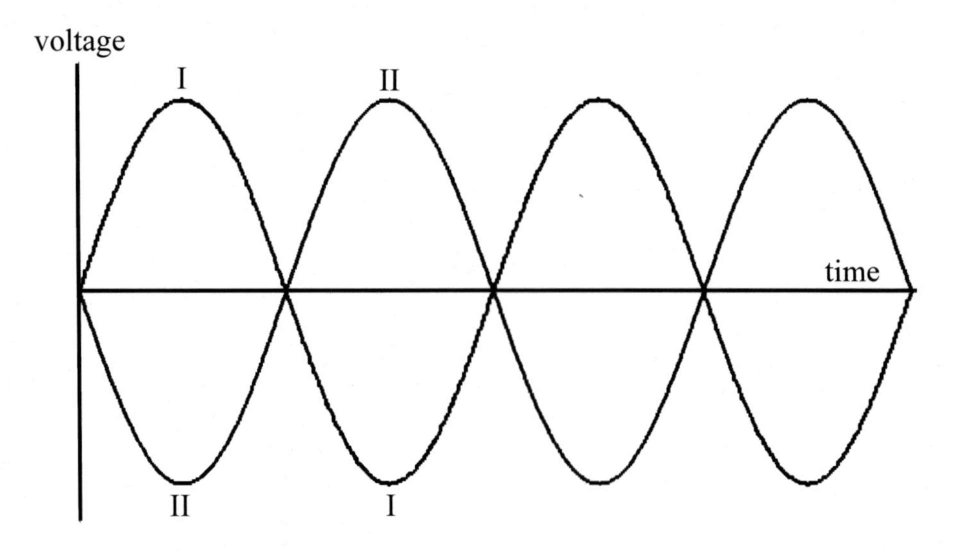

Figure 8.6d: Bi-phasic AC voltages

Most household appliances are designed to accept single phase voltage. Some heavy duty household appliances, such as some air-conditioning units, are designed for two phase voltage. Many power outlets and electrical devices have an additional terminal, called the ground terminal. This terminal is for safety, and its role is explained in a forthcoming section.

A transformer is a device that changes the voltage provided by a given AC source. A transformer can increase that voltage or reduce it (step up or step down transformers). A transformer consists of two conducting coils, the primary and the secondary, wrapped around a common core, usually made of iron. The two ends of the primary coil serve as input terminals, and those of the secondary coil serve as output terminals. The source is connected to the primary terminals, and the load to the secondary terminals. Some transformers produce several output voltages from one input voltage. Those transformers have, accordingly, more output terminals. A transformer does not generate electrical energy; it just changes the voltage that the power generator generates. The power relayed by the transformer is still given by $P=I\cdot V$ ([8.6]), but V is the output voltage of the transformer, and I is the output current of the transformer. A transformer that increases the voltage of the source delivers less current. Similarly, a transformer that reduces the voltage delivers more current. An ideal transformer that is not connected to a load does not draw energy from the power source to which it is connected, because it does not deliver any current. Although the primary and secondary coils are separated from each other, electrical energy can be transmitted through space from the primary to the secondary coil by the process of electromagnetic induction. A transformer cannot change the voltage of a DC source.

Electricity generated in power plants goes through a sequence of step-up and step-down transformers before it reaches the customer. A step-up transformer causes the electrical energy to be transmitted at a higher voltage and at a lower current. Lower currents are more economical to transmit, but on the flip side, their associated higher voltages are more hazardous to humans, as will be explained in the following. Therefore, step up transformers are used to raise the voltage on its way from the power plant to population centers, before it is distributed to the customers. Step-down transformers, located in populated areas, relay safer lower voltage to households.

We have said earlier that the power and current that a battery of a given voltage delivers depend on the resistance R of the load, subject to limitations that depend on the battery. Similar limitations apply also to real AC sources and to equations [8.9]. In the case of AC wall outlets, those limitations are not the result of the amount of current and energy that the power plant can produce (as is the case with batteries). A power plant can produce more power and current than any household could ever need. The bottleneck is in the wires that bring the current to the load. The current that the load uses (I_{load}) has to pass through an electrical wire, which has its own small resistance (R_{wire}). The power consumed by the wire is therefore $P_{wire}=I_{load}\cdot R_{wire}$. That power is converted in the wire into heat. If the wire heats too much, its wrapping and objects around it may catch fire. There are safety measures in every household electrical network for averting that hazard. Those measures, which include fuses and circuit breakers, cut off the current in a loop that draws too much current. The resistance of the wires that carry the current to the load has to be small enough, so that the current would not heat the wires excessively on its way to the load. This can be accomplished by using wires of the appropriate thickness. The resistance of a wire is inversely proportional to its cross sectional area (formula [8.7]). However, thicker wires cost more. Building codes specify the minimum wire thickness that is needed for the anticipated power consumption and the type of fuse or circuit breaker that comes with it.

Note: In the previous discussion we dealt with a resistor connected to an AC source, and said that the current through the resistor changes in phase with the voltage. Both voltage and current oscillate at the same frequency and at the same phase: When the voltage is zero, so is the current. When the voltage is at its maximum so is the current. There are circuit elements, such as capacitors and inductors, that when connected to an AC source, the current through them is not in phase with the voltage. Formulas [8.9] do not apply to those elements.

Many electronic systems require DC current. A rectifier is a device that changes AC voltage into DC voltage. If the rectified voltage is not at the level that the device needs, the original AC voltage is first adjusted by a transformer, and only then it is rectified. The combination of a transformer and a rectifier is often called a DC adapter.

8.2.7 Wiring inside the house

The power company provides households with a connection to the power network. The power company keeps the voltage supplied to that connection at the standard value (e.g. 120V for a single phase household connection in the U.S.). From that connection, the customer has to distribute the power to all the loads that need it. All the loads are designed to operate at the standard voltage supplied by the power company. (Some devices, though, have also built-in auxiliary inputs for other power sources). To operate, each load has to become a part of a closed loop, whose power source has the standard voltage. If a number of loads have to be connected to one power supply, so that each load gets the full voltage of that power supply, the loads have to be connected in a certain way, called connection in parallel. This is the standard way of connecting loads to the power in houses. Figures 8.7 and 8.8 illustrate how such connections are made.

When a number of loads are connected to one power source, this power source causes currents to flow in each of the loads. The power source and each of the loads obey Ohm's Law (V=IR). For the source, V is its own voltage, I is the current that it delivers to all the loads, and R is the equivalent resistance of all the loads. For each load, R is its own resistance, I is the current that flows through it, and V is the voltage across its own terminals. Usually, the voltage of the power source and the resistances of the individual loads are known. Everything else (the equivalent resistance, the current that the source delivers, the current through each resistor, and the voltage across each resistor) depend on how the loads are connected. There are formulas that relate all those parameters. In the following, we'll deal with formulas for connecting resistors in parallel ([8.10]) and in series ([8.11]).

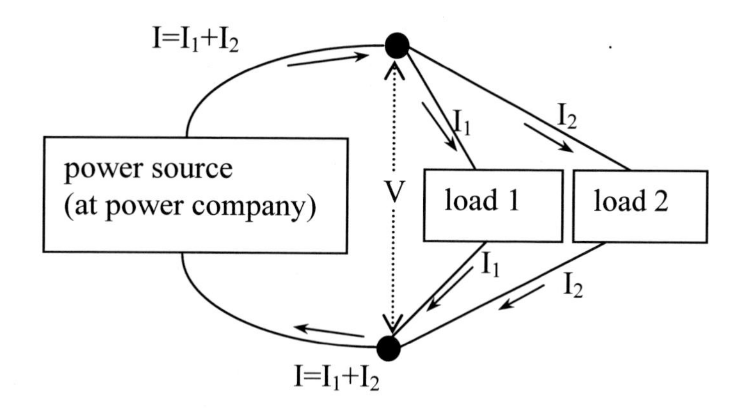

Figure 8.7: Two loads are connected in parallel to the terminals of the power source (full circles) that are provided by the power company. Each load "sees" the full voltage V of the source, and forms a closed loop with it. Each load draws its current according to Ohm's Law, as if the other loads did not exist. The current provided by the power source (I) is equal to the sum of the currents that are drawn by the loads.

The relationships between the parameters of the individual loads that are connected in parallel and those of the equivalent load, as "seen" by the source are:

$$V = V_1 = V_2$$
$$I = I_1 + I_2$$
$$\frac{1}{R} = \frac{1}{R_1} + \frac{1}{R_2} \qquad \text{(Parallel connection)} \qquad [8.10]$$
$$P = P_1 + P_2$$

where V indicates voltage, I current, R resistance, and P power. Indices 1 and 2 indicate parameters of loads 1 and 2. V, I, R, and P that belong to the same load or to the source obey also equations [8.5] and [8.6].

Example
Two loads; a 75W light bulb and a 1.5kW microwave oven are connected in parallel to a 120V power source, as illustrated in figure 8.7. (a) What is the current drawn by each load? (b) What is the current supplied by the source? (c) What is the power supplied by the source?

Solution
Let an index 1 denote the light bulb and an index 2 the oven. Since the loads are connected in parallel, $V_1=V_2=120V$. $P_1=75W$, $P_2=1,500W$.
Substituting in [8.6] ($I=P/V$) yields $I_1=0.625A$, $I_2=12.5A$.
Based on [8.10] $I=I_1+I_2= 13.125A$
Based on [8.10] $P=P_1+P_2= 1,575W$

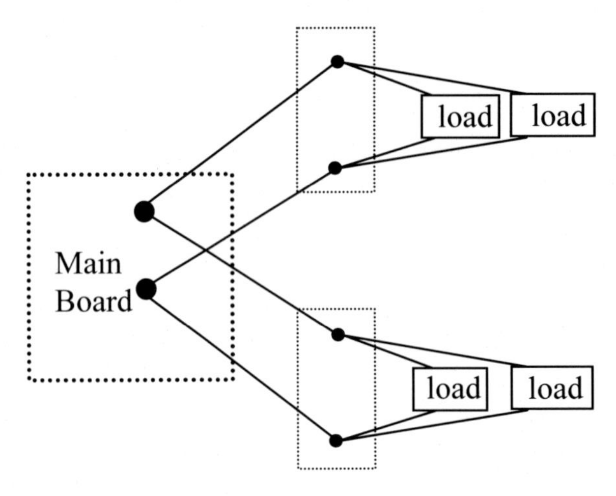

Figure 8.8:. Schematics of a power distribution wiring in a house. Two local boxes are connected to the terminals of the main board. All loads are connected in parallel to the source.

The entire electrical distribution wiring in a house is connected to the terminals on the main power board. Groups of adjacent rooms usually have a common box, to which the loads of those rooms are connected. All those boxes are connected to the terminals on the main power board, which are provided by the power company. Overall all the loads are connected to those terminals in parallel, as illustrated in figure 8.8.

Although the electrical wiring in a house provides parallel connections to the loads (figure 8.8), another connection scheme–series connection–is quite common. It is always present when wires connect a load to the power source. The wires themselves are resistors, and they are connected in series to the load. Figure 8.9 illustrates two general loads connected in series. When dealing with wires that connect a device to the source, load 1 represents the resistance of the connecting wires, and load 2 represents the device that uses the electricity.

In series connection, the current that leaves the power source passes through one load, then through the other, and then returns to the source. The voltage that each load gets (V_1 and V_2 in figure 8.9) is only a part of the voltage V of the source. The parameters of the individual loads and the equivalent load are related through:

$$\begin{aligned} I &= I_1 = I_2 \\ V &= V_1 + V_2 \\ R &= R_1 + R_2 \\ P &= P_1 + P_2 \end{aligned} \qquad \text{(Series connection) [8.11]}$$

where V indicates voltage, I current, R resistance, and P power. V, I, R, and P that belong to the same load obey also equations [8.5] and [8.6].

The implications of [8.11] are important when considering the wires that are used. First, because of the wires, the load does not get the entire voltage of the source. Some of that voltage "is taken" by the wires. If too much voltage is taken by the wires, the load may not get the voltage that it needs to operate properly. The voltage is distributed between the wires and the load according to the proportion of their resistances. Therefore, the resistance of the wires has to be as small as possible.

In addition to that, the current that flows through the load is equal to the current through the wires. That current heats the wires. The resistance of the wires should be kept small, to minimize the wasted heat, and to prevent over-heating of the wires.

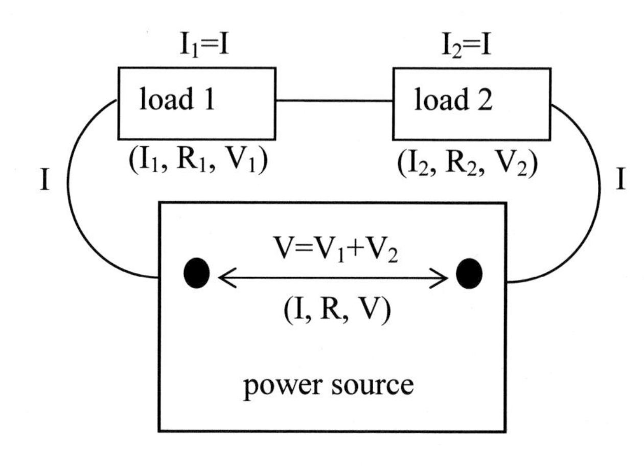

Figure 8.9: Two loads are connected in series to a power source

Example

A toaster of 0.75kW is designed to operate at 120V. (a) What is its resistance? (b) Assume that its resistance is constant, what would be the voltage across the toaster if the resistance of the wires connecting it to the 120V source is 2Ω ? (c) How much power it would consume when connected by those wires?

Solution

According to [8.6], $R=V^2/P$, and in this case, the resistance of the toaster is $R=120^2/750=19.2\Omega$.

Let index 1 denote the toaster, and index 2 the connecting wires. Since the toaster and the wires are connected in series, their equivalent resistance R, as "seen" by the source is $R=R_1+R_2=19.2+2=21.2\Omega$. The current I that the source delivers to that equivalent resistance is I=120/21.2=5.66A (I=V/R, Ohm's Law). Because the wires and the toaster are connected in series, the current through the toaster is $I_1=I_2=I=5.7$A ([8.11]). Based on Ohm's Law (V=IR), the voltage across the toaster is 5.66x19.2=108.67 volts.

The power consumed by the toaster is (P=IV) 5.66x108.67=615watt= 0.615kW.

Conclusion: Because of the resistance of the wires, the toaster does not receive the current that it needs, which causes it to operate not at its full capacity. In order for the toaster to receive more voltage, it has to be connected to the source by wires of smaller resistance.

Excessive resistance of the wires and of the connections between wires degrades the voltage that reaches the loads. When designing the electrical wiring of a house, wires with appropriate resistances have to be used, in order to keep the degrading of the voltage within acceptable limits.

8.2.8 Electrical safety

Due to the electrolytes contained in all body fluids, the human body is a conductor. When the human body comes in contact with a power source and closes a loop, it draws current. The physiological effects of the current would depend on the amounts of current that pass through various organs and tissues. The typical resistance of the human body is approximately $10^5\Omega$. When touching the electrodes of a 12 V car battery, the current that passes through the body is too small to cause any damage. However, the 120V household outlet can deliver fatal amounts of currents, because the resistance of the body would draw them, and the source is capable of delivering them. In those two cases, V, the current, and the resistance of the body obey Ohm's Law [8.5]. On the other hand, accidentally having a contact with the 20,000 volts across the spark plugs of a car is not fatal, albeit unpleasant. The car's ignition system cannot deliver a high enough current to be fatal. The delivered current is much less than what would be expected according to Ohm's Law. The danger in electrical currents is due to their interference with critical physiological processes, e.g. function of the nerve cells. Such damages occur even if the current is too small to cause burns. During surgery, even a very low voltage that accidentally contacts sensitive organs could deliver small amounts of current that are dangerous.

Most household lightings and appliances use single-phase voltage. Larger motors and heavy machinery use two and three phase sources. In addition to the power terminals to which the load has to be connected to form a closed loop, many electrical outlets and appliances have an additional terminal: the ground terminal. For example, a standard power cord of a refrigerator has three prongs: two for the voltage source that provides the power, and one for the ground. The power company connects one of its power terminals to the ground. This terminal is sometimes called the neutral terminal, and the other is called the "hot" terminal. The customer connects the loads to the power terminals (the "neutral" and the "hot"). It is up to the customer to also connect the load itself to the ground ("to ground the load"). This is done through the ground (third) terminal of the household electrical receptacle. This terminal is connected by a metal wire to a metal water pipeline or to a metal rod pushed into the ground. Since the ground is a conductor, any load connected to that ground terminal becomes electrically connected to the ground of the power company.

If something goes wrong, and the "hot" wire touches the envelope of the load (for example, the metal surface of a refrigerator), that surface becomes an extension of the "hot" wire, and acts as the "hot" electrode of the power source. If that refrigerator is not grounded and a person touches it, and if another part of that person touches the ground or a conductor connected to the ground (e.g. a water faucet), that person becomes a part of a closed electrical loop, which includes also the "hot" wire and the ground. An electrical shock is delivered to that person (figure 8.10). However, if the refrigerator is grounded, as required, that risk is averted. Once the loose wire touches the envelope of the refrigerator, an electrical loop is closed through the ground wire of the refrigerator. Because the resistance of the envelope of the refrigerator is small, the current through that loop will be large [8.6], and will trip the circuit breaker. That would disconnect the power to the refrigerator.

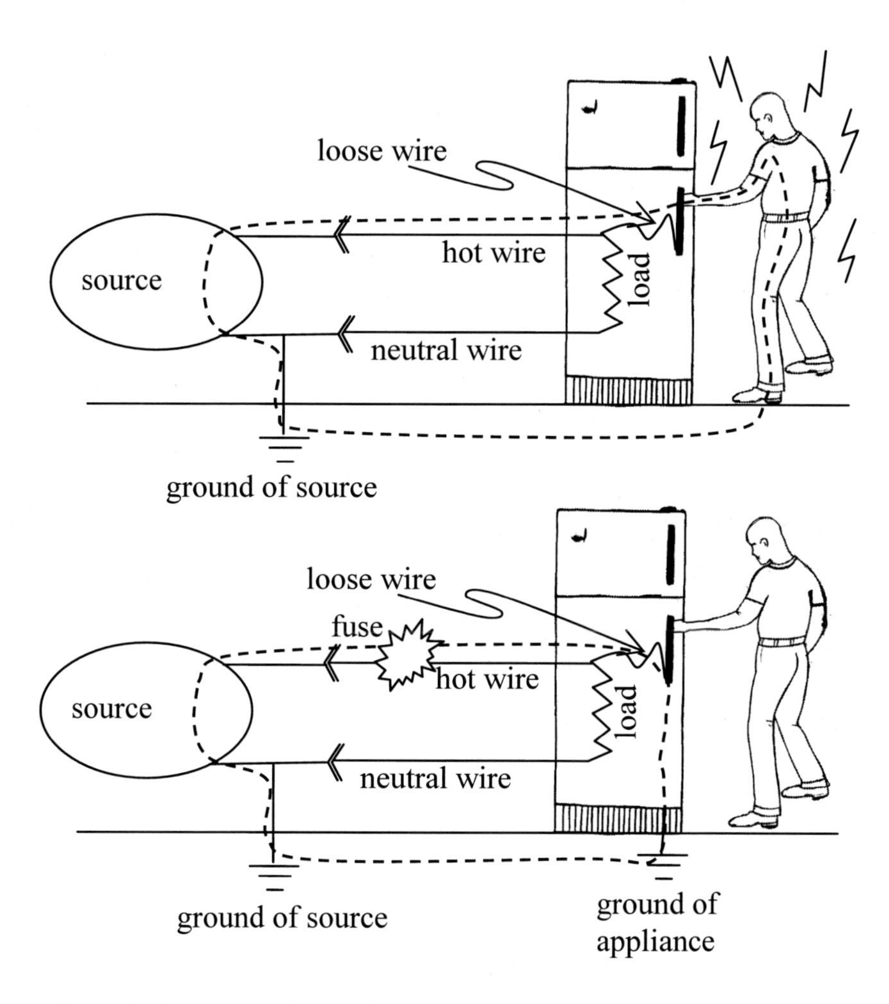

Figure 8.10: Top: The "hot" wire in an ungrounded refrigerator becomes loose and touches the door. A person that touches the door closes a loop and gets an electric sock. Bottom: If the same refrigerator is grounded, a closed loop is formed as soon as the loose wire touches the door. A large current flows through that loop because it bypasses the load, and blows the fuse. The refrigerator is thus disconnected from the power.

PROBLEMS

Section 8.1

1. A point charge of +2μC (+2 micro coulombs) is placed at the origin of the x-axis. A second point charge of -3μC is placed to its right at x=4m. A third point charge of -5μC is place at x=-4m.
(a) Find the magnitude and direction of the force that the second charge exerts on the first.
(b) Find the magnitude and direction of the force that the third charge exerts on the first.
(c) Find the magnitude and direction of the total force that the second and third charges together exert on the first.

2. Two identical point charges, separated by 3 meters, exert a force of 6.5×10^{-9}N on each other. What are their charges?

3. Two point charges exert a force of 16N when separated by 3m. What would be the force that they exert on each other when they are moved (a) 6 m apart? (b) 9 m apart?

Sections 8.2.2- 8.2.3

4. A battery can generate a current of 2 milliamperes for 8 hours. What is the total charge that passes through one of its electrodes?

5. What is the energy provided by a battery of 1.5 volts when it moves a total charge of 8μC from one electrode to the other?

6. What is the power that can be provided by a source whose specs are 110V, 2A? Express your answer in watts and in horsepower.

7. A toaster whose resistance is 15Ω is connected to an outlet of 110V.
(a) How much power does it consume?
(b) What is the current that flows through it?
(c) How much power will this toaster withdraw if connected to a source of 12V? How will this affect its operation?
(d) How much power will it withdraw if connected to a source of 150V? How might this affect its operation?

Sections 8.2.4-8.2.5

8. Find the resistance of an aluminum wire, 1 mile long and a diameter of 6mm.

9. An aluminum wire and a copper wire of the same length have the same resistance. The diameter of the copper wire is 2mm. What is the diameter of the aluminum wire?

10. The specification of a battery says that it can deliver 200 ampere hour at 12 volts. How much energy can this battery deliver?

11. What is the current drawn by
(a) 2.5kW space heater. (b) 1kW microwave oven. (c) 100W light bulb, when connected to a 120V AC source?

12. What are the current and power drawn by a 100,000Ω resistance when connected to a 120V AC source.

13. An iron of 0.5kW is designed to operate at 120V. How much power will it consume when connected to a source of 240V (assuming that its resistance won't change)?

14. What is the resistance of a 45W 120V light bulb, when operating as designed? Based on the answer, what should be the resistance of a 90W 120V light bulb?

Sections 8.2.7- 8.2.8

15. A load of 100 Ω and a load of 200 Ω are connected in parallel to a source of 120V. Draw the circuit diagram and find the current through each load and the total power consumed by the loads.

16. A load of 100 Ω and a load of 200 Ω are connected in series to a source of 120V. Draw the circuit diagram and find the current through each load and the voltage across its ends.

17. A load of 100,000 Ω and a second load are connected in series to a power source of 450 V. What is the resistance of the second load if the voltage across the 100,000 Ω is only 25 volts?

18. The specs of an electric water heater say 2kW, 120V.
(a) What is its resistance?
(b) The wires connecting it to the power source have a resistance of 0.5 Ω. What is the actual voltage on the water heater? How much power is wasted in the wires and how much power does the water heater actually consume?

19. A bedroom is designed to have three electrical outlets, each should be able to deliver up to 1,200 watt. All those outlets have to be connected to one AC room's source of 120V. The room's source should be connected to the main source, which is provided by the power company.
(a) Draw a schematic diagram showing the electrical wiring in this room and the connection to the main source.
(b) Indicate the current that would flow in each wire should it be used to its maximum intended capacity.
(c) You have to choose one circuit breaker that would cut off the power to the room if more than 3,600 watts are used in it. In your diagram, show where you

would place the circuit breaker, and the current that should trigger it. Specify the current limits that each of the wires from the room's source to the outlets and from the main-source to the room's source should be able to carry safely.

9. OPTICS

And sweet is the light, and a pleasant thing is for the eyes to behold the sun. (Ecclesiastes 11, 7)

9.1 The nature of light

It takes time for a light beam to propagate from one point to another. The speed of light, which is commonly denoted by c, is very large: $c=2.99792458\text{x}10^8$ m/s, or approximately $c=3\text{x}10^8$m/s. A pulse of light can travel the distance between the Earth and the moon in approximately one second. This value of c is for light traveling in vacuum or in air. In transparent media, such as glass or water, light propagates slower. The speed of light in transparent media will be discussed in a later section.

9.1.1 Particles or waves?

Waves are quite a common phenomenon. We are all familiar with water waves that propagate in a pond after we throw in a pebble. The wavy nature of other phenomena, such as sound, may not be so obvious to our senses, and we need special instruments to recognize it. Waves can have many shapes. One common shape is sinusoidal waves. These waves can be described by a sine function. Figure 9.1 shows a snapshot of a sinusoidal wave. The x-axis describes the x-coordinate of a point of the wave at a certain instant. The y-coordinate describes the entity that propagates like a wave. In water waves, the y-coordinate describes the height of a point of the wave with respect to calm water. In sound waves the y-coordinate describes the change in the local air density of the air with respect to quiet air.

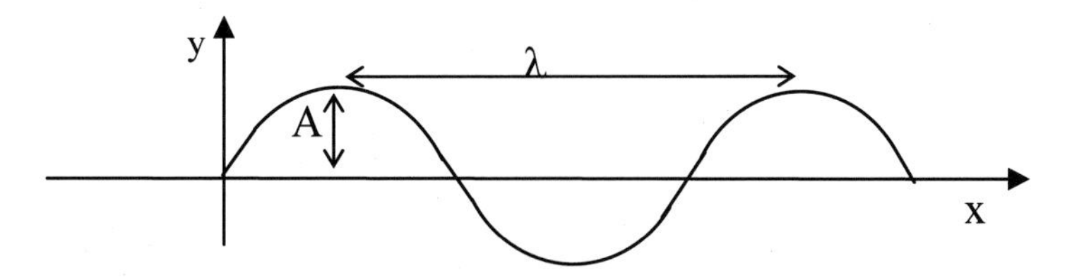

Figure 9.1: Sinusoidal wave

The distance between two consecutive crests of a wave is called the wavelength, and is denoted here by λ (figure 9.1). If you watch a cork floating in a pond when waves propagate in it, you'll notice that the cork moves up and down as the waves pass through. The frequency of these oscillations is the frequency of the wave (f). In general, the frequency of a wave is the frequency at which a point at a given location oscillates. The velocity of the wave (c) is the velocity at which the wave's crest is moving. λ, f, and c are related by the formula:

$$c = \lambda \cdot f \quad [9.1]$$

The amplitude of a wave (A in figure 9.1) is its largest deviation from the quiet state of the medium in which the wave propagates. Waves that are not sinusoidal can be constructed by adding sinusoidal waves of different wavelengths. In our discussion we will concentrate on sinusoidal waves.

We can hear sounds, which are air vibrations that propagate in the form of waves. We can feel showers and rain, which are streams of particles. We can sense light. What is light? Is it something like sound, which propagates in the form of waves, or is it a stream of some particles, like the rain?

We cannot see what light is made of. It may consist of large number of tiny particles that we cannot see, or it may be made of waves that we cannot discern, like sound waves. In order to answer this question, we have to investigate phenomena that can tell apart waves from beams of streaming particles.

When a stream of particles hits an obstacle with an opening, e.g. when rain falls on a roof with a hole, the stream of particles is divided into two parts. The part that hits the roof is blocked, and the part that passes through the hole continues straight on, as if the roof did not exist. When a wave encounters the same obstacle, e.g. when a radio is playing on the roof, the sound waves propagate through the hole and fill the room underneath, even if the roof itself is sound proof. The trajectory of a particle in a stream is not affected by an obstacle that it does not hit, while the entire wave is affected by an obstacle that blocks part of its path. Waves go around obstacles. This characteristic property of waves is called diffraction.

There is another difference between streams of particles and waves. When we throw pebbles into a pond, each pebble starts its own water wave. After a while, these waves propagate and interfere with each other. Some of the interfering waves are stronger than the original individual waves, while others are weaker or completely extinguished. This extinction will never happen with streams of particles. Two merging water jets will never leave dry points where they meet. We say that waves can undergo constructive interference (two waves reinforcing each other) and destructive interference (two waves weakening or annihilating each other), while beams of particles cannot interfere destructively. Constructive and destructive interference of waves are direct consequences of the basic characteristics of waves: wavelength (λ), frequency (f), and wave velocity (c). There are formulas that relate these wave characteristics and the patterns of interference of waves.

The total energy of a beam of particles is the sum of the energies of the individual particles. For example, the total energy of a jet of water droplets is the sum of the kinetic and potential energies of all the droplets that make up the jet. The energy of a wave depends on its amplitude. The larger the amplitude, the more energy is carried by the wave. For example, one-foot waves on the beach are pleasant, while fifteen-foot waves can cause severe damage, since they are more energetic.

So, waves differ from beams of particles by their ability to diffract and to interfere, and by the way that they carry their energy. With these differences in mind, experiments were conducted in order to determine whether a beam of light behaves like a wave or like a stream of particles.

Young's double slit experiment

In Young's (Thomas Young, 1773-1829) double slit experiment, a beam of light passes through two narrow slits in a screen, which are very close to each other. After passing the slits, the light hits a second screen. If light is a stream of particles, the two slits should create two bright spots of light on the second screen, much like two jets of water that pass through two windows (figure 9.2 left). However, in Young's experiment, instead of two spots of light, an entire pattern of bright and dark fringes is obtained (figure 9.2 right). Careful measurements and analysis have confirmed that these patterns can be attributed to diffraction and interference of waves. The bright fringes are formed by constructive interference of diffracted waves, and the dark fringes are formed by destructive interference. The wavelength of the waves can be determined from the spacing between the fringes, the distance between the two screens, and the separation between the slits. Hence, light must be a wave.

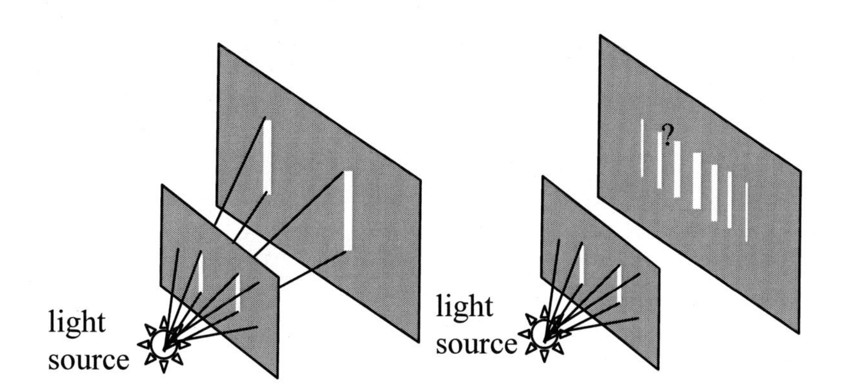

Figure 9.2: Young's double slit experiment. Left: what would be expected from beams of particles. Right: the actual observed fringes.

Using double slits and similar devices it was found that different wavelengths of light waves correspond to different beam colors. It is customary to characterize light colors by their wavelength. For example, a certain yellow beam is characterized by a wavelength of 580 nm. ($1nm=10^{-9}m$). Based on equation 9.1, it is also possible to characterize each color by a frequency $f = c / \lambda$. For example, light whose wavelength is 580 nm has the frequency of $3x10^8$[m/s]/ 580 x 10^{-9}[m] = 5.17x10 ^{14} Hz. Table 9.1 summarizes the ranges of wavelengths and frequencies for visible light. Transitions from color to color are gradual, and the numbers in the table indicate the regions where the gradual transitions occur.

Color	*wavelength range (nm)*	*frequency range (x10^{14} Hz)*
Red	750-640	4.0-4.7
Orange	640-590	4.7-5.1
Yellow	590-570	5.1-5.3
Green	570-480	5.3-6.3
Blue	480-430	6.3-7.0
Violet	430-400	7.0-7.5

Table 9.1: Relationships between color, wavelength, and frequency

The photoelectric effect

Some materials exhibit the photoelectric effect. When hit by light, they release electrons, hence the name photoelectric. The released electrons can activate specifically designed electronic circuits, which indicate the presence or absence of an incident light. Photoelectric materials are used in many safety and security devices in which an object that crosses a beam of light triggers an alarm. Photoelectric materials are also used in remote control units, such as those that control TV sets.

Normally, the electrons are tied to the other particles that constitute the photoelectric material. By breaking those ties, the incident light causes the release of the electrons. It was found that whether or not electrons are released depends on the wavelength of the incident light. For any given photoelectric material, only light whose wavelength is shorter than a certain critical value can release electrons. It does not matter what total energy is carried by the light beam (how bright is the light). A faint beam with a short wavelength will release electrons, while even a very energetic (bright) beam whose wavelength is too long will not release any electron. However, the intensity of the incident light affects the number of released electron, if the wavelength was short enough to release them in the first place. Bright light will release more electrons than dim light. These intricate features of the photoelectric effect were explained by Albert Einstein (1879-1955) in the following way. A beam of light in the photoelectric effect may be viewed as made up of a stream of particles, called **photons**. The photons are mass less, and they move at the speed of light $c = 3 \times 10^8$ m/s. The energy E of a photon in a beam depends on the wavelength λ, or on the frequency f, of that beam. (The values of λ and f could be determined by passing the beam through a double slit or by some similar experiment). The energy of a photon is proportional to the frequency f of the light, and, hence, it is also inversely proportional to the wavelength λ of the light. The energy of a photon is given by:

$$E = h\frac{c}{\lambda} = h \cdot f \qquad [9.2]$$

where $h=6.63 \times 10^{-34}$J·s is Plank's constant.

So, for example, red photons are less energetic than blue photons (see table 9.1).

The tie of an electron to the photoelectric material may be broken when the electron is hit by a photon. However, for this to happen, the photon has to carry enough energy. That explains why a beam whose wavelength is too long does not release any electrons. Its photons are too weak. When the energy of the photon is higher than what is needed for breaking the tie, the excess energy is given to the released electron in the form of kinetic energy.

The second part of the photoelectric effect has to do with the number of released electrons. The brightness of a light beam is determined by the number of its photons; the more photons it has, the brighter is the beam. When a brighter beam hits a photoelectric material, more photons collide with electrons, and consequently more electrons are released, compared to a faint beam of the same wavelength. All this could happen, of course, only with beams whose individual photons carry enough energy to break the ties of the electrons with which they collide.

A dilemma and its solution

We are having here a dilemma. On one hand, as in the case of the photoelectric effect, a beam of light behaves like a beam of particles (photons). On the other hand, that same beam of light behaves like a wave, e.g. in Young's experiment. The only way out of this dilemma has been to conclude that a beam of light from a given source might behave in certain circumstances like a wave, with characteristic amplitude, wavelength, and frequency. That same beam in different circumstances might behave like a stream of particles–photons–each having its own trajectory and energy. This is called the **duality principle**. The same physical object may behave as a particle in some circumstances, and as a wave in others. It was found that it applies not only to photons but also to other particles, and it is most noticeable in processes that involve atomic and subatomic particles. All the phenomena that we will discuss in the following could be explained by considering light as a stream of photons, and that is the approach that we will use.

In our approach, photons are characterized by their energy, which is associated with a wavelength and a frequency according to equation [9.2]. When a photon enters our eye, it creates a sensation of color commensurate with its associated wavelength and frequency, as listed in table 9.1. So, photons are those tiny peculiar energetic particles that trigger in us the sensation of light and color.

The energy of a light-beam

A monochromatic beam of light consists of photons that have the same frequency (and wavelength). A Polychromatic beam is made up of a mixture of photons of various frequencies. Since the energy of a single photon is given by $E=hf$, the energy of N photons that make up a monochromatic beam is given by:

$$E_f = N_f \cdot h \cdot f \qquad [9.3]$$

where E_f is the total energy of the monochromatic beam consisting of N_f photons of frequency f. The total energy of a polychromatic beam is the sum of the energies of its constituting monochromatic beams, as expressed by equation [9.3].

The spectrum of a beam is a list of the wavelengths of its photons and the relative energy of each wavelength in the mixture. Because of the one-to-one correspondence between wavelength and frequency (9.1), a spectrum may also be organized by frequency. It is common to present spectra in a graphical way.

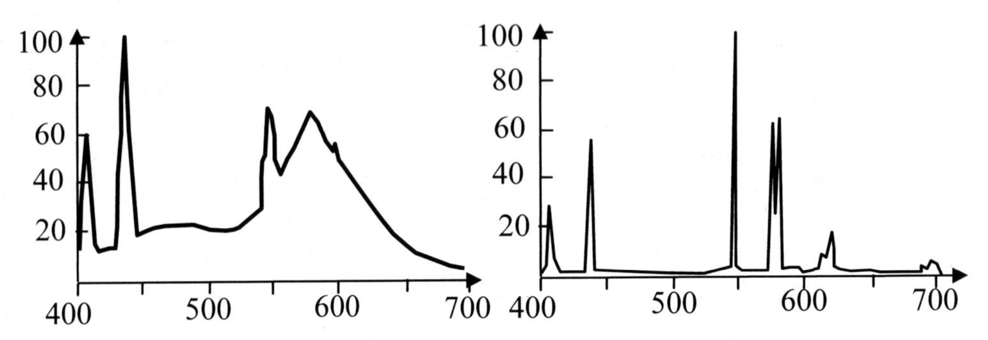

Figure 9.3: Spectra (relative energy vs. wavelength in nm) of a common incandescent white light source (left) and a common mercury vapor lamp (right).

9.1.2 *Generation and annihilation of light*

A book that falls and hits the floor loses energy. Before the fall it had potential energy (mgh), and after hitting the floor it lost this energy (h=0). We can hear the sound of the book hitting the floor because some of the lost potential energy is transformed into acoustic energy, expresses as sound waves that reach our ears. The rest of the lost potential energy is transformed into other energy forms, such as heat. Light is generated in certain processes where electrically charged particles, usually electrons, lose their energy. In an atom, electrons circulate in orbits around the nucleus. If an electron falls from one orbit to an orbit closer to the nucleus, the electron loses some of its energy, similar to the book that fell to the ground. The lost energy of the electron is converted into light energy, which is realized by a photon. The energy of the photon, E=hf, is equal to the energy lost by the electron.

In order to lose energy, one has to have some. A book cannot lose potential energy if it is already on the ground. It has to be lifted above the ground level. When it is lifted, energy is invested in it by an outside source (the lifting person). The same is true for the electron. In order to be able to emit a photon by falling closer to the nucleus, the electron has to be in a higher energy level. It can get the energy to be raised to a higher level through a collision with another particle. An electron can also be bumped into a higher energy level by being hit by a photon. In such cases the projectile photon loses its energy and vanishes. A photon may give away its energy and vanish also in other processes. For example it may add kinetic energy to its target, by making it, as a whole, move at a greater speed. So, photons are generated when electrons or other charged particles lose energy, and are annihilated by giving away all their energy. The photon's lost energy is transferred to other particles.

The generation of light is governed by the laws of quantum physics, which apply to processes that take place in atoms and other small-scale systems. A major difference between the principles of classical physics and the principles of quantum physics is that in classical physics an object can get any amount of energy. (A book can be raised to any level above the ground level). In quantum physics, not all energy levels are permissible. An electron can circulate only at certain orbits around the nucleus. The orbit with the lowest energy is called ground level or ground state, and all other orbits are called excited levels or excited states. These states determine the energy of the released photons $E = h f$. Other objects, such as molecules and crystals, also have ground states and excited states. When they absorb energy, they get into an excited state. When they later fall back to a lower state, they release energy in the form of photons.

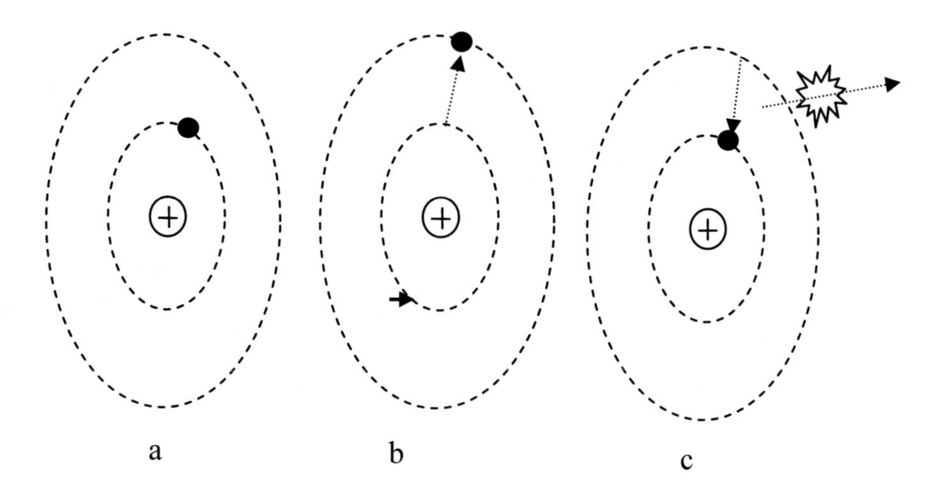

Figure 9.4: Creation of a photon: (a) An electron (dark dot) orbits around a nucleus (+). The electron is orbitting in its ground level. (b) As a result of a collision with another particle (not shown), the electron gains energy and is bumped to a higher orbit. (c) The electron falls back to its lower orbit, giving away its gained energy to a newlly created photon, which leaves the atom.

Some of these processes generate photons that cannot be detected by the human eye, but they still have all the properties of visible photons. They move at the speed of light, they have wavelength and frequency, and their energy is $E=hf$.

Photons whose wavelengths are in the range 800-10^5nm are called infrared (IR) radiation. They are invisible to the human eye, but eyes of some animals, such as snake, can detect them. All objects emit infrared radiation. The spectrum of the infrared radiation depends on the temperature of the emitting object. This fact is utilized in the visual system of snakes and in some night vision goggles to distinguish between hot objects, such as prey or engines, from their colder background.

Photons whose wavelengths are in the range 50-380nm approximately are called ultraviolet (UV) radiation. UV light is invisible to the eye. UV photons are present in sunlight, and they are partially blocked from reaching the surface of the Earth by ozone layers in the upper atmosphere. These photons are important to the photosynthesis process, in which plants generate organic material from inorganic gasses and water. When these photons interact with skin cells, the skin gets tanned. However, excess exposure to UV light may cause skin cancer.

The term electromagnetic radiation refers to photons of any energy. It includes Gamma rays, X rays, ultra-violet light, visible light, infrared radiation, microwave, TV waves, and radio waves. Electromagnetic radiation in the range of TV waves and radio waves are generated by large-scale oscillations of electrons in antennas, and they are usually treated as waves rather than as photons. Figure 9.5 shows the wavelengths of the various kinds of electromagnetic radiation.

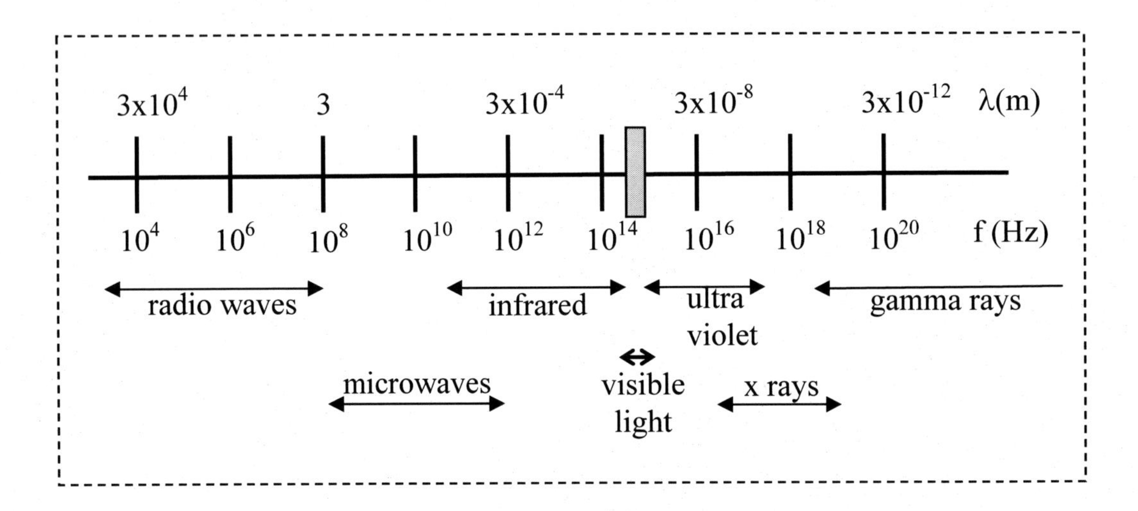

Figure 9.5: The electromagnetic spectrum. The wavelength (λ) and frequency (f) ranges of major types of radiation.

9.2 The perception of color

9.2.1 Photon-eye-brain interactions

Rays of light from objects that we look at reach our eyes. The lens of the eye focuses the rays on the retina, at the back of the eye, and forms there an image of the objects. At the retina, the photons are absorbed by two kinds of cells: rods and cones. The absorbed photons trigger electrical pulses that propagate from the retinal cells to the brain, where they are processed. The perception of color is due to inputs that the brain receives from the cone-cells. There are three kinds of cone cells–the so-called red, green, and blue cone-cells. Each cone-cell absorbs and responds to the entire spectrum of the visible light. However, "blue" cones are more sensitive and relay stronger electrical signals when illuminated by photons of wavelengths around 420 nm, "green" cones have a sensitivity peak around 530 nm, and "red" cones have a sensitivity peak around 600 nm. The sensitivity of the rod-cells, too, depends on the wavelength of the illuminating photons, but rods are not involved in color perception. Figure 9.6 shows how the sensitivity of the cones and of the rods depends on the wavelength of the sensed photons.

Different mixtures of photons that reach a small region in the retina can create the perceptions of thousands of different colors. However, there is no one-to-one correspondence between mixtures of incoming photons and the perception of the color that they create. Different mixtures may end up creating the perception of the same color. For example, a certain mixture of red, green, and blue photons creates the perception of white. However, mixing red green and blue photons of other wavelengths may also create the perception of white. Daylight, which contains photons of the entire spectrum of the rainbow, also creates the perception of white. Mixing red and green photons in certain proportion creates the perception of yellow, the same perception that is created by yellow photons. So, by knowing the wavelength of a photon we can determine the perceived color of the beam, but by knowing the perceived color of a beam we cannot usually determine its spectrum.

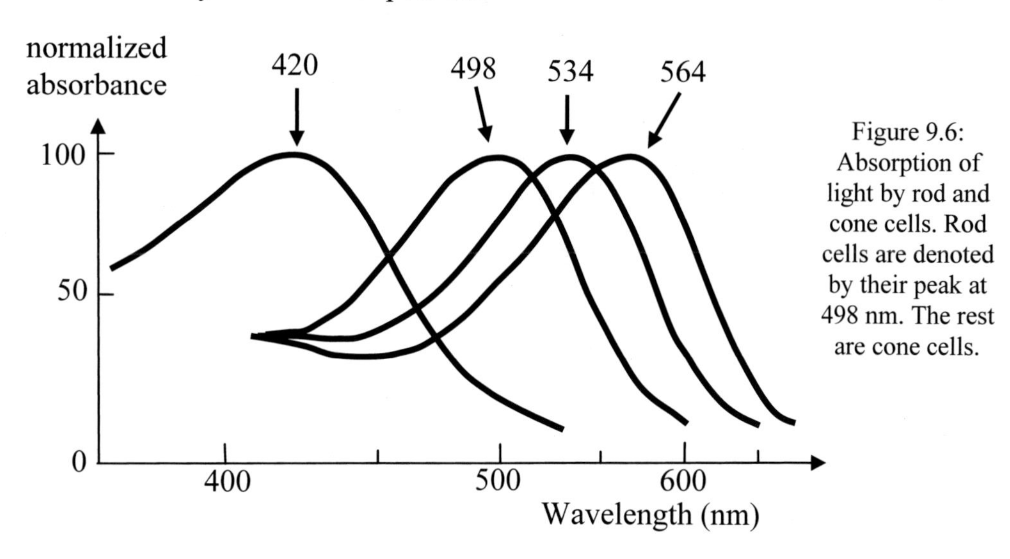

Figure 9.6: Absorption of light by rod and cone cells. Rod cells are denoted by their peak at 498 nm. The rest are cone cells.

By mixing three beams of different colors we can get many colors that are different from the colors of those individual beams. Such three colors are called primary colors. If the primary beams are certain red, green and blue, we can mix from them almost any perceived color. This fact is used in color TV's. The screen of a color TV is made up of thousands of small elements called pixels. Each pixel consists of three dots: red, green, and blue (RGB). Each dot can be made to emit photons of its corresponding color. (In reality, each dot emits also smaller amounts of photons of other colors). The number of photons that a dot emits is controlled by a beam of electrons, which is controlled by the electronic circuits of the TV set. Thus, by controlling the proportions of red, green and blue photons that are emitted by the three dots of a pixel, the pixel could be perceived as having any intended color. This is done to each of the screen's pixels, thus generating the entire color picture from its pixels.

In design and industry it is sometimes needed to specify colors by numbers, so that they could be manufactured and applied in precise ways. The RGB system is one way of doing that. The wavelengths of the three primary sources, whose luminous power output can be controlled, are specified first. (Luminous power is the energy delivered by the beam (formula [9.3]) per unit time). The maximum power of the sources is chosen such that when all three beams are at their maximum, the resulting color of their mixture is white. When the powers of the beams are reduced gradually from maximum to zero, while keeping those powers equal to each other, the resulting mixture turns gradually from white to gray to black.

In order to specify the perceived color of a specimen, it is compared with a mixture of those three beams. The powers of the beams are adjusted until the color of their mixture matches the color of the specimen. The relative powers of the three primary beams are then used to specify the color of the specimen. The values of these relative powers are called the **RGB components** of the perceived color of the specimen. For example, R=0.3, G=0.7, B=0.65 means that the match was obtained when the power of the red beam (R) was 30% of its maximum value, the power of the green beam (G) was 70% of its maximum, and the power of the blue beam (B) was 65% of its maximum. Other normalizations of the primary beams are quite common. For example, the range 0-255 instead of 0-1 is used for specifying the RGB components of pixels of computer monitors. Different schemes may choose also different primary beams.

A disadvantage of the RGB specification system is that small differences in the RGB components sometimes correspond to big differences in how the perceived color is described in words.

When only one RGB component is different from zero, the name of the color is the name of the non-zero component, i.e. red, green, or blue. Such a color also gets the attribute **saturated**.

When two RGB components are different from zero and the third is zero, the perceived color belongs to one of three hues: mixtures of red and green constitute the yellow hue group, mixtures of green and blue constitute the cyan hue group, and mixtures of blue and red constitute the magenta hue group. When one of the RGB components is zero, the hue is called **saturated**. The ratio of the two primary RGB components defines the specific hue within the group. For example, equal powers of red and green create the perception of yellow. When the ratio of the powers of red to green is 3/1 the hue is called orange. (The blue component in these cases is zero).

When all three RGB components are different from zero, the perceived color is that of a mixture of a saturated hue and white (or gray). For example, R=0.5, G=0.3, B=0.2 is a combination of (R=0.3, G=0.1, B=0) + (R=0.2, G=0.2, B=0.2). The first color is saturated orange because the B component is zero, and the ratio of red to green is 3/1, which is orange. The second color is gray because R=G=B=0.2. Overall, R=0.5, G=0.3, B=0.2 is perceived as a mixture of saturated orange and white, and is called unsaturated orange. The degree of saturation is determined by the ratio of the power of the saturated hue to the total power of the unsaturated hue. In the last example the saturation level is (0.3+0.1)/(0.5+0.3+0.2)=40%.

The total luminous power of a mixture is the sum of the powers of its components. The luminous power of R=0.5, G=0.3, B=0.2 is 1. The luminous power of pure white in this normalization scheme is 1+1+1=3.

There are other color specification systems such as Munsell's, and a number of CIE systems. They are based on similar principles and emphasize various aspects of the perceived color.

9.2.2 Perceiving reflected light

There are objects, such as the sun, light bulbs, and TV screens, that generate their own light, but the vast majority of the objects around us can be seen due to the light that they reflect. In the process of reflection, an object will selectively reflect photons of certain wavelengths, and absorb others. The reflected light that reaches our eyes creates the perception of the color of the illuminated object. Ideal black objects completely absorb all the incoming wavelengths. Ideal white objects completely reflect all the incoming wavelengths.

The perceived color of a reflecting object will depend on the spectrum of the illuminating light and on the reflective properties of the object. For example, some yellow dyes completely reflect yellow photons, partially reflect red and green photons, and completely absorb blue photons. Consider a white paper with a dot of this particular yellow die on it. In plain daylight the dot will look yellow, and the paper will look white. Now imagine taking it into a dark room. Because of the lack of light, the paper and the dot will not reflect any light and we won't be able to see them. When illuminated by blue photons, the paper will look blue (white paper completely reflects all colors), but the yellow dot will look black (this yellow dot absorbs blue photons). When illuminated by yellow photons the paper and the dot will look yellow, because both reflect yellow photons. When illuminated by red photons, the dot and the paper will look red, with different intensities. The dot will not reflect red photons as effectively as the paper.

One property of light sources is their **color rendering** ability. The perceived colors of an illuminated object are determined by the photons that the object reflects to our eyes. These, in turn, are determined by the spectrum of the illuminating light, and by the absorption and reflection properties of the object. Low-pressure sodium lamps are an extreme example. These light sources emit only yellow photons. When illuminated by this light, an object, which is colorful in daylight, will have various intensities of yellow, with no other color.

The principles that govern color perception of light that is transmitted through mediums are similar. A light-transmitting medium, such as glass, selectively absorbs some wavelengths and transmits others. The perceived color of the transmitted light depends on the spectrum of the source and on the absorption properties of the medium.

9.2.3 Eye accommodation and glare

We can see clearly scenes that have various illumination levels. The eye can adapt itself to perceive details in bright scenes, whose illumination is 10^{12} times stronger than that of the dimmest scenes that can still be accommodated by the eye. However, when we look at any one given scene, the ratio between the intensities of the brightest and the dimmest parts of that scene that we can perceive is only 1000:1. The eye adapts itself to the average illumination of the entire scene. If the range of illuminations of regions within the scene is greater than 1000:1, the eye won't be able to distinguish details in the brightest and darkest regions.

We feel the sensation of glare when a part of a scene is illuminated much stronger than the rest of it. Glare may be annoying, such as the light that comes from an exposed light bulb, or it may be incapacitating, such as the light from an incoming high beam of a car in the dark. The same high beam during the day may not cause any discomfort, even when aimed directly at us. What matters is not the absolute intensity of the light, but its intensity relative to the rest of the scene. Glare may be also caused by an abrupt change in illumination, such as turning on the light in a dark room or leaving a strongly lit tunnel into the dark night. Gradual change in illumination has to be provided to allow time for the accommodation of the eye.

9.2.4 The 'Temperature' of color.

Black objects absorb all the light that falls on them. An example of an object that meets this definition is a hole of an electric socket. No matter how strong is the light that reaches such a hole, once it gets inside, it won't come out, and the hole will look black. In general, a hole in the envelope of a cavity behaves like a black object. Consider now an oven, such as those used in pottery, that has a small hole in it. This hole acts like a black object. When the oven is heated up, its inside and the pottery in it glow. Some of the glowing light will come out through the small hole. This is an example of a black body radiation or cavity radiation. Measurements were taken of such spectra. It was found that the spectra did not depend on the material of the oven, only on its temperature. For any given temperature, the spectrum was continuous and had one peak. As the temperature of the oven increases, the maximum shifts to shorter wavelengths– from red to blue (figure 9.7). Max Planck (1858-1947) derived the exact mathematical expression of these spectra by assuming that light is made out of photons, whose energies are given by [9.2].

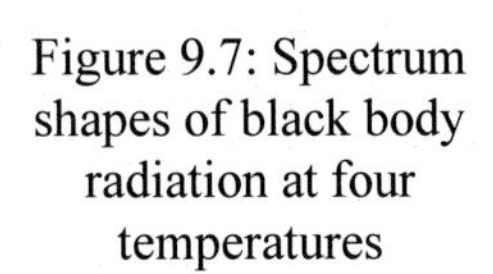
Figure 9.7: Spectrum shapes of black body radiation at four temperatures

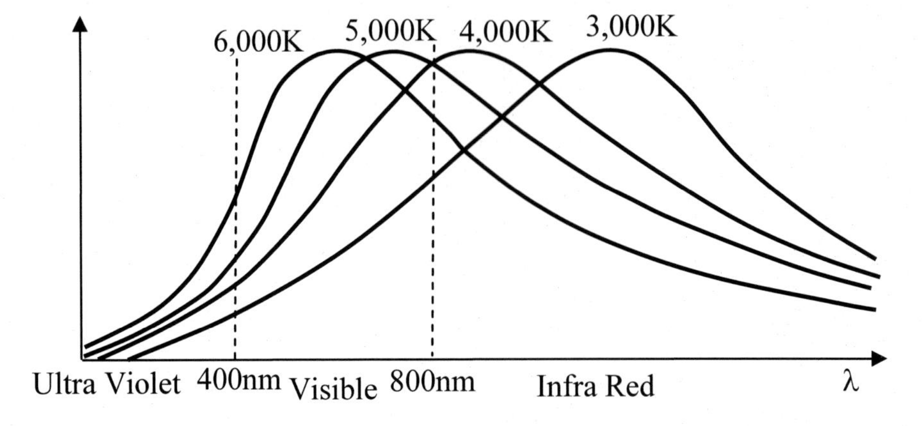

Some light sources have spectra that resemble those of black body radiation at a certain temperature, even though the light source itself may not be at that temperature. The temperature of the black body is then attributed to the light source. For example, the spectrum of blue sky is similar to that of black body at 12,000K to 25,000K. Therefore, it is said that the temperature of blue sky is from 12,000 K to 25,000 K. (The letter K indicates degrees Kelvin). This is not the real temperature of the sky. The color temperature of a candle is 1,900 K, and that of a gas filled tungsten light is below 3,000 K. The actual temperature of a tungsten filament is around 2400 K.

9.3 Photometry

9.3.1 Radiant power and luminous flux

A beam of light consists of many photons. A monochromatic beam consists of photons that have the same wavelength. A polychromatic beam consists of a mixture of photons of different wavelengths.

The radiant power of a light source is the energy that its photons carry per unit time (power = energy/time). If a monochromatic light source emits N_f photons during a time interval t, and if each photon has the frequency f, the power output P_f of this light source is given according to [9.2] by:

$$P_f = \frac{N_f \cdot h \cdot f}{t} \qquad [9.4]$$

Example

How many photons per second are emitted by a light source whose wavelength is 680 nm and whose radiant power is 50 W?

Solution

Substituting the data into [9.4] gives: P_f=50W, N_f=?, $h = 6.63 \cdot 10^{-34} J \cdot s$, f=$3x10^8$(m/s)/$680x10^{-9}$(m) (based on 9.1), and t=1s. The answer is N_f =$1.7x10^{20}$ photons/s.

The concept of the radiant power of a light source, as presented by [9.4], does not take into consideration the human factor. The sensitivity of the eye depends on the photon's wavelength. Two identical light sources, that have the same radiant power in watts, but one emits violet photons and the other emits yellow photons, will not be perceived by the eye as having the same brightness. The yellow source will look brighter. It was found that for an average eye, one watt of light at 555 nm appears as bright as two watts of light of 610 nm, and as four watts of 490 nm. In order to determine the luminous efficiency of a given light source, we compare it with a yellow light source of wavelength 555 nm that appears to have the same brightness. The ratio between the radiant power of that yellow source to the radiant power of the given source is called the **luminous**

efficiency of the given source. Figure 9.8 gives the luminous efficiency as function of wavelength.

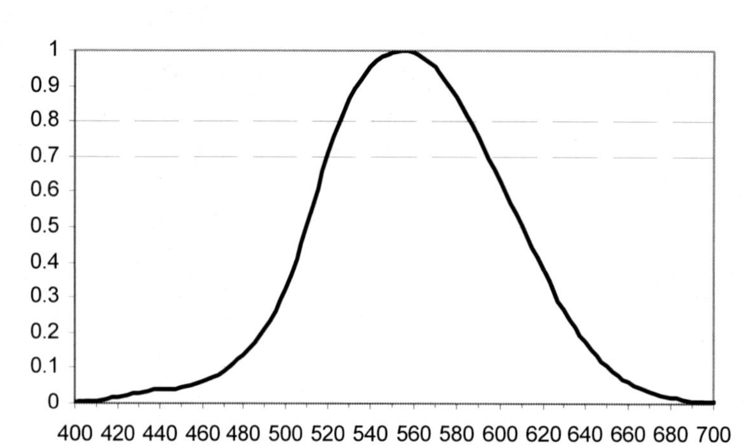

Figure 9.8 The luminous efficiency as a function of wavelength (nanometers)

The lumen (lm) is the SI unit that specifies the total luminous output of a source, or its **luminous flux.** The lumen incorporates the sensitivity of an average eye with the spectrum of the source. A lumen is defined as the luminous flux of a source of green-yellow light (555 nm) whose total power output is 0.00146 watt. The luminous flux of any other source, whose spectrum is known, can be calculated based on this definition and on the measured luminous efficiency of each participating wavelength. The power of each wavelength is first multiplied by its luminous efficiency. The sum of all these weighted powers is the luminous flux of the source, expressed in lumens. Similar sources of different colors that have the same luminous flux (lumens) appear to have the same brightness.

Example
How many photons per second are emitted by a light source of wavelength 555 nm and luminous flux of 1 lumen?

Solution
Based on the definition of the lumen, the output power of this source is 0.00146 watt. The frequency of a 555 nm photon is 5.4×10^{14} Hz. Let N be the number of such photons. Based on [9.4],
$0.00146 = N \times 6.63 \times 10^{-34} \times 5.4 \times 10^{14}/1$
from which we get $N = 4.08 \times 10^{15}$ photons.

Example
What is the luminous flux of a source whose spectrum consists of 0.0824 watt of light at 555 nm and 0.1237 watts of light at 610 nm.

Solution
The luminous efficiency of the yellow light is 1, and that of the 610nm light is 0.5. Therefore, 0.00146 watt of the yellow light provides 1 lumen, and

0.00146 watts of the 610 nm light provide 0.5 lumens. The total luminous flux of the source is
1.0·(0.0824/0.00146) + 0.5·(0.1237/0.00146) =98.8 lumen.

9.3.2 Illuminance and luminance intensity

The attribute '**luminous**' is assigned to light sources. The attribute '**illuminance**' is assigned to objects that receive light.

A typical light bulb that consumes 100 watts of electric power emits light whose luminous flux is 1200 lumen. These 1200 lumens are carried by photons that leave the light bulb in all directions. If we want to read a book in this light, we cannot be too far from the light bulb, because not enough photons will reach the book and illuminate it. What matters as far as being able to see clearly the book is how many lumens reach the book and the area of a page that these lumens are illuminating.

The **illuminance** measures the luminous flux that reaches a spot divided by the area of that spot:

$$E=F/A \qquad [9.5]$$

Where E is the illuminance, F is the luminous flux at the spot, and A is the area of the spot. In the SI system, the unit of illuminance is the lux (lx).
1 lux=1 lumen / m^2.

Other non-SI units of illuminance are the Phot (ph), which is 10^4lux, and the lumen per square foot (also called foot-candle (fc)), which is 10.76 lux.

A lighting designer has to choose light sources and to plan their placement, so that they provide the appropriate illuminance levels where needed. The required illuminance depends on the activity types in the illuminated area and on other factors such as duration of the activity, reflectance and contrast of the illuminated area, and age of the occupants. For example, values for an office may be: 750 lux on task areas, 250 lux in general areas, and 80 lux in the corridor. Detailed guidelines vary from country to country; commonly used guidelines in the US are those of IESNA. The illuminance (lux) at a given surface due to a given light source will depend on the nature of the source, its luminance (lumen), the distance between the source and the illuminated surface, and the angle of incidence between the beams of photons and the surface.

Let's consider now the relationship between some common light sources and their induced illuminance on illuminated surfaces.

Illuminating by a parallel beam.

In a parallel beam, all the photons move on parallel lines. A sunbeam is an example of a parallel beam. Another example is a very narrow beam, because the rays that make it are almost parallel to each other.

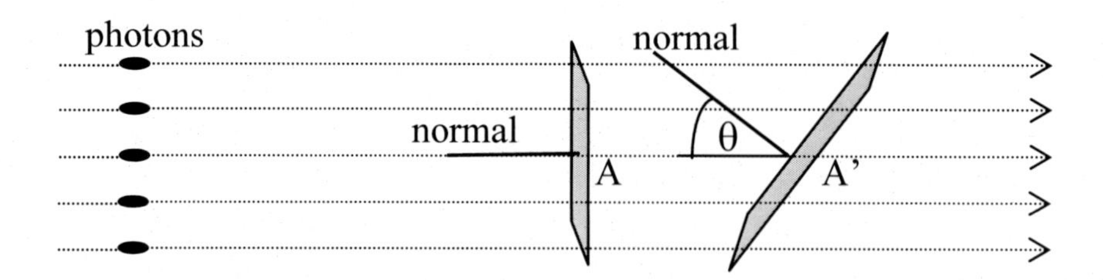

Figure 9.9: Effetcs of a parallel beam on two imaginary planes.

Figure 9.9 illustrates photons moving in a parallel beam. They first encounter an imaginary plane of area A. The rays of the beam are perpendicular to the plane, or in other words, they hit the plane in the direction of the normal. If the luminous flux of the photons is F lumens, the illuminance of area A is given by $E_A = \frac{F}{A}$ lux. Then, the same photons move on and hit another imaginary plane, A'. A' is tilted with respect to A by the angle θ, and the rays make an angle θ with the normal to A'. The exact same photons that hit A hit also A'. So, A and A' get the same luminous flux of F lumens. However, since the area of A' is greater than the area of A, the same F lumens are now distributed over a larger area, and the illuminance of A' ($E_{A'}$ lux) is less than the illuminance of A. It is given by: $E_{A'} = \frac{F}{A'}$.

Since A'=A/cosθ, we get that

$$E_{A'} = \frac{F}{A}\cos\theta = E_A \cos\theta \qquad [9.6]$$

That means that the largest illumination of a given surface at a given point is obtained when the incident beam is in the direction of the normal to that surface ($\cos(0^0)=1$). The illuminance of the surface under these conditions (F/A) is equal to the flux density of the beam, which is also called the illuminance of the beam.

Example

The illuminance of sunbeams that reach the ground depends on the location, the date and time, and the weather. At a certain point and time, when the sun was 60^0 above the horizon, the illuminance of its sunbeams was 80,000 lux. Calculate the illuminance induced by these beams (a) on the ground. (b) on a vertical wall.

Solution

The angle that the beam makes with the normal to the ground is 30^0. Therefore, based on [9.6], the illuminance on the ground would be 80,000·cos30=69,282 lux.

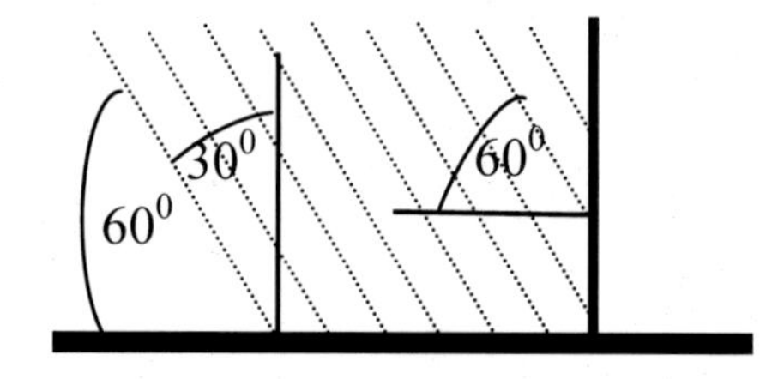

The angle that the beam makes with the normal to the wall is 60^0. Therefore, the illuminance on the wall would be 80,000·cos60=40,000 lux.

Illumination by a point source

A point light source is a source whose dimensions are small compared with its distance to the area that it illuminates. The photons that leave a point source propagate in all directions. Figure 9.10 illustrates a uniform point source. The photons (black dots) propagate uniformly in all directions. As those photons propagate outwards, they first pass through an imaginary sphere of radius r. All the photons that left the point source reach that sphere and pass it. They all continue and reach a second imaginary sphere of radius R and pass through it.

The luminous flux of F lumens, which was emitted by the point source, is carried by the photons and reaches the first sphere, and then the second sphere. Both imaginary spheres receive the same amount of lumens. However, the illuminations of the two spheres are not equal to each other, because of their different areas. Since the area of the outer sphere is greater than the area of the inner sphere, those F lumens are spread over a larger area. The illumination E, which is the luminous flux F divided by the illuminated area is $E=F/(4\pi r^2)$ lux for the inner sphere, and $E=F/(4\pi R^2)$ lux for the outer sphere. Since we are dealing with a uniform source, and since the photons hit the spheres in the normal direction, the illumination of any small spot of the inner sphere would be $F/(4\pi r^2)$ lux, and the illumination of any small spot of the outer sphere would be $F/(4\pi R^2)$ lux.

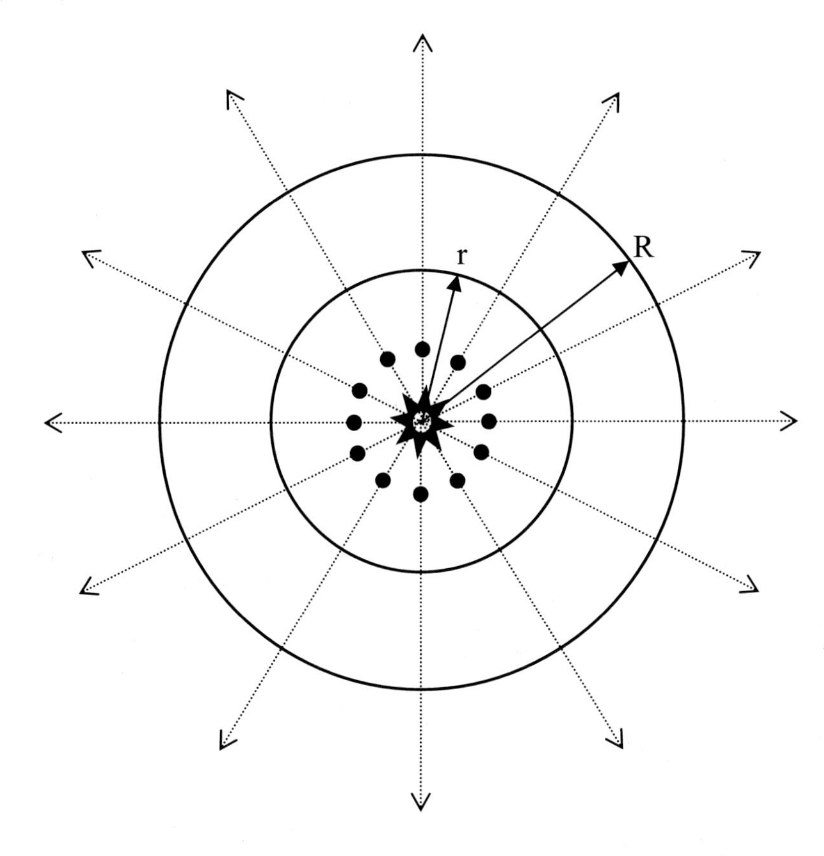

Figure 9.10: Photons leaving a point source and passing through two imaginary spheres.

In general, a point source whose luminous flux is F lumens, which distributes light equally in all directions, creates an illumination of E lux on a surface that is perpendicular to the incident beam and at a distance r meters from the source, according to:

$$E = \frac{F}{4\pi r^2} \qquad [9.7a]$$

If photons from a point source hit as small surface at a distance r away, and if their angle of incidence with the normal to the surface is θ (figure 9.10.a), the illumination is affected not only by the distance r, but also by the angle θ, similarly to equation [9.6]. The illumination E would be:

$$E = \frac{F}{4\pi r^2}\cos\theta \qquad [9.7b]$$

Equation [9.7b] becomes [9.7a] when the incident light is perpendicular to the illuminated surface ($\theta=0^0$, $\cos(0^0)=1$).

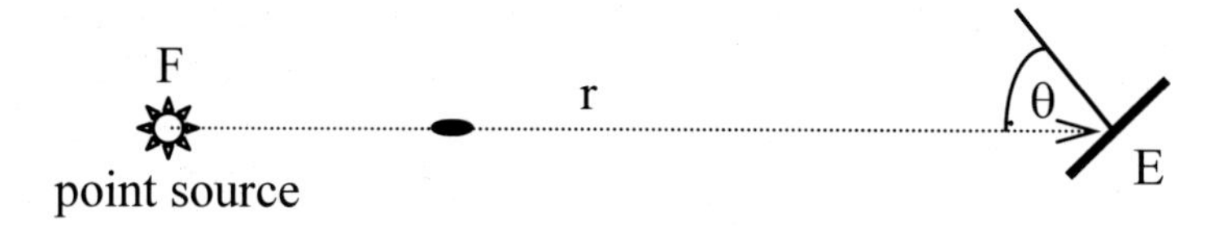

Figure 9.10a: The illumination due to a point source

Figure 9.10 described a point source that emits photons uniformly in all directions. There are point sources that emit their photons in a non-uniform way. For example, the reflector behind the light bulb in a car's headlight redirects photons that are initially emitted backwards, and spreads them in front of the car. In such cases, different directions receive different amounts of light. We need an entity to describe how the light emitted by a point source is distributed in different directions.

Luminance intensity, I, is defined as the luminous flux (lumen) which is directed by a point source into a solid angle, divided by the size of that solid angle (measured in steradian). All the points of a straight line that originates at a point source receive the same luminance intensity, no matter how far they are from that source. The SI unit of the luminance intensity is the **candela** (cd), also called **candlepower**.

If a source of F lumens spreads its light uniformly in all directions, these F lumens are directed into a solid angle of 4π, which is the solid angle in steradian subtended by the entire sphere. For this source, the luminance intensity in candela, I, in any direction would be

$$I = F / 4\pi \qquad [9.8]$$

A point source of 1 lumen that spreads light uniformly in all directions has luminance intensity of $1/4\pi$ candela in any direction.

Equation [9.8] can be interpreted also as follows: Assume that a non-uniform point source generates luminance intensity I at a certain point. The same luminance intensity would have been generated at that point by a uniform source of F lumens, where F and I are related by [9.8]. Because of that, we can replace the $F/4\pi$ in equations [9.7] by the luminance intensity I at the illuminated point, and get:

$$E = \frac{I}{r^2}\cos\theta \qquad [9.9]$$

where E is the illuminance (lux) of the illuminated small surface element, and I is the luminance intensity (candela or candlepower) directed by the source to that illuminated surface element.

Many point-light-sources emit light in a non-uniform manner, e.g. a luminaire consisting of a light bulb and a reflector. Although it has final dimensions, a luminaire may be treated as a point source if the illuminated object is not too close to the source. A rule of thumb states that if the distance between the luminaire and the illuminated point is at least five times greater than the dimension of the luminaire, the luminaire may be considered a point source. If we know the luminance intensity I in a given direction, we can find E, the illuminance at a distance r from the source in this direction, by using equations [9.9]. Photometric data of luminaires are often provided in polar candela-distribution diagrams or tables. These diagrams and tables provide the I in any direction from the source. This I is then substituted in equation [9.9], according the situation at hand.

θ (degree)	0	10	20	30	40	50	60	70	80	90
I (candela)	980	910	850	820	800	720	510	300	80	0

Table 9.3: Distribution of illuminance intensity of a certain lamp

Example

A desk in a library is illuminated by two identical lamps, whose polar luminance distribution is given by table 9.3. The angles in the table are with respect to the main axis of the lamp, which points down. The lamps hang 2.5 m above the floor–one right over the desk, and the other to the side, as shown

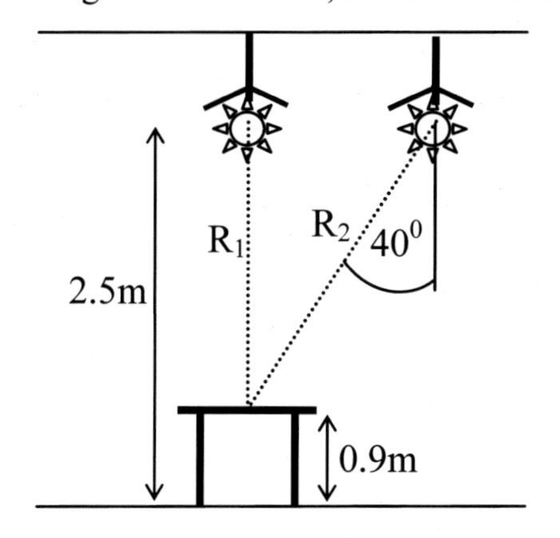

Figure 9.11

in figure 9.11. The desk's surface is 0.9 m above the floor. Find the illuminance at the center of desk's surface due to the two lamps.

Solution

The illuminance at the center of the desk is the sum of the illuminances due to the lamps, as given by [9.9]. We have to find r and θ for each lamp. For the left lamp, $r=R_1=2.5-0.9=1.6m$, and $\theta=0$. The luminance intensity in this direction is given in the table, and it is 980 cd. By substituting in [9.9] we get $E_1=\frac{980}{1.6^2}\cos 0^0=382.8lux$. For the right lamp, $r=R_2=R_1/\cos 40^0=2.1m$. The angle with the normal $\theta=40^0$, and the luminance intensity is 800 cd (table 9.3). From [9.9] we get: $E_2=\frac{800}{2.1^2}\cos 40^0=236.8lux$. The total illuminance at the center of the table is $E=E_1+E_2=382.8+236.8=619.6$ lux.

Luminance

You may have noticed that some small light sources, such as Christmas lights, may look brighter than similar larger sources, such as regular light bulbs, when they are at the same distance from the observer. Both lights can be considered point sources. The difference is that the size of the Christmas bulb is much smaller than that of the regular bulb. What determines the perception of brightness is the 'density of the light source' or the amount of light that it emits in a certain direction divided by its surface area, as seen from that direction. The concept of luminance is used to express this feature of the point source. It is defined as the luminance intensity of a point (very small) source divided by its area that is facing the observer. The unit of luminance is candela per square meter (cd/m^2), sometimes called nit (nt).

Non-SI units of luminance: The stilb is 10^4 cd/m^2. The candela per square foot is 10.76 cd/m^2. The lambert (L) is $10^4/\pi$ cd/m^2, the foot Lambert (fL) is $1/\pi$ candela per square foot, the apostilb (asb) is $1/\pi$ cd/m^2. And, yes, the candela per square inch is 452 foot Lambert.

Long line and other sources

So far we have dealt with point sources. If we need to figure out the illumination by a large source, such as large square or rectangular sources, it is possible to first divide the source into elements, each small enough to be considered a point source. After calculating the illumination by each of these elements, their contributions are added together. This is the basis for computer programs that calculate and display lighted areas in various designs.

There are few cases of large sources whose illumination can be described by simple formulas. One of these cases is the **infinite straight-line source**. To a worker working under a long line of fluorescent lamps in a large factory hall, the light source looks like an 'infinite straight-line'. If the worker is not too close to an end of the hall, the illumination E is given by (figure 9.12):

$$E = \frac{L}{2\pi r}\cos\theta \qquad [9.10]$$

Where L is the luminance of the light source, r is the distance from the illuminated point to the line source, and θ is the angle that r makes with the normal to the illuminated surface. The luminance L is the luminous flux of the light source (lumen) divided by its length (m). If there are no gaps between the lamps, L is the luminous flux of a single lamp divided by its length. If there are gaps between the lamps, L can be approximated by the total luminous flux of the line (number of lamps times the lumens of each) divided by the length of the line.

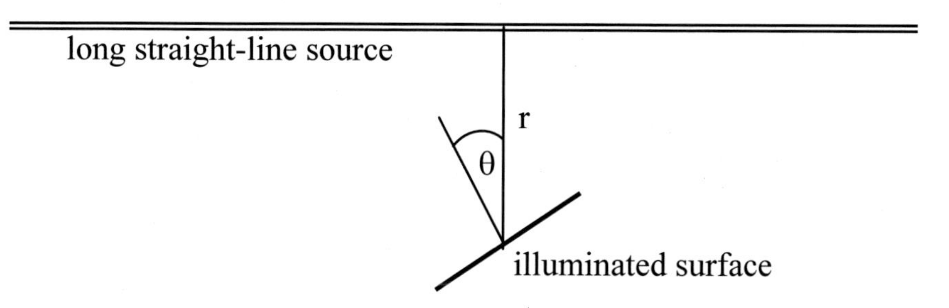

Figure 9.12: Illumination by an infinite straight-line source (Formula [9.10])

When the illumination of indoor spaces has to be calculated, reflected light from the walls, the ceiling and the floor has to be considered. There are practical procedures, semi-empirical formulas, and computer programs that are used in such cases, but they are beyond the scope of this book.

Luminous efficacy and quantity of light

An artificial light source needs energy in order to generate light. The ratio between the lumens obtained and the watts invested is called (overall) Luminous efficacy. The overall invested watts should include the power consumed by the light source itself, and the power consumed by supporting components, such as ballasts. Light sources with wavelength 555 nm have high luminous efficacy, because they emit light to which the eye is most sensitive. The highest luminous efficacy possible is 683 lumen/watt, which is obtained when all the invested energy is transformed into light energy at 555 nm.

The **Quantity of light** emitted by a source over a period of time is measured in lumen-second.

Example
What is the overall luminous efficacy of a fluorescent lamp of 25 W, 1400 lumen, that uses a ballast of 10 W?

Solution
The overall luminous efficacy is 1400/(25+10) = 40 lm/W.

Concepts Summary:
Table 9.4 summarizes the various concepts discussed in this section.

Concept	Symbol	Meaning	common unit (and abbreviation)
Radiant power	P	Total power of electromagnetic energy. Eye sensitivity to different wavelengths **is not** considered.	watt (W)=joule/sec
Luminous flux	F	Total quantity of visible light. Eye sensitivity to different wavelengths **is** considered.	lumen (lm).
Illuminance (illumination)	E	Luminous flux per unit area	lux=lm/m^2 footcandle=lm/ft^2
Luminous intensity	I	Brightness of a small source. Total quantity of visible light emitted in a given direction per unit of solid angle.	candela (cd) lumen/steradian (lm/sr)
Luminance		Luminous intensity per unit area of the source.	cd/m^2
Luminous efficacy		Emitted luminous flux per input power.	lm/W

Table 9.4: Summary of concepts

9.4 Daylight

9.4.1 Sunlight

The energy of the photons that the sun emits is generated in nuclear reactions that take place inside the sun. Hydrogen nuclei fuse together and form helium nuclei, releasing energy in the process. This energy undergoes several transformations, as it spreads inside the sun to its surface. This energy flow maintains the surface of the sun at an average temperature of 6000 degrees K. Like any hot object, the sun glows. The observed spectrum of the sunlight is consistent with theoretical calculations of the spectrum emitted by a black body at 6000 K.

The light from the sun propagates through space in all directions. Those beams that reach the Earth consist of parallel rays. They lose some of their photons by diffusion to the atmosphere. These diffused photons scatter in the atmosphere in all directions, and eventually reach the ground from all directions. The rays of the sunbeam itself reach the ground in a well-defined orientation, which changes during the day. The illumination created by the sunlight depends also on the prevailing atmospheric conditions.

The typical value of the total sun energy that reaches the atmosphere is about 1350 W/m^2. The illumination intensity of direct sunlight on horizontal surfaces in tropical regions amounts to approximately 10^5 lux, while the illumination intensity of the diffused light, which comes from all directions, is approximately 10^4lux. The diffused illumination on a vertical surface is only half of that on a horizontal surface, because a vertical surface ‘sees’ only half of the sky. The luminance of the sun is 1.5x10^9 cd/m^2, and that of clear blue sky is about 5000 cd/m^2.

Direct light (like sunbeams) creates sharp shadows. Diffused light, which comes from all directions, creates soft shadows, which may not be noticeable in many cases. When an object is illuminated by a direct parallel beam, its shadowed area does not get any light, while the illuminated areas get uninterrupted light. On the other hand, an object that is illuminated from all direction cannot block all the light from reaching any area next to it. This is the condition of soft shadows.

9.4.2 Light and the atmosphere

The blue color of the sky is the result of resonant scattering. Air molecules in the atmosphere scatter different wavelengths with different effectiveness. Violet and blue photons are scattered most effectively, and the effectiveness decreases gradually for green, yellow, orange, and finally red photons. Blue photons that originally were moving with the other photons of the sunbeams are scattered many times in the atmosphere and reach the ground from all directions. If there were no atmosphere around the Earth, the sun would look like a big bright star on a background of black sky.

Scattering is a probabilistic process. The longer a photon travels, the greater are its chances to hit an air molecule and to be scattered by it. The shortest path of a sunbeam through the atmosphere is when the sun is high in the sky. Along this path, large portions of the initial blue photons are being scattered from the beam into the surroundings. As result, the sun itself appears yellow, which is white light less some of its blue photons. The scattered photons undergo many more scattering in the air, till the entire hemisphere looks blue.

In sunset and sunrise, sunbeams traverse longer paths through the atmosphere than they do in midday. As a result, not only violet and blue photons are absorbed in noticeable amounts, but also green and yellow. The sun itself looks red, which is white light less large portions of violet, blue, green and yellow photons. The remaining red sunbeams reach observers on the ground directly, hence the sun's red color. These beams reach also clouds and mountaintops, which reflect them to observers on the ground. That is why these objects look red in sunset and sunrise. Clouds and mountaintops may also be illuminated by beams that have more orange and yellow photons in them than the beams that directly reach ground observers. By reflecting those beams, the clouds will look orange and yellow, while the sun will look red.

The spectrum of the sunlight contains, in addition to visible light, light in the **infrared** region (IR), whose wavelengths are longer than that of red light, and light in the **ultraviolet** region (UV), whose wavelengths are shorter than that of the violet light. When absorbed, infrared radiation warms up the skin. When absorbed, ultraviolet radiation causes changes in various molecules. One result of absorption of UV by the skin is tanning. However, excessive exposure to UV can modify normal skin cells into cancerous ones. Ultraviolet radiation is absorbed in the upper atmosphere by ozone molecules. Depletion of ozone will allow more ultraviolet radiation to reach the surface of the earth, raising the number of skin cancer occurrences. Ultraviolet light is important in the photosynthesis process in plants, which is essential for plants' growth.

9.5 Light in various mediums.

9.5.1 Types of mediums

Light can propagate in various mediums. A ray of light retains it direction in transparent mediums. When light propagates through translucent mediums it diffuses or scatters. Opaque mediums completely absorb light. Transparent and translucent mediums absorb some of the light that propagates through them. The amount of light being affected by a medium through absorption and scattering will depend on the length of the ray's path through the medium, and on the wavelength of the involved photons.

There are processes in which light photons are absorbed by a substance, and that causes the substance to emit other photons. **Fluorescent** materials absorb photons of high frequency, such as ultraviolet photons, and emit photons of lower frequency, such as visible light. The high frequency photon excites the material. The material then falls back to a lower excited state, emitting a photon of lesser energy than the original one. The excited material may cascade down all the way to the ground state, emitting photons of less energy than the original one.

In the process of **resonant scattering**, an atom (or a molecule) absorbs a photon of a certain wavelength from a beam of light. The atom becomes excited, and immediately falls back to its ground state, emitting a photon. The emitted photon has the same wavelength as the absorbed one, but it is emitted in a random direction. As a result, the beam will have less photons of this wavelength, while the surroundings of the beam will get all these scattered photons.

9.5.2 Attenuation

A ray of light that propagates in a transparent uniform medium, such as glass or water, moves on a straight line. As the photons of the ray hit molecules of the medium, some of them may be deflected, some may be absorbed, and the rest will continue to move unaffected. The end result is that the ray will lose some of its original intensity. These processes are probabilistic in nature, and for any given medium they depend on the wavelength of the involved photons. If the ray is polychromatic, its various wavelengths may be absorbed in different amounts. Thus, the perceived color of a polychromatic beam may change as it passes through a transparent medium.

Attenuation is defined as the ratio of the intensity of the light after it has passed through an absorbing medium, divided by the intensity of the light that entered this medium. For example, attenuation of 0.8 means that 80 percent of the incident light passed the medium. Attenuation depends on the wavelength of the light, the material of the medium, and the thickness of the medium.

If the attenuation of a certain pane of glass is 0.8, the attenuation of a pane twice as thick would be $0.8^2=0.64$: Eighty percent of the light passed the first half of the thick pane, and entered the second half. Eighty percent out of that light passed the second half, and emerged out. Overall, eighty percent of the eighty percent of the original light, which is sixty four percent, passed the double pane.

The phenomenon of attenuation is described by the formula:

$$I = I_0 \exp(-\mu L) \qquad [9.11]$$

Where I_0 is the intensity of the incident beam, L is the length of the path that the beam passed in the medium, I is the intensity of the emerging beam, and μ is the linear absorption coefficient. The linear absorption coefficient depends on the material of the medium and on the wavelength of the beam.

Not only visible light is attenuated according to [9.11]. This formula is valid also for photons of other regions of the electromagnetic spectrum, such as x-rays and gamma rays, and for radioactive beams that consist of alpha and beta particles. Those radiations can penetrate and pass through many materials that are opaque to visible light.

Example

The linear absorption coefficient of lead to γ rays emitted by a radiation therapy machine is 0.77 cm^{-1}. What percentage of those photons that reach a 10 cm lead wall will pass through it?

Solution

According to formula [9.11]: $\mu=0.77\ cm^{-1}$, L=10 cm, $I/I_0=?$ Substitution yields $I/I_0=4.5x10^{-4}=0.045\%$.

9.5.3 Reflection

When a ray of light reaches the boundary between two mediums, such as air and glass, part of the light is reflected and another part, the refracted, moves into the second medium. The reflected and refracted rays continue to propagate in straight lines. The energy carried by the incident ray is split between the reflected and the refracted rays. If the reflecting medium is opaque, the refracted ray will be absorbed practically at the boundary. When conditions of total reflection (to be discussed shortly) are met, there is no refracted ray. In such cases, the reflected ray carries all the energy of the incident ray. The law of reflection deals with the relationship between the incident ray and the reflected ray. It holds no matter what happens to the refracted ray.

The Law of Reflection.

The Law of Reflection states: "The angle of reflection is equal to the angle of incidence. The incident ray, the reflected ray, and the normal to the reflecting surface are in one plane". Figure 9.13 illustrates the law of reflection. It is customary to measure the incidence angle (θ) and the reflected angle (θ') with respect to the normal at the reflecting point. In these notations, the law of reflection is formulated as: θ=θ'.

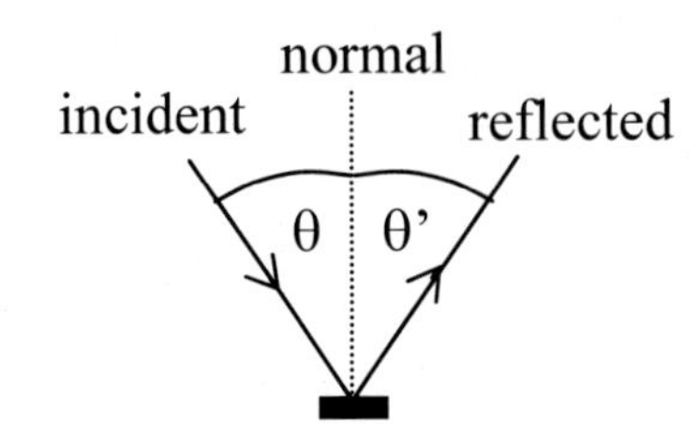

Figure 9.13: reflection of light

When a parallel beam falls on a smooth reflecting boundary, all the incident photons hit the surface with the same incidence angle. They will all be reflected at the same angle too, thus the reflected beam will also be parallel. This is called **specular reflection**. If the reflecting boundary is not smooth, the incident rays will be reflected in different directions, according to the incident angle of each individual ray. The reflected beam will diffuse. Although the beam as a whole diffused into many directions, each of its individual rays is reflected by its own reflecting surface-element according to the law of reflection.

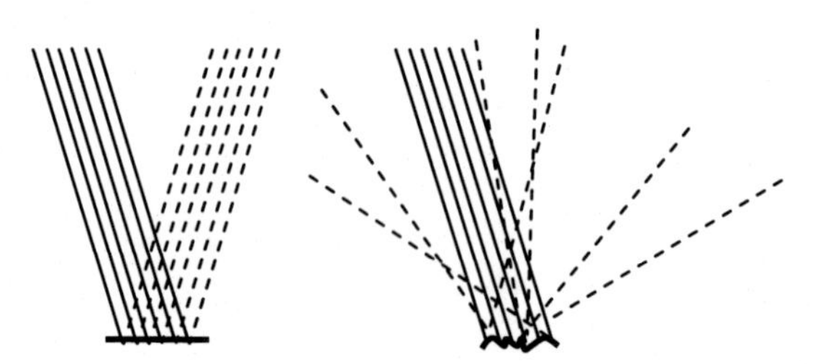

Figure 9.14: A parallel beam (full lines) is reflected into a parallel beam by a smooth polished surface (left) and into a diffused beam by a rough surface (right)

The **reflectance** of a surface is the ratio between the energy of the reflected beam and the energy of the incident beam. Generally, matte surfaces (such as brick) have lower reflectance than glossy surfaces (such as polished silver). But this is not always the case. For example, a flat white wall may have higher reflectance than black tinted glass. The reflectance of a surface may depend also on the angle of incidence and on the wavelength of the light.

In some design problems it is necessary to trace a beam and to determine how it spreads after being reflected by a flat surface, like a wall or a mirror. A beam consists of many rays, and by tracing a sample of these rays, including rays at the boundary of the beam, it is possible to determine the shape of the reflected beam. Each individual ray is reflected according to the law of reflection (reflected angle = incident angle). Based on this law, a beam of parallel rays will remain parallel after being reflection by a smooth flat surface, such as a mirror, and its cross section will not change (figure 9.15, left).

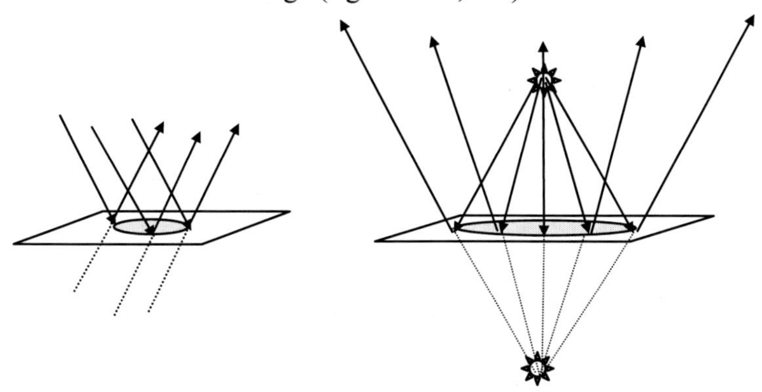

Figure 9.15: Tracing the reflection of a parallel beam (left) and of a beam that originates at a point (right). The reflections of the beam on the right illustrate why the image of the source appears to be in the dark side of the mirror, symmetrically to the source. After reflection, the cone of rays that has been emitted by the point source (upper star) appears as if it is emitted by the virtual image (bottom star).

The rays of a beam that start at a point source will hit a mirror at different angles, and will be reflected at different angles. The beam will continue to spread after being reflected (figure 9.15, right). Let's trace now a cone of rays that start at the point-source and hit the mirror. The reflected rays will spread as if they are a cone, originating at a point on the other side of the mirror (dotted lines). When these rays reach our eye, the eye cannot tell if the rays started at a real object, or whether their imaginary continuations (the dotted lines in figure 9.15) intersect at a point. This is why the image of the point source appears as if it is located in the dark side of the mirror, symmetrically to the real point source.

In order to trace a region that is illuminated by a beam, it is enough to trace the outermost rays of the beam, as illustrated in figure 9.15. All the other rays will lie within this boundary.

Let's find out now what is the illumination intensity at a surface that is illuminated by a point source, such as a small lamp, placed in front of a mirror. This illuminated surface receives direct light from the source and reflected light

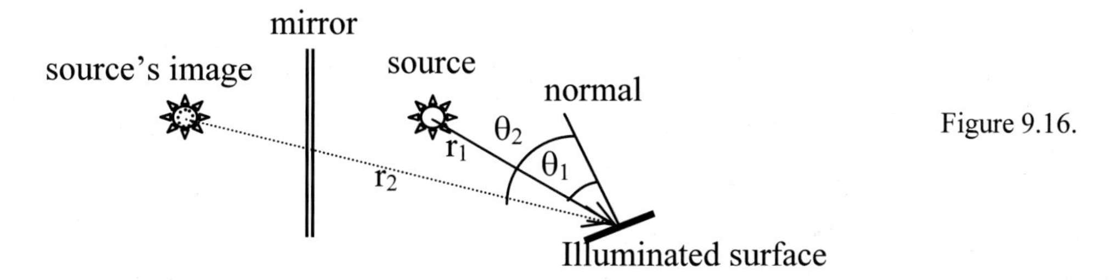

Figure 9.16.

from the mirror. The contribution from the mirror is equivalent to the contribution that would have come from an imaginary point source at the location of the image (figure 9.16). Replacing the contribution of the mirror by the contribution of an equivalent point source, which is an image of the original source, is called **the method of images**. The illumination intensity at a surface element due to a source of luminous flux F (lumen) and the light reflected by the mirror is given, based on [9.7], by:

$$E = \frac{F}{4\pi r_1^2}\cos\theta_1 + \frac{\rho F}{4\pi r_2^2}\cos\theta_2 \qquad [9.12]$$

The first term on the right hand side is the contribution of the source, and the second term is the contribution of the mirror. The ρ is the reflectance of the mirror. If the mirror is a perfect reflector, $\rho=1$. This formula applies only to polished surfaces, which cause specular reflection. It does not apply to diffused reflection.

9.5.4 Refraction of light

When a block of glass partially covers a straight line drawn on a piece of paper, that line appears "broken" (figure 9.17). The reason is that the rays of light that started at the line and passed through the glass changed their direction as they traversed into air. In general, when a ray of light passes from one transparent medium to another it changes its direction, or refracts.

Figure 9.17: Refraction of light that passes through a glass pane.

The amount of refraction depends on the angle of incidence at the boundary between the two mediums, on the wavelength, and on materials of the mediums.

The index of refraction

The speed of light depends on the medium in which it propagates. In vacuum, the speed of light is approximately $c=3\text{x}10^8$ m/s. In mediums other than vacuum, the speed of light is less than that, and it depends on the medium and on the frequency of the light.

The index of refraction n of a certain medium is defined as the ratio between the speed of light in air and the speed of that light in that medium:

$$n_{\text{of a medium}} = \frac{c}{\text{speed of light in the medium}} \qquad [9.13]$$

Since the speed of light in any medium is less that the speed of light in vacuum, n of any medium (other than vacuum) is greater than one. The index of refraction depends on the wavelength of the incident light.

Substance	Index of refraction, n
Air	1.000293
Carbon dioxide	1.00045
Water	1.333
Glycerin	1.473
Fused quartz	1.458
Glass (crown)	1.52
Glass (flint)	1.6-1.9
Diamond	2.419

Table 9.5: Index of refraction (n) of different substances for yellow light (λ=589 nm)

Wavelength in vacuum (nm)	400	500	600	700
Index of refraction (n) (quartz)	1.470	1.462	1.458	1.456

Table 9.6: Index of refraction of fused quartz as a function of wavelength

The frequency of light does not change as it passes from one medium to another, but the wavelength λ changes, in accordance with the change in the speed:

$$\lambda_{\text{in a medium}} = \frac{\lambda_{\text{in vacuum}}}{n_{\text{of the medium}}} \qquad [9.13a]$$

The Law of Refraction

Consider a ray of light that propagates in a transparent medium whose index of refraction is n_1. It hits the boundary of the medium at an angle θ_1 with respect to the normal. As the ray passes the boundary into the adjacent medium it changes its direction (refracts). Its new direction in the second medium, whose index of refraction is n_2, is at an angle θ_2 with respect to the normal (figure 9.18). The relationship between the incident and the refracted rays are given by the Law of

Figurer 9.18: Refraction of light

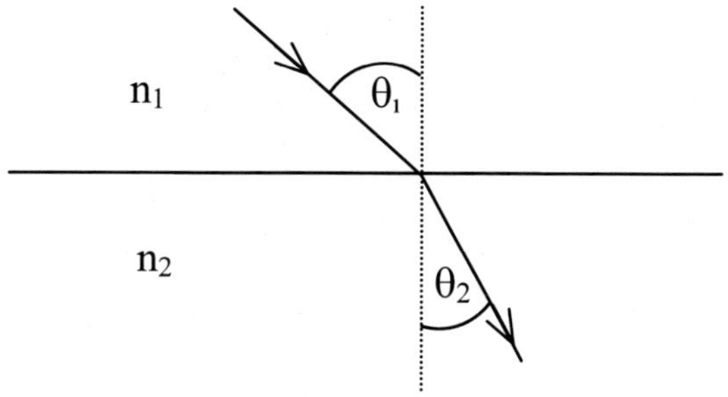

n_1 θ_1 n_2 θ_2

Refraction (Snell's Law):

$$n_1 \sin\theta_1 = n_2 \sin\theta_2 \quad [9.14]$$

Both rays and the normal are found on the same plane.

According to Snell's Law, if the incident ray hits the boundary in the normal direction ($\theta_1=0$), the refracted ray will continue in the same direction ($\theta_2=0$).

If we reverse the direction of the refracted ray, it becomes an incident ray (with the angle θ_2). According to Snell's Law, as this ray refracts into medium 1, it will trace exactly the path of the original incident ray.

When a ray passes into a denser medium (larger n), it moves closer to the normal, and when it passes into a less dense medium (smaller n), it moves farther away from the normal.

When a white beam hits the boundary between two transparent mediums at an angle other than normal, its various wavelengths refract in different angles. The eye is now able to notice the different colors of those separated beams. This is the basic process in the formation of rainbows, in which white light from the sun is separated into beams of different colors by droplets of water. This is also the process that makes diamonds and chandeliers glitter. Rays of different colors that make up the incident white beam are refracted in different directions. Some of those rays reach our eyes, depending on the relative orientation of the refracting surfaces with respect to our eyes, and create the perception of their colors.

Example

Two rays, one of $\lambda=400$ nm and the second of $\lambda=700$ nm, move in air and are refracted into quartz. The incident angle of the two rays is 60^0. What would be the refraction angle of each beam?

Solution

The refraction process follows Snell's Law [9.14] $n_1 \sin\theta_1=n_2 \sin\theta_2$. Using these notations, we have for the first ray ($\lambda=400$nm): $n_1=1$, $\theta_1=60^0$, $n_2=1.47$ (table 9.6), and $\theta_2=?$ Substituting yields

$$\theta_2 = \sin^{-1}\left(\frac{1\cdot \sin 60^0}{1.47}\right) = 36.1^0 .$$

For the second ray ($\lambda=700$nm) we have: $n_1=1$, $\theta_1=60^0$, $n_2=1.456$ (table 9.6), and $\theta_2=?$ Substituting yields: $\theta_2 = \sin^{-1}\left(\frac{1\cdot \sin 60^0}{1.456}\right) = 36.5^0$.

Total reflection

Consider now a beam that moves in a dense medium (e.g. water), and reaches a boundary with a less dense medium (e.g. air). In such cases $n_1>n_2$. By rearranging [9.14] we get:

$$\sin\theta_2 = \frac{n_1}{n_2}\sin\theta_1 \quad [9.14a]$$

Since $n_1/n_2>1$, the refraction angle θ_2 will be greater than the incident angle θ_1. When the angle of incidence θ_1 satisfies the condition $\sin\theta_1 = \frac{n_2}{n_1}$, the right hand side of [9.14a] becomes 1, and consequently $\sin\theta_2=1$, or $\theta_2=90^0$. That means that the refracted beam will not penetrate the second medium. It will move along the boundary between the two mediums (figure 9.19). If the incidence angle θ_1 is further increased, the sine of the refracted angle θ_2 should become, according to [9.14a], greater than 1. However, the sine function can vary only between 1 and -1. This is how equation [9.14a] tells us that there will not be any refracted beam (for those θ_1's). When

$$n_1>n_2 \quad \text{and} \quad \theta_1 \geq \sin^{-1}\left(\frac{n_2}{n_1}\right) \qquad [9.14b]$$

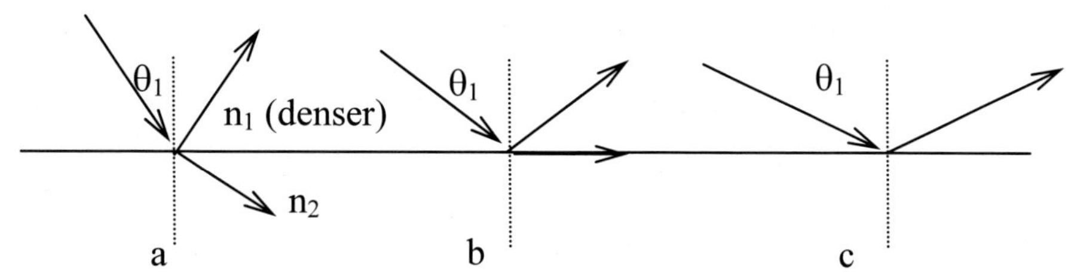

Figure 9.19: Beams in a medium denser than its adjacent ($n_1>n_2$). (a) When $\sin\theta_1 < \frac{n_2}{n_1}$, the refracted beam is further away from the normal than the incident beam. (b) At the critical angle ($\sin\theta_1 = \frac{n_2}{n_1}$) the refracted beam moves on the boundary between the two mediums. (c) In total reflection, ($\sin\theta_1 > \frac{n_2}{n_1}$), there is no refracted beam. In all cases there is a reflected beam, satisfying the Law of Reflection.

there will not be a refracted beam into the less dense medium (medium 2). This situation is called **total reflection**, because the incident beam generates only a reflected beam. In all other cases, the incident beam will generate both reflected and refracted beams. The critical angle of total reflection satisfies the equality part of the greater or equal sign in [9.14b]. Total reflection can only happen when the incident beam propagates in a medium that is denser than its adjacent medium, and only if the angle of incidence is greater than the critical angle.

The energy of the incident beam is divided between the refracted and reflected beams. In total reflection, all the energy of the incident beam is transferred to the reflected beam.

Total reflection enables the transmission of light signals in glass cables (fiber optics). Pulses of a narrow light beam carry coded messages in a glass fiber. When a beam hits the surface of the fiber, it does so at an angle greater than the critical angle. Therefore, it does not escape from the fiber through refraction; it is completely reflected back into the fiber.

Example
What is the critical angle of total reflection for light of wavelength λ=589 nm that moves from water to air? What does it mean?

Solution
Using the notations of [9.14b] we get: n_1=1.333 (table 9.5). n_2=1, which yields: $\theta_1 = \sin^{-1}\left(\frac{1}{1.333}\right) = 48.6^0$. That means that at incidence angles that are greater than 48.6^0 there will not be refraction from the water into the adjacent air, and for incidence angles smaller than 48.6^0 there will be refraction into the air.

9.5.5 Lighting control

In lighting design, light from selected light sources has to be distributed in a given space, according to practical and esthetical requirements. Reflection and refraction are two main tools that are used to accomplish those requirements. In simple cases, the laws of reflection and refraction could be implemented manually and provide insights on how reflection from surfaces and refraction by transparent elements would re-distribute the illumination within the given space. In more complex situations, mathematical formulas and computer programs, which are based on these fundamental laws, provide the insights.

We feel different sensations of light when the space that we are in is illuminated diffusely or sharply. In many situations, sharp light that is provided by the sources has to be diffused. Reflection and refraction by polished flat surfaces maintain the sharpness of the original beam. A smooth polished reflector reflects a sharp beam in a way that preserves most of the spatial relationships between its rays. Rays that move next to each other in the incident beam will move in similar directions also in the reflected beam. The same is true for smooth polished refracting surfaces. On the other hand, rough, etched, wavy, and similar textured surfaces diffuse the sharpness of a beam. Rays that move in similar directions in the incident beam move in completely different directions after interacting with those surfaces. Matte walls diffuse light that falls on them. Transparent materials with textured surfaces diffuse the light that passes through them. Precise tracing of diffused beams can be done by using special formulas or computer programs.

Quite often, it is required to direct light from a source to a specific region of the illuminated space. Polished reflectors are commonly used for redirecting light from light bulbs. Reflectors have a variety of shapes, depending on what is required of the reflected light. By tracing a large number of rays emanating from the source, and employing the Law of Reflection for each of them as it hits the surface of the reflector, it is possible to find out how dispersed or focused will the reflected beam be.

There are mathematical formulas for calculating the properties of the light reflected from simple surfaces such as spherical, parabolic, and elliptical mirrors. These reflections can also be described in general terms. An imaginary point called 'focus' characterizes each such mirror. The focal length of a mirror is the

distance between the center of the mirror and the focus. The position of the light source with respect to the focus determines the properties of the reflected beam. Figure 9.20 shows reflected beams from spherical mirrors. Parts a, b, and c trace rays reflected by a concave spherical mirror, and part d traces the rays from a convex spherical mirror. If the light source is placed at the focus of a concave mirror (a), the reflected rays form a parallel beam. If the light source is placed between that mirror and the focus (b), the reflected rays form a diverging beam. If the light source is placed further away than the focus (c), the reflected rays converge at a point and then diverge. In a convex mirror (d), the reflected rays

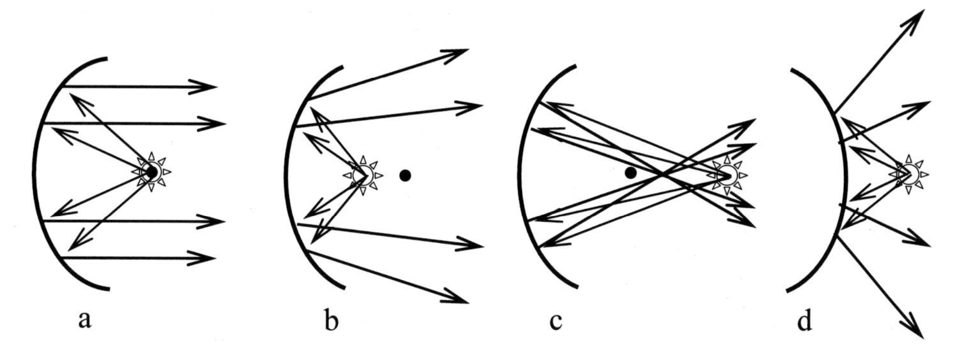

Figure 9.20. Reflected rays by concave spherical mirrors (a,b, and c), and a convex spherical mirror (d). The shape of the reflected beam depends on the position of the source (star) with respect to the focus (full dot).

always diverge. So, it is possible to manipulate the shape of the reflected beam by varying the position of the light source with respect to the mirror.

Another way of controlling the shape of beams is by lenses. Lenses are made of transparent materials. Rays of light enter them through one surface, refract, and leave the lens through the opposite surface. Quite often the entry and exit surfaces are spherical. Combinations of flat and spherical surfaces are also common. There are two kinds of lenses: converging and diverging. The center of a converging lens is thicker than its periphery, while the center of a diverging lens is thinner than its periphery. Like mirrors, each lens is characterized by an imaginary point called focus. The distance between the focus and the center of the lens is called the focal length. There are formulas that enable us to calculate the shape of a beam after it has passed through a lens. The general characteristics of such beams can also be described in a graphical way.

Figure 9.21 a, b, and c shows how light rays that are emitted by a light source are refracted by a converging lens. Figure 9.21 d shows how a diverging lens refracts rays. These patterns are similar to those of lenses, as shown in figure 9.20.

If the surfaces of lenses and mirrors are textured, the light that they reflect or refract loses all or some of its sharpness, and becomes diffused.

Mirrors and lenses are also used in a variety of optical devices such as eyeglasses, cameras, microscopes, telescopes, etc. They form images of sources of light. Each point of the source has its corresponding image point. A cone of rays emitted by a point of the source produces a cone of rays that seems to be emitted by the corresponding point of the image. The same process that is illustrated in figure 9.14 for a flat mirror is happening also in curved mirrors and lenses. Because of the curved geometry of the latter, the images may appear larger or smaller than the original source, upright or upside down, and in various positions with respect to the mirror or the lens.

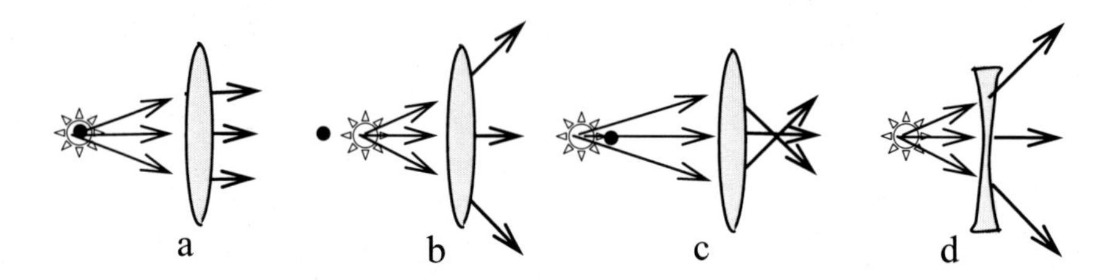

Figure 9.21: Refraction of a beam emitted by a point source through converging lenses (a, b, and c) and through a diverging lens (d). Dots indicate the position of the focus, stars the position of the source.

PROBLEMS

Section 9.1

1. How long would it take a pulse of light to travel a distance equal to the circumference of the equator (24,910 miles)?

2. What properties distinguish between a beam of particles and a wave?

3. What is the frequency of an electromagnetic wave whose wavelength is 10^{-2}m? What are such waves used for?

4. How much energy is carried by a photon whose wavelength is 0.5 nanometers? What is the frequency of that wave?

5. How many red photons (wavelength 700 nanometers) carry a total energy of 1 joule? (b) How many yellow photons (wavelength 580 nm) carry that total energy?

Section 9.2

6. What are the hues and saturations of the following RGB combinations?
(a) R=0.7, G=0.3, B=0; (b) R=0., G=0.5, B=0.5; (c) R=0.8, G=0., B=1.

7. What are the hues and saturations of the following RGB combinations?
(a) R=0.9, G=0.5, B=0.2; (b) R=0.4, G=0.6, B=0.8; (c) R=0.8, G=.3, B=1.

Section 9.3

8. How many photons per second are emitted by a radiant source of 75W that emits photons whose wavelength is 550 nm?

9. How many photons per second are emitted by a radiant source of luminous flux of 1200 lumen that emits photons whose wavelength is 550 nm?

10. What is the luminous flux of a source whose spectrum consists of 0.013 watt of light at 550 nm and 0.09 watts of light at 610 nm.

11. A light fixture, consisting of a standard light bulb and a reflector, hangs 40 cm over the center of the working area of a jeweler. The effect of the reflector is to double the amount of light that falls on the work area (compared to the light of the light bulb alone). The following are the lumens provided by available light bulbs: 120, 300, 600, 900, 1,200, and 1,800. Which is the smallest light bulb that would satisfy the requirements of 1,000 lux at the center of the working area?

12. A 1,800 lumen light bulb hangs from a tree limb 4 m above the ground. (a) What is the luminance intensity of that light bulb? (b) What is the illuminance on the ground at the brightest point (right under the light bulb)? (c) What is the illuminance on the ground, two meters away from the brightest point?

13. Two identical light sources, whose illuminance intensity distribution is given in table 9.3, hang 3 m above the ground, 4m away from each other. What is the illuminance on the ground at the mid point between the two?

14. A warehouse is illuminated by long lines of fluorescent lamps. Each line consists of straight fluorescent lamps, with very narrow gaps between them. The length of each lamp is 1.1 m, and its luminous output is 800 lumens. The effect of the reflector above the lamp is to double the amount of the light that falls on the floor (compared to the light of the fluorescent lamp alone). The lines are 4m above the floor, and the distance between lines is 3 m. Find the illuminance on the floor due to three consecutive lines of lamps, at a point right under the middle line of lamps.

Section 9.5

15. In order to shield the patients waiting outside an x-ray room, the walls are covered by a sheet of lead 0.5 cm thick. What percentage of the x-ray photons that hit the wall are stopped by it, if the linear absorption coefficient is 1.55 cm^{-1}?

16. A 1,200 lumen light bulb is placed 40 cm in front of a mirrored wall, whose reflectance is 0.92. What would be the illuminance at the brightest spot on the opposite wall, 3 m away from the mirrored wall?

17. What is the speed of yellow light in (a) water, (b) diamond?

18. What is the wavelength in fused quartz of a photon whose wavelength in air is 600 nm? What is the frequency of that photon in air and in fused quartz?

19. A beam of light travels in air and hits glass at 40^0 with respect to the normal. What is the angle of the refracted beam?

20. A beam of light travels inside an aquarium full with water and hits the glass at an angle of 20^0 with respect to the normal. (a) What would be the angle of the refracted beam in the glass? (b) What would be the angle of the beam that emerges from the glass to the air?

21. (a) What is the angle of total reflection for water-air interface?
(b) What is the angle of total reflection for diamond-water interface?

10. ACOUSTICS

10.1 Propagation of sound

When someone talks to us, the movement of her lips seems synchronized with what we hear. However, we always hear the thunder lagging after the lightning. When, at a distance, a heavy hammer is driving a pole into the ground, we first see the hammer hitting the pole, and after a while we hear the thump. These and similar phenomena illustrate that it takes time for the sound to travel from the source to the observer. Measurements have shown that at sea level and at a temperature of 20^0C the speed of sound in air is 343 m/s. The speed of sound increases when the temperature of the air increases and when the density decreases.

Example
A lightning flash and its thunder are generated in a cloud at the same time. How long will it take the lightning and the thunder to reach the ground, if they start in a cloud 3 km above ground? (Neglect changes in the speed of sound due to changes in air density and temperature. The speed of light is $3x10^8$ m/s. This is approximately a million times faster than the speed of sound.)

Solution
We use the formula for motion at constant velocity: $x=v \cdot t$. For the lightning and the thunder, x=3km=3,000m. For the lightning $v=3x10^8$m/s, yielding $t=10^{-5}$sec, which is 10 millionths of a second. For the thunder, v=343m/s, yielding t=8.7sec. Practically, it could be said that the lightning reached the ground instantly, and the thunder arrived 8.7 seconds later. By measuring the time delay between seeing the lightning and hearing the thunder (or estimating it e.g. by counting the seconds, such as one Mississippi, two Mississippi, ...), it is possible to find the distance to the cloud where they originated.

Even in calm conditions, air molecules are constantly in random motion, colliding and bouncing from each other. As a result, the molecules become distributed evenly in a state of equilibrium. When a vibrating element, such as a string of a violin, is activated, its vibrations cause adjacent molecules of air to vibrate. That disturbs the equilibrium condition. Compression and rarefaction regions are formed, where the density and pressure of the air are higher and lower, respectively, than those at the equilibrium state. Those patterns of compression and rarefaction propagate in the air, away from the vibrating source, forming what are called sound waves. As such a pattern reaches an observer, the sequences of high and low pressure cause membranes inside the ear to vibrate.

Those vibrations are sensed by specialized nerve cells that relay that information to the brain, where the perception of sound is formed. Each sound, e.g. a word, a honk, etc. has its own pressure pattern. However, all patterns travel through air at the same speed, which is the speed of sound. In the following, when we speak about the pressure pattern of a sound wave we mean the pattern of the pressure-difference between the actual pressure and the average equilibrium pressure of calm air. Figure 10.1 illustrates a sound wave propagating in air.

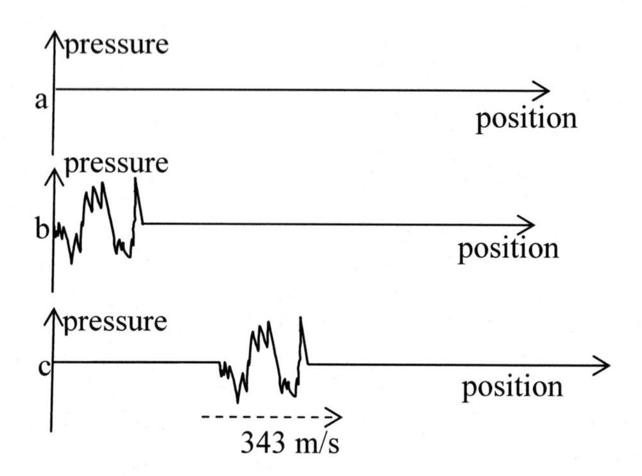

Figure 10.1: Propagation of sound in air. a: Air at equilibrium - uniform pressure and density. b: Sound is generated. A pattern of pressure is formed at the source. c: The pattern propagates at the speed of sound. The pattern is also called a sound wave.

The pressure pattern of a sound wave is formed because of shifts of molecules from their equilibrium state. Those shifts also create density patterns and velocity patterns that are associated with the pressure pattern. The shapes of the pressure and density patterns are similar. The velocity pattern looks different from the pressure and the density patterns.

It should be noted that while the patterns travel away from the source, the air molecules stay pretty much in their original neighborhoods. The situation is similar to a "wave" made by spectators in a stadium. They raise and lower their hands in a coordinated way, creating the impression of a wave traveling around the stadium. However, the hands of the spectators are moving only near their seats. In a wave in a stadium, the hands of the spectators are moving up and down while the wave is traveling horizontally around the stadium. This is an example of a transverse wave, in which the oscillations that make up the wave are perpendicular to the wave's direction of propagation. When sound propagates, the molecules that make up the pressure patterns oscillate along the direction of the wave's propagation. Sound waves are examples of longitudinal waves.

Although longitudinal waves and transverse waves differ from each other in the directions of their oscillations, they are described graphically in the same way. Figure 10.1 shows how deviation of the disturbance that makes up the wave depends on position. This deviation may be longitudinal, as in the case of sound waves, or transverse, as in the case of water waves.

Sound can also propagate in media other than air. We can hear sounds through a wall or when we dive under water. The velocity of sound depends on the medium in which it propagates. The speed of sound in water is 1440 m/s and in steel it is approximately 5,940 m/s. Sound cannot propagate in a vacuum, because there are no molecules there to vibrate.

10.2 Sound waves

When two stones hit a pond at the same time at different points, they initiate two sets of waves that eventually meet each other. The wave pattern at the meeting region is a combination of the patterns of the two individual waves. Meeting waves interfere with each other. According to the **superposition principle**, the interference pattern is the sum of the patterns of the individual waves. The superposition principle applies to all kinds of waves, including sound waves. In some cases, the superposition principle holds only for small waves.

If, after interfering, the two waves continue to two separate regions, each wave retains its original shape, unaffected by the previous interference. (figure 10.2).

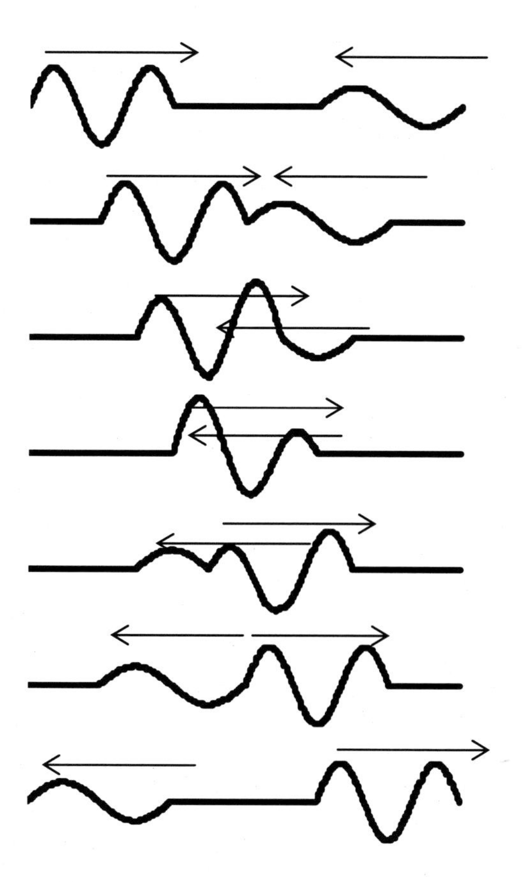

Figure 10.2: Snap shots of two waves, traveling in opposite directions (top to bottom). When they overlap with each other, the waves interfere. The instantaneous pattern of the interfered wave is the sum of the instantaneous patterns of the individual waves. As the waves re-emerge (bottom two patterns), each wave regains its original shape and properties.

Sinusoidal waves constitute a family of waves that are easy to describe and to analyze. It is simple to figure out how these waves interfere with each other and how they interact with other objects. A sinusoidal wave is a wave whose spatial pattern looks like the sine function (figure 10.3). (In the following, the term sinusoidal wave will include also sinusoidal waves that start at values other than zero. So, waves whose shape is like a cosine function will also belong to the family of sinusoidal waves. This family of waves is also called harmonic waves). In sinusoidal sound waves, the pressure as a function of position has the shape of a sine function. The largest value of the wave (with respect to the calm state) is called the amplitude (A). The distance between two consecutive crests of the wave is called the wavelength (λ). This is also the distance between two consecutive troughs, or as a matter of fact, between any two consecutive points that have the same phase (angle) of the sine function. Imagine now a sinusoidal water wave propagating in a pond where a cork is floating. As the wave passes, the cork will oscillate up and down. The frequency of those oscillations is called the frequency of the wave (f). (Frequency is defined as the number of cycles per unit time). The frequency is the same all along the wave. The pitch of a sinusoidal sound wave is related to its frequency; the higher the frequency the higher the pitch that we hear. The human ear can hear sinusoidal waves in the frequency range of 20 to 20,000 Hz. However, the ear is most sensitive in the frequency range of 100 to 5,000 Hz.

The speed of the wave (c) is defined as the distance traveled by its crest per unit time. It can be shown that for any wave:

$$c=f\lambda \qquad [10.1]$$

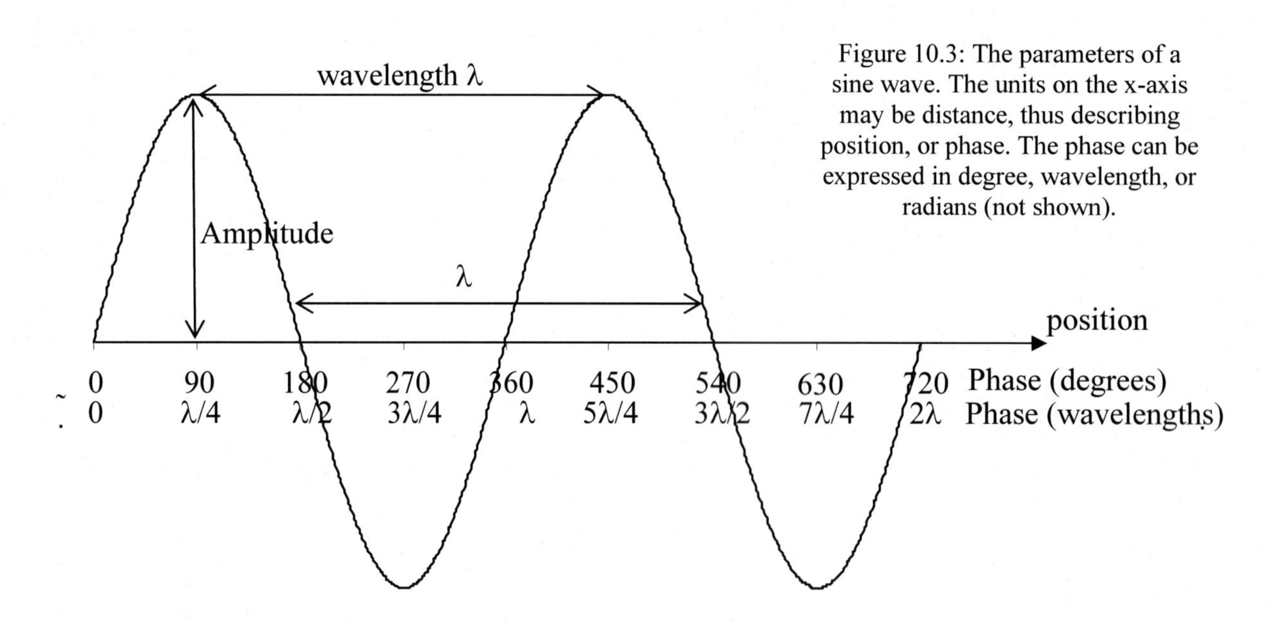

Figure 10.3: The parameters of a sine wave. The units on the x-axis may be distance, thus describing position, or phase. The phase can be expressed in degree, wavelength, or radians (not shown).

All sound waves in a given medium propagate at the speed of sound c, which is typical to that medium (e.g. c=343 m/s in air at 20^0C; c=1440 m/s in water, c=5,940 m/s in steel). Based on [10.1], sinusoidal waves of lower frequency have longer wavelength than those of higher frequency. For example, in air, when f=125 Hz, λ=2.75m; when f=4,000 Hz, λ=0.086m. In the following, when we talk about the frequency of a sound wave, we imply that the sound wave is sinusoidal.

One of the outcomes of a mathematical theory developed by Fourier (Fourier, Jean Baptiste Joseph, 1768-1830) is that a wave of any shape can be expressed as a superposition of sinusoidal waves of various amplitudes and frequencies. The sinusoidal waves that are used to express that arbitrary wave are called its Fourier components. Figures 10.4a and 10.4b illustrate a wave and its Fourier components. When those components are superimposed, the original wave is reconstructed. The list of the intensities of each Fourier component that make up a given sound is called the spectrum of that sound.

The frequency range of human speech is from 100 to 5,000 Hz. That means that any sound wave of human speech can be expressed as a superposition of sinusoidal waves whose frequencies are in the range of 100 to 5,000 Hz.

The human ear can hear sinusoidal waves whose frequencies are in the range of 20 to 20,000 Hz (the audible range). If a certain sound wave has Fourier components whose frequencies are outside the audible range, the ear simply does not hear them. Only those Fourier components that are within the audible range contribute to the superposition that is heard by the ear.

10.2.1 Distortion and manipulation of sound waves

The detailed pattern of a sound wave determines how that sound is perceived by our brain. Any sound pattern can be expressed as a superposition of Fourier

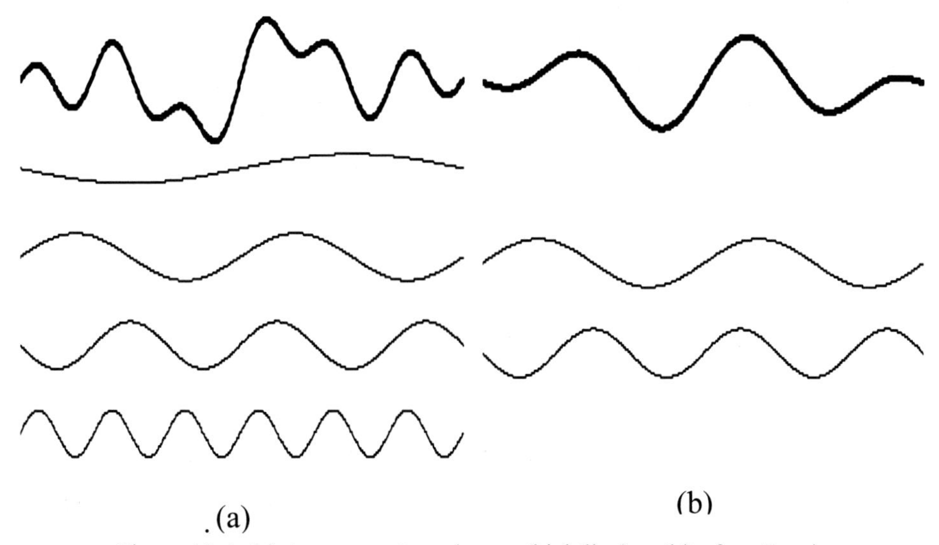

(a) (b)

Figure 10.4: (a) A wave pattern (upper thick line) and its four Fourier components. (b) When two of the Fourier components (with the longest and shortest wavelengths) are eliminated from the superposition, the resulting signal (upper thick line) is a distortion of the original one.

components. When the sound pattern interacts with objects, e.g. reflected or absorbed by walls, that interaction may affect different Fourier components in different amounts. As those affected components are superimposed, their superposition pattern may be different from the original pattern. This is perceived by the ear as a distortion of the original sound. In order not to distort a sound wave, all its Fourier components have to be affected by the same amount. In some applications, though, distortions of the original sound are introduced intentionally in order to achieve specific effects, so distortion, in spite its generally negative connotation, is not always undesirable Many spaces for music used the distorting properties of materials to emphasize or deemphasize aspects of sound waves in beneficial ways.

In architectural acoustics, it is common to group the Fourier components of sound waves into bands. This simplifies analysis by providing accurate enough information to make design decisions while keeping the data set small and easy to manage. The frequency ranges of bands used depends on the application at hand. One set of bands is the octave bands. The centers of those bands are at the following frequencies: 63, 125, 250, 500, 1,000, 2,000, 4,000, 8,000, and 16,000 Hz. As you can see, each frequency in this sequence is twice its predecessor. If f denotes the center frequency of an octave band, then the lower limit of that band would be at $f/\sqrt{2}$, and the upper limit would be at $f \cdot \sqrt{2}$. For example, the lower limit of the band whose center is at f=1,000 Hz is at $1{,}000/\sqrt{2} = 707\text{Hz}$. The upper limit of that band is at $1{,}000 \cdot \sqrt{2} = 1{,}414\text{Hz}$. That means that sinusoidal waves whose frequencies are between 707 and 1,414 Hz belong to the band of 1,000 Hz. It should be noted that in some cases, such as the non-typical acoustic criteria of an experimental music space, full spectrum data is used.

10.3 Sound intensity

It takes energy to generate a sound wave. That energy is usually provided by vibrating elements of the sound source. The vibrating energy is relayed to the air molecules and it is stored as acoustic energy by the pressure pattern of the wave and by the moving molecules. This is similar to a vibrating spring that stores elastic and kinetic energy. As the sound wave travels, it carries that energy. When it encounters an object, the wave may transfer some or all of its energy to that object. For example, sound waves that hit a window make it vibrate; some of the acoustic energy of the wave has been converted into mechanical energy of the window. This is an important concept in architectural acoustics, since windows reflect sound but also absorb and transmit it through this mechanical vibration. Massive, laminated, double paned windows are frequently used to minimize transmission. For example, a 6 mm thick pane of glass of certain dimensions will vibrate very well at about 2,000 Hz and will therefore transmit sound relatively easily at that frequency. Using a 6mm laminated pane (instead of a monolithic one) will dampen some of the vibration and therefore reduce transmission. Further, if a second pane of glass of a different thickness is added, then the problematic frequencies of the panes will not match and the entire system will become very effective at stopping sound.

The amount of energy per unit time that a source provides to the sound-wave is the effective power of the source. This is also the total power carried by the wave. As the wave propagates away from the source, that power is distributed over the entire wave front. The intensity (I) of a wave at a point is defined as the power carried by an area element of the wave front at that point, divided by the size of that area element. The unit of sound-intensity in the SI system is watt/m^2. The intensity I is proportional to the square of the pressure that constitutes the sound wave at that point.

In general, the intensity of the sound at a given point depends on the distance of that point from the source of the sound, and on the type of the source. Consider, as a first example, a small (point) sound source that transmits to the surrounding air sound energy uniformly in all directions. The sound wave spreads out as a spherical shell, whose center is the source. The original power of the source is distributed evenly over the entire spherical front of the wave. Let P (watt) be the power provided by the source. When the front of the wave has propagated R meters away from the source, those P watts are distributed evenly over the area of the sphere, which is $4\pi R^2$. The intensity I of the wave, or the power density of the wave, which is the power per unit area (watt/m^2), is given by:

$$I = \frac{P}{4\pi R^2} \qquad [10.2a]$$

Consider now, as a second example, a sound source that has the shape of a long straight line, such as a busy highway. The sound waves spread out of a line source as a cylinder whose center is that line. The sound energy is now distributed over the surface of the cylinder, and the relationship between P_{line}, which is the power provided by a unit length of the source (watt/m in the SI system) and I, the intensity of the sound (watt/m^2) at a distance R from the source would be:

$$I = \frac{P_{line}}{2\pi R} \qquad [10.2b]$$

Please note that the units of P_{line} in [10.2b] differ from P in [10.2a] since P_{line} has a length associated with it.

Equation [10.2b] holds when the line source is surrounded by air. In the case of a highway, corrections have to be introduced to account for the sound reflected and absorbed by the ground.

Example

What would be the sound intensity 5 m away (a) from a 10^{-5} watt point source? (b) from a 10^{-5} watt/m line source.

Solution

According to [10.2a] we get: $P=10^{-5}$ watt, R=5m, I=?

Substituting yields $I=3.2 \times 10^{-8}$watt/m^2.

According to [10.2b] we get $P=10^{-5}$ watt/m, R=5m, I=?

Substituting yields $I=3.2x10^{-7}$watt/m^2.

	Intensity (I) (watt/m^2)	Intensity Level (IL) (dB)
Threshold of hearing	10^{-12}=0.000,000,000,001	0
Human breathing	10^{-11}=0.000,000,000,01	10
Whisper	10^{-10}=0.000,000,000,1	20
Quiet urban area at night	10^{-8}=0.000,000,01	40
Conversation at 1m (3f)	10^{-6}=0.000,001	60
Busy office	10^{-5}=0.000,01	70
Car without muffler	10^{-2}=0.01	100
Threshold of pain	10	130

Table 10.1: Intensities and intensity levels of typical sounds

Table 10.1 lists typical intensities of common sounds. As you can see, the human ear can hear sounds having a huge range of intensities (second column). Sometimes it is cumbersome to handle this range of numbers. It is common to describe the loudness of sound by using the decibel (dB) scale (third column), which will be explained in full detail in following sections.

10.3.1 Sound loudness

It was found that when people say that two sounds of different frequencies appear to have the same loudness, the intensities (in watt/m^2) of those two sounds may be quite different. The ear and the brain perceive the loudness of a sound based on both its frequency and its intensity. The unit ***phon*** has been introduced in order to measure the perception of loudness. Two sounds that have the same number of phons have the same loudness. One phon is defined as the loudness of the sound whose intensity level is 1 dB, and whose frequency is 1,000 Hz. The loudness in phons of a sound whose frequency is 1,000 Hz is equal to the intensity level of that sound expressed in dB. For example, the loudness of a 1,000 Hz sound whose intensity level is 80 dB is 80 phons. However, if the frequency of the sound is different from 1,000 Hz, its loudness in phons may be different from its intensity in dB. Figure 10.5 shows the relationships between loudness in phons and intensity level in dB for frequencies in the audible range. It could be seen that the ear is less sensitive to low frequency sounds of small intensity level. If we listen to music on the radio at a normal sound level, and then we lower the volume, all the Fourier components will be lowered by the same number of decibels. (That is what the regular volume knob does). Since our ears are less sensitive to the low frequency components, their loudness (in phons) would be reduced by more than medium and high frequency components. For example, if the original sound consists of two components: 50 Hz and 500 Hz, both at 80 dB, and we reduce the volume by 40 dB, both new intensities would be 40 dB. However (see Figure 10.5), the loudness of the 50 Hz component will fall below the threshold of hearing, while the loudness of the 500 Hz component would be about 40 phons, which could be heard easily. That would distort the reduced sound. This distortion can be corrected by selectively amplifying the low frequency components.

The use of the decibel scale, named after the inventor of the telephone Alexander Graham Bell (1846-1922), relies on the mathematical function log(x) and its properties. Before the advent of computers and hand held calculators, tables of log(x) were widely available and were used to simplify the calculation of multiplications and divisions of numbers. Nowadays, values of log(x) are readily available in most scientific calculators, or online by typing the equation into the Google search engine, and even by text message from your mobile phone by texting Google at 466453 (GOOGLE). The following review lists some useful properties of log(x).

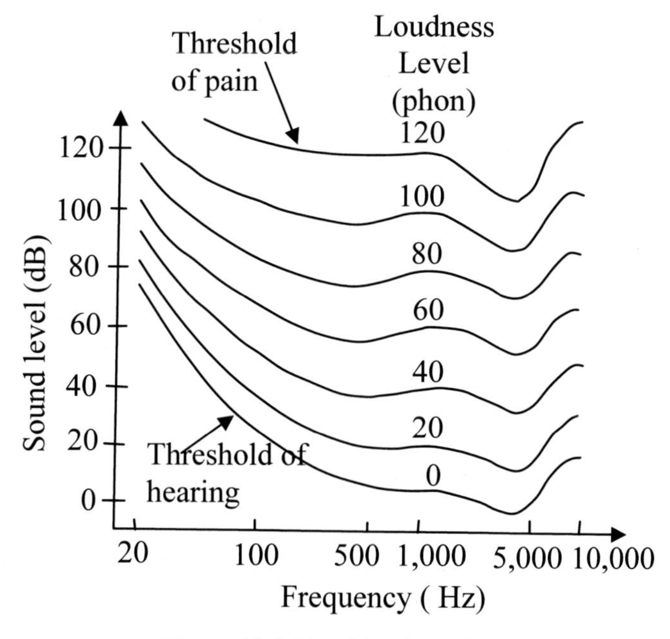

Figure 10.5: Equal loudness-level curves.

10.3.2 The log(x) function (review)

1) log(x·y)=log(x)+log(y), meaning that the log of the product of two numbers, x and y, is equal to the sum of their logs.

2) log(x/y)=log(x)-log(y) , meaning that the log of the ratio of two numbers, x and y, is equal to the difference of their logs.

3) log(10)=1

Based on 1) and 3) we get that $\log(100)=\log(10\cdot 10)=1+1=2$. $\log(1000)=3$, and in general $\log(10^n)=n$ for n=1,2,3,...

Based on 2) and 3) we get that log(1)=log(10/10)=1-1=0.

log(0.1)=log(1/10)=0-1=-1. log(0.01)=log(1/100)=0-2=-2, and in general $\log(10^{-n})=-n$.

Table 10.2 lists the values of log(x), accurate to two significant digits, for some numbers in the range 1 to 10. Based on these values and on the properties of log(x) mentioned above, log of numbers outside that range, as well as numbers within that range could be calculated. For example:

$\log(20)=\log(2\cdot 10)=\log(2)+\log(10)=0.30+1=1.30$.

log(1.5)=log(3/2)=log(3)-log(2)=0.48-0.30=0.18

x	1	2	3	4	5	6	7	8	9	10
log(x)	0	0.30	0.48	0.60	0.70	0.78	0.85	0.90	0.95	1

Table 10.2

The inverse function of (log-of...) is (10-to-the-power-of...). This is expressed by the identity $x=10^{\log(x)}$.

If we know that y=log(x), where y is given, and we want to find the value of x, we calculate $x=10^y$. For example, if log(x)=3, we find x by $x=10^3=1000$.

10.3.3 The decibel scale

Let I be the intensity of a certain sound (expressed in watt/m^2 in the SI system). The intensity level (IL) of I is defined by:

$$IL = 10 \cdot \log\left(\frac{I}{I_0}\right) \qquad [10.3a]$$

where I_0 is the intensity of the weakest audible sound. The unit of IL is called decibel (dB). The IL of I_0 is 0 dB. ($I_0=10^{-12}$watt/m^2). The third column of table 10.1 is calculated from the second column by using equation [10.3a].

When equation [10.3a] is solved for I we get:

$$I = I_0 \cdot 10^{IL/10} \qquad [10.3b]$$

We use [10.3a] to find the intensity level IL (in dB) when the intensity I (in watt/m^2) is given. We use [10.3b] to find I when IL is given.

A difference of up to 1dB is perceived as basically the same level. A difference of 3dB is barely perceived as different. A difference of 5dB is clearly different, and a difference of 10dB is substantially different. This is true for any intensity level.

Example
The sound level inside a car is 45dB. What is the sound intensity?

Solution
Using the notations of [10.3] we get: IL=45dB, $I_0=10^{-12}W/m^2$, I=? Substituting we get: $45 = 10 \cdot \log\left(\frac{I}{10^{-12}}\right)$, or $4.5=\log(I)-\log(10^{-12})$, which gives $\log(I)=4.5+(-12)=-7.5$. Since $I=10^{\log(I)}$ we get $I=10^{-7.5}=3.2x10^{-8}\ W/m^2$.

In general, if the intensity level is given as IL (in dB), the intensity I (in $watts/m^2$) would be:

$$I = 10^{\log(0.1 \cdot IL - 12)} \qquad [10.3c]$$

Example
One car at the street creates a sound level of 60 dB at a window high above. What would be the sound level at that window generated by two identical cars at the street?

Solution
Let I denote the sound intensity that one car creates at that window. Sound intensity is proportional to the power of the sound source. The power of two identical sources is twice the power of one source. Therefore, the sound intensity created by those two cars would be $2 \cdot I$. Based on [10.3] the sound level of two cars (IL_2) would be: $IL_2 = 10\log\left(\frac{2 \cdot I}{I_0}\right)$. Based on property 1 of log(x) we have: $IL_2 = 10\log(2) + 10\log\left(\frac{I}{I_0}\right)$. The second term in the right hand side is the sound level of one car, which is 60dB. Therefore, based on table 10.3, $IL_2=10 \cdot 0.3+60=63dB$. An increase of 3dB in the sound level corresponds to doubling the power of the source. This general result is true not only for 60dB, but also for any sound level. A decrease of 3dB corresponds to halving the power of the original source.

Example
Show that an increase of 20 dB in the sound level corresponds to increasing the sound intensity 100 times.

Solution
Let I_1 denote the initial intensity, whose sound level is IL_1, and I_2 denote the final intensity, whose sound level is IL_2. Based on [10.3]:

$$IL_1 = 10\log\left(\frac{I_1}{I_0}\right),$$

$$IL_2 = IL_1 + 20 = IL_1 + 2\cdot 10 = 10(\log\left(\frac{I_1}{I_0}\right) + 2) =$$

$$10(\log\left(\frac{I_1}{I_0}\right) + \log(100)) = 10\log\left(\frac{I_1 \cdot 100}{I_0}\right)$$

(We have used 2=log(100), and the sum of the logs is the log of the product). Meaning that sound level IL_2 corresponds to sound intensity of $100 \cdot I_1$.

Example
If at a distance R_1 from a point sound source the sound intensity is I_1 and the sound level is IL_1, what would be the sound intensity I_2 and the sound level IL_2 at a distance R_2, which is twice R_1?

Solution
Based on [10.2a], $I_1 = \frac{P}{4\pi R_1^2}$ and

$I_2 = \frac{P}{4\pi R_2^2} = \frac{P}{4\pi(2R_1)^2} = \frac{P}{4\pi R_1^2 \cdot 4} = \frac{I_1}{4}$. So, the sound intensity at R_2 is one fourth of that at R_1.

Based on [10.3], $IL_1 = 10\log\left(\frac{I_1}{I_0}\right)$

$$IL_2 = 10\log\left(\frac{I_2}{I_0}\right) = 10\log\left(\frac{I_1/4}{I_0}\right) = 10(\log\left(\frac{I_1}{I_0}\right) - \log(4)) = IL_1 - 10\cdot\log(4)$$

$$IL_1 - 10\cdot 0.60 = IL_1 - 6$$

So, the sound level at R_2 is 6dB less than that at R_1.

The decibel scale and common intuition

Due to the wide range of sound intensities that can be detected by the ear and processed by the brain, common intuition developed from experiences may be misleading when dealing with sound phenomena.

Consider first how common intuition works. Let's say that a pound of grapes costs one dollar, and that we ate all the grapes but one (say we ate 99% of the grapes). The remaining amount of one grape is insignificant compared to the original one pound, and its value to us is also insignificant. By generalization, common intuition would lead us to believe that when the amount of something is reduced by 99%, it looses most of its significance to us.

Now, let's deal with a similar situation that involves sound. The sound intensity level that reaches a worker in an office through an open door is at a normal level of 70 dB. When that door is closed, it completely blocks all the sound. If the door is left slightly open, say to only 1% of its total area, 1% of the sound intensity passes through that crack and reaches the worker in the office. The other 99% are blocked by the door. If we apply the common intuition based on the grapes experience, we may conclude that the 1% of the sound energy that reaches the worker is insignificant. Let's see how many dB's correspond to this 1%.

Let I_{70} denote the intensity (in watt/m^2) of the sound that comes through the open door. Based on [10.3a]

$$70 = 10\log\left(\frac{I_{70}}{I_0}\right).$$

The intensity of the sound that comes through the open crack I_1 is one percent of what comes through the open door: $I_1 = \left(\frac{I_{70}}{100}\right)$. Based on [10.3a], the intensity level of the sound that comes through the crack IL_1 would be:

$$IL_1 = 10\log\left(\frac{I_1}{I_0}\right) = 10\log\left(\frac{I_{70}}{I_0 \cdot 100}\right) = 10(\log\left(\frac{I_{70}}{I_0}\right) - \log(100)) =$$

$$= 10\log\left(\frac{I_{70}}{I_0}\right) - 10\log(100) = 70 - 20 = 50\text{dB}.$$

So, the intensity level of the sound that comes through the crack is 50dB. Referring to table 10.1, this is not insignificant. The worker can easily listen through the crack to what is said in the other room.

The same conclusion, of course, would be reached if the analysis was done using sound intensities. Based on table 10.1, (or formula [10.3b]) 70 dB corresponds to 10^{-5} watt/m^2. If only 1% of it passes through the crack in the door, the intensity of the passed sound is $10^{-5}/100=10^{-7}$ watt/m^2. Based on table 10.1, that intensity is in the significant range, and could be heard.

(Note: In this example we assumed that the sound spreads through the open door in the same pattern that it spreads through the crack.)

Doors which are designed to stop sound can be very complicated and expensive for this reason. They also must be very carefully installed. Full perimeter seals must be used and full engagement of the seals with the head, jamb, and threshold is necessary to maximize the door's ability to stop sound. Mass and damping play significant roles too, but a door without fully engaged perimeter seals is sure to fail, even if 99% of the open area is covered.

The decibel scale helps to emphasize that even a small fraction of the original intensity may be significant. It makes it possible to express a wide range of intensities by numbers in the range of 0-120. However, everything that is computed in the logarithmic-based decibel scale can be expressed by the conventional linear scale.

10.4 Reflection, absorption, refraction, and transmission

10.4.1 Echoes and reflection of sound

An echo happens when a sound reflected by an obstruction reaches our ears slightly after the original sound. For the brain to be able to distinguish the echo, the reflected sound has to arrive after a minimal time delay. If the reflected wave reaches the observer sooner, the brain considers the original sound and its reflection as one sound. That minimum delay depends on the relative intensities of the original and reflected waves, and it varies from 40 to 100 milliseconds. A representative minimal delay time is 50 milliseconds, which corresponds to a difference of 17m (56 feet) between the distance traveled by the original wave, known in architectural acoustics as the direct sound, and the distance traveled by the reflected wave, before they reached the observer. So, in order to create an echo, one has to be at least 17/2=8.5m in front of a reflecting wall. In a typical living room we don't hear echoes, even though the walls are reflecting sound, because the distances involved in a typically-sized living room are much too small to meet this minimal delay requirement. However, short distances aren't the only way to prevent echoes. Even if the distance from a wall is greater than 8.5m, diffusion, which spreads the reflected sound, and absorption can be used to weaken the reflection to a level which cannot be distinguished as a clear echo.

Only part of the energy of the incident wave is reflected by an obstruction. Another part, which is called the refracted wave, moves into the obstruction. The direction of the refracted wave and its energy depend on the material of the obstruction and on the angle of incidence, and the direction of propagation of the refracted wave is different from that of the incident wave, except when the incident wave hits the surface of the obstruction at 90^0. As the refracted wave passes through the obstruction, it looses some of its energy (it is attenuated). It may eventually emerge from the other side of the obstruction. This emerging wave is called a transmitted wave.

In this context, any increase in the resistance of a medium to the passage of sound constitutes an obstruction. For example, warm air attenuates sound more than cold air. Therefore, a layer of warm air above a layer of cold air is also an obstruction. In atmospheric inversion, the air close to the ground is colder than the air above it. Sound from sources on the ground propagates upwards, hits the layers of warm air, and some of it is reflected back to the ground while some of it is refracted.

Different physical processes are responsible for reflecting sound by the surface of an obstruction and for transmitting it through the medium of the obstruction. An obstruction that reflects sound effectively may also be a good transmitter and an obstruction that is not an effective reflector may not be a good transmitter. This is especially true when considering a wide range of frequencies because different frequencies of sound affect materials in different ways.

The **reflection coefficient** ρ of a surface (e.g. a wall) is the ratio of the intensity of the reflected wave to the intensity of the incident wave. In architectural acoustics, when dealing with acoustical processes that take place inside a room and not considering the effect of sound on adjacent spaces, sound that leaves the room can be considered absorbed sound. Absorbed sound includes sound refracted into walls, ceiling, and floor, and sound that left the room through openings. In this context, an open window is considered to be the perfect absorber because it does not reflect any sound. The **absorption coefficient** α of an element of the surface of the room (e.g. a wall) is the ratio between the intensity of the sound that it does not reflect (due to refraction or free passage through openings) and the intensity of the incident sound. These two coefficients are related by:

$$\alpha=1-\rho \qquad [10.4]$$

meaning that $I_{reflected}=\rho \cdot I_{incident}=(1-\alpha) \cdot I_{incident}$, where I stands for sound intensity. So, an absorption coefficient $\alpha=0.8$ means that 20% of the sound generated in the room was reflected back into it. Figure 10.6 illustrates that terminology.

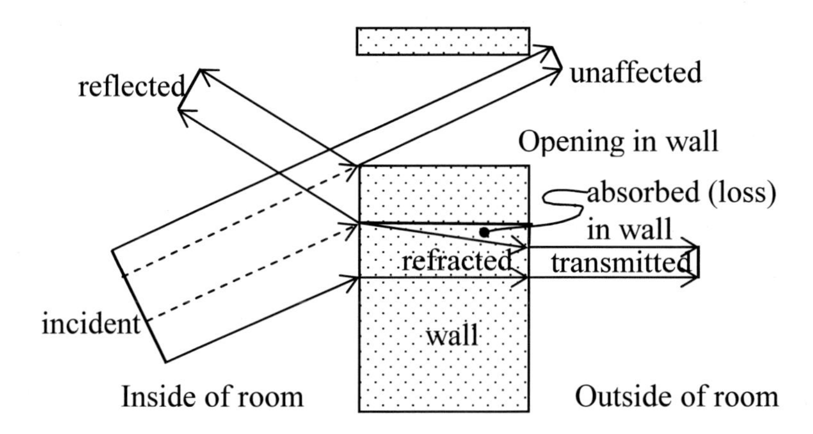

Figure 10.6: Terminology for cases where sound hits a wall. For an observer and sound source inside a room: absorbed=incident-reflected=refracted+unaffected. For an observer on one side of the wall and source on the other side: absorbed or loss (in wall)=entered-transmitted. In the latter case, the refracted part of the beam at the point of incidence is what enters the wall

The reflection and absorption coefficients depend on the material of the wall and the angle of incidence. Even though many architecture applications are characterized by sound incident on materials at specific angles, measurement standards for the absorptive properties of materials give absorption coefficients which are averaged over all angles of incidence. This is done for simplicity since listing data for all angles of incidence would amount to a daunting set of data. Table 10.3 gives absorption properties of some materials. (Please be aware that some manufacturers market materials as having absorption coefficients higher than 1.00. This is theoretically impossible based on the definition of the absorption coefficient (nothing can absorb more energy than there is to absorb), but manufacturers are not necessarily lying. If the absorbing material is thick, it absorbs also sound that falls on its sides, thus increasing the effective absorption area.)

Material\frequency (Hz)	125	250	500	1,000	2,000	4,000
Brick wall	0.02	0.02	0.02	0.03	0.04	0.05
Concrete masonry/coarse	0.36	0.44	0.31	0.29	0.39	0.25
½" gyp. bd, 3-5/8" studs,½" gyp. bd.	0.27	0.1	0.05	0.04	0.03	0.03
½" gyp. bd. ceiling	0.11	0.11	0.05	0.06	0.04	0.05
Marble or glazed tile	0.01	0.01	0.01	0.01	0.02	0.02
Glass window	0.35	0.25	0.18	0.12	0.07	0.04
Water surface (e.g. swimming pools)	0.01	0.01	0.01	0.01	0.02	0.02
Heavy carpet, foam backed	0.08	0.24	0.57	0.69	0.71	0.73
Fabric curtain 14 oz. velour, 50% full	0.07	0.31	0.49	0.75	0.70	0.60
Audience on heavy upholstery	0.76	0.83	0.88	0.91	0.91	0.89
Fiberglass, 2" 1 pcf, mount. type A	0.22	0.67	0.98	1.0	0.98	1.00
Metal roof deck, plain	0.40	0.30	0.15	0.10	0.04	0.12

Table 10.3: Sound absorption coefficients

10.4.2 Tracing reflected sound

When a sound beam hits an obstruction, part of it is reflected and another part is refracted. In order to trace the reflected wave, the incident beam may be divided into narrow rays, each of which is reflected by the surface of the obstruction. The properties of the entire reflected beam may be inferred by tracing a sample of those rays. The Law of Reflection states that the angle of the reflected ray is equal to the angle of the incident ray. The angles are measured between the rays and the normal to the reflecting surface. The incident ray, the normal, and the reflected ray are found in the same plane (figure 10.7).

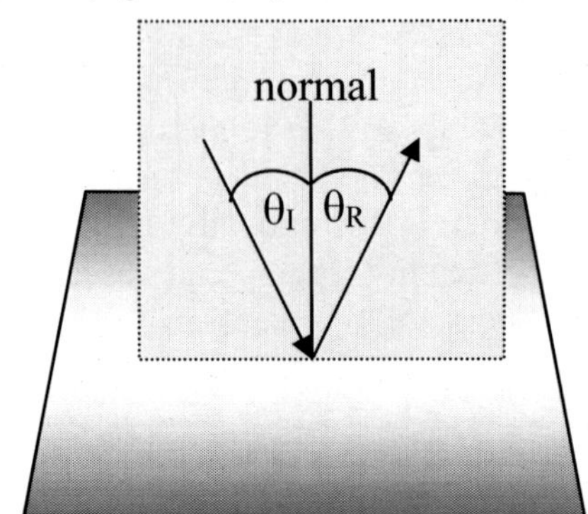

Figure 10.7: Reflection of sound. The angle of incidence (θ_I) is equal to the reflection angle (θ_R)

The reflection patterns from some common surfaces are illustrated in figures 10.8. Those reflection behaviors occur when the wavelength of the sound is smaller than the typical sizes of the reflecting surfaces; typically the wavelength should be at most a fifth of the size of the reflecting element in order for the Law of Reflection to apply. If that is not the case, another phenomenon, diffraction, enters the game. Diffraction is discussed later in this chapter. All the patterns of figures 10.8 are obtained by applying the Law of Reflection (figure 10.7) to each of the rays in the sample.

The method of images: The intensity of the sound reflected from a flat surface may be calculated based upon equation [10.2a]. In such cases the distance R is the distance between the observer and the image of the source.

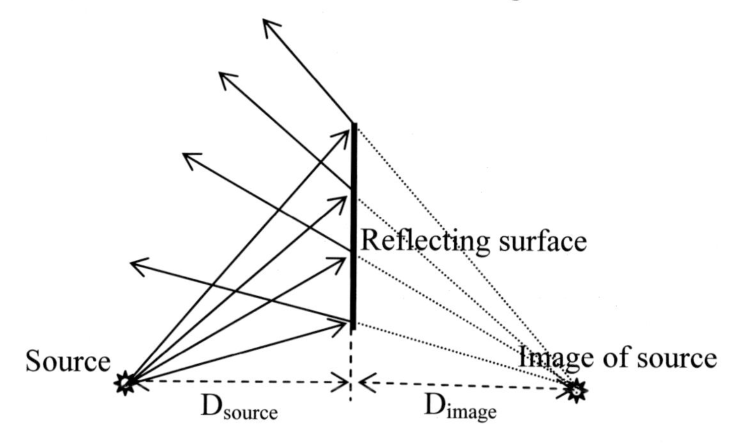

Figure 10.8a: Reflection by a flat surface. The reflected rays appear as if they are coming from a mirror-image of the original source. The distance of the image from the surface (D_{image}) is equal to the distance of the source from the surface (D_{source}).

Since only a fraction ρ of the incident sound is reflected, where ρ is the reflection coefficient, the reflected intensity would be:

$$I = \frac{\rho \cdot P}{4\pi R^2} \quad [10.2c]$$

The time that it takes the sound to travel from the source to the reflector and from there to the observer is equal to the travel time from the image source directly to the observer. Therefore, it appears as if the image source emits the reflected sound at the same moment as the real source. The boundary of the region of space that is reached by the reflected sound can be found by tracing the rays that are reflected from the edges of the reflector (figure 10.8a). This is the same as considering the image point as a light source, and the reflector as an obstruction to that light. The shade that would have been created by the reflector is where the reflected sound is found. Plane reflecting surfaces are commonly hung from ceilings of music and lecture halls to redirect sound to regions where it needs to be strengthened.

Example
The effects of sound reflection from the ground
Consider a sound point source, situated 1.5 m above a marble floor, and emitting sound at $2x10^{-6}$ watts. (a) Find the sound intensity level due to that source at a listening point 1 m above the floor and 6 m away. (b) Find the sound intensity level at that point when the reflection from the floor is taken into account.

Solution
First, we find the sound intensity, using formula [10.2a]. We have $P=2x10^{-6}$, $R^2=36.25\ m^2$ (see figure), which yields $I_1=4.39x10^{-9}$ watt/m^2. This sound intensity corresponds to sound intensity level of

$$IL_1= 10\cdot\log\frac{4.39x10^{-9}}{10^{-12}}=36.4dB.$$

The contribution of the reflected sound can be found by considering the image source (see figure) and using formula [10.2c].
In this case, $P=2x10^{-6}$, $R^2=42.25\ m^2$, and $\rho=0.99$ (based on table 10.3). Substituting in [10.2c] yields: $I_2=3.72x10^{-9}$ watt/m^2. The total intensity due to the direct plus reflected sounds is: $I=I_1+I_2=8.11x10^{-9}$watt/m^2. The corresponding sound intensity level is I=39dB.

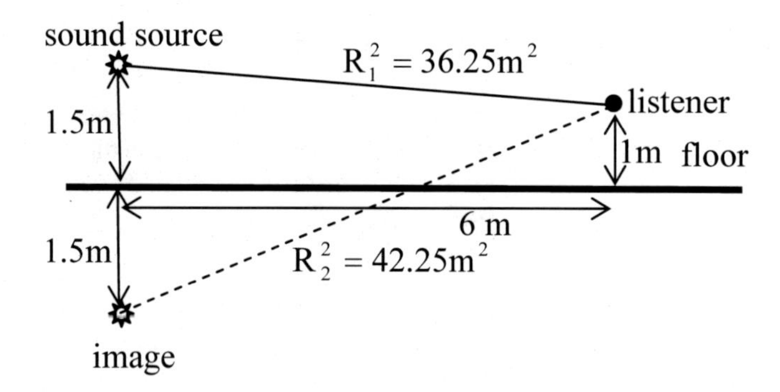

Comment. The procedure shown in the example above provides an exact way for calculating the effects of the reflected sound. In general, the sound reflected from the ground reinforces the direct sound. Very close to a ground with high reflective coefficient, the effect of the reflected sound is to double the direct sound. Above the ground, when the listener is far from the source, R_1 becomes very close to R_2. If the ground has high reflection coefficient, the outcome of the reflection is doubling the intensity of the direct sound (or increasing the sound intensity level by 3 dB).

In various applications, curved surfaces are used to reflect sound. There are two general types of curved surfaces: convex and concave. Like in flat surfaces, the spatial pattern of the reflected sound can be found by ray tracing. Figure 10.8b traces the reflection from a convex surface. The reflected rays always diverge from a convex surface, regardless of the location of the source. Convex surfaces are widely used when the sound has to be spread around.

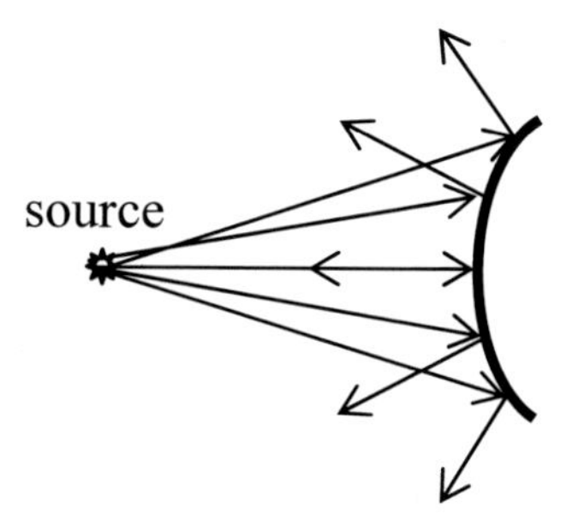

Figure 10.8b: Reflection from a convex surface. All the reflected sound diverges.

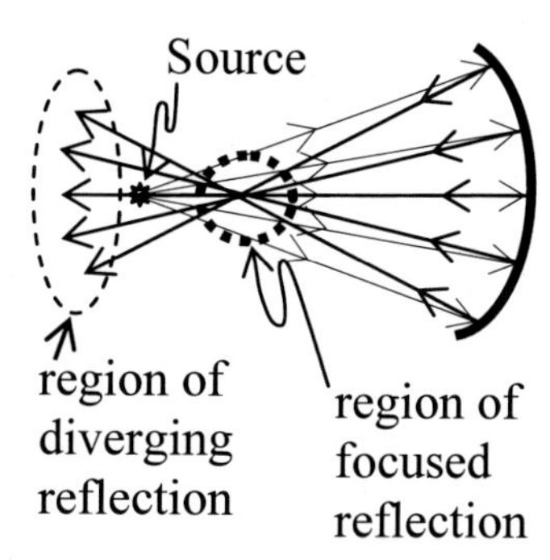

Figure 10.8c: Reflection of rays of a distant source by a concave surface. The reflected sound first converges, and then continues on and diverges.

The shape of the pattern of the rays reflected by a concave surface depends on the depth of the concavity and on the location of the source with respect to the surface. If the source is close to a shallow concave surface, all the reflected rays diverge, similar to flat or convex surfaces (figures 10.8a and b). If the source is far enough from the concave surface, the reflected rays first converge, and as they continue-on, they diverge (Figure 10.8c). The pattern of the rays reflected from any concave surface can be found by ray-tracing, similar to figure 10.8c.

Example: Trace the reflected rays from a concave hemi-spherical reflector of radius R, when the source is placed at a distance of R/2 from the reflector

Solution: Figure 10.8d illustrates the situation; the source is marked by a star. According to the law of reflection, the angle of incidence is equal to the angle of reflection. Angles are determined with respect to the normal of the surface at the point of incidence. The normal at a point on a sphere is the radius to that point. Figure 10.8d shows a group of incident rays, marked by open arrows; normals at the points of incidence, marked by dotted lines; and the reflected rays, marked by solid arrows. The reflected rays first converge, creating regions of concentrated reflection surrounded by regions devoid of reflection. Then the rays diverge. A carefully drawn ray-tracing diagram, from any reflecting surface, can show the boundaries of those regions.

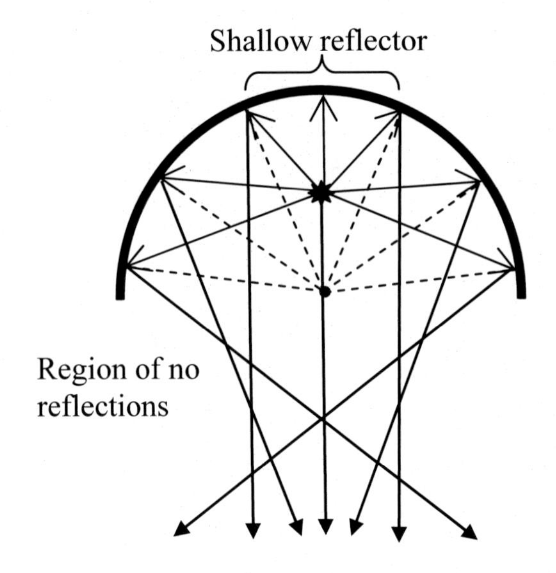

Figure 10.8d Reflection by a hemi-spherical surface (thick line). The source, denoted by a star, is midway between the center of the sphere and the center of the reflector. Dotted lines denote the normals.

Some surfaces have formulas that describe the spread of their reflected rays. One such group is the shallow spherical reflectors. A shallow spherical reflector is a small part of a much larger sphere of a radius R. The opening of a shallow reflector is much smaller than the R. In Figure 10.8d, the area next to the center of the reflector is a shallow reflector. A shallow reflector reflects all the rays that come from a very far source to one point, called the **focus,** located at a distance of $f=R/2$ from the reflector. After passing through the focus, the reflected rays diverge, similar to figure 10.8c. The location of a source with respect to the focus determines the general characteristics of its reflected rays. When the source is between the reflector and the focus, the reflected rays diverge. When the source is farther than the focus, the reflected rays first converge and then continue-on and diverge, as in figure 10.8c. If the source is exactly at the focus, all the reflected rays travel as a parallel beam. This can be seen in figure 10.8d.

In general, when the source is far from a concave surface, there are pockets of strong reflections, pockets deprived of reflection, and areas of diverging reflection. Therefore, reflections from concave surfaces require detailed planning.

Concave surfaces shaped as a parabola also have a focal point. The characteristics of their reflected rays depend on the location of the source with respect to the focus, similar to those of shallow spherical reflectors. However, those characteristics are not limited to shallow parabolas; they are the same also for deep parabolas.

It is quite common to have a sound-reflecting-wall behind a stage. Such a wall redirects part of the sound, which otherwise would be lost, back to the audience. That reinforces the sound that reaches the audience. That idea was taken to its utmost in the Hill Auditorium at the University of Michigan in Ann Arbor, which was designed by Albert Kahn and Associates and opened in 1913.

The interior shaping of the hall is that of a rotated parabola. This geometry makes for splendid acoustics, throughout the entire hall, for solo vocalists standing at the focal point of the parabola. But problems can arise as performers spread out beyond the focal point. As with all venues for music, trade offs exist, making some venues well known for performances that take advantage of their best traits.

Figure 10.8e shows how sound is reflected by an irregular surface. The reflection pattern is a combination of reflections by concave and convex surfaces.

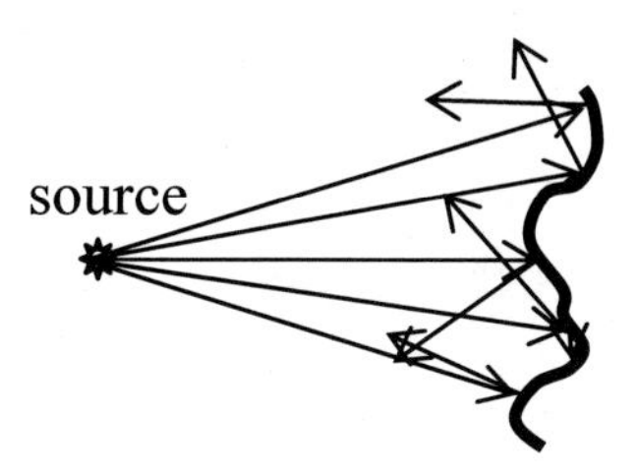

Figure 10.8e: Reflection from an irregular surface. The reflected sound is diffused in all directions.

A few words of caution about the use of plane, concave, and convex reflectors in critical acoustic spaces: all applications of this type must be done very carefully, with consideration of the frequency dependent effects, the aggregate effects of all treatments when considered together, the special implications of all treatments, and the additional more complicated effects such as diffraction (to be discussed later in this chapter).

10.4.3 Reverberations

When we are indoors, a sound signal that reaches us directly from a source is followed by a train of its reflections from the walls, floor, ceiling, and other objects. The wave patterns of those reflected signals are similar to the original one, but their intensities are smaller and they usually decrease with the delay of their arrival (Figure 10.9). When the reflected signals arrive close to each other, the brain perceives all of them as one continuous decaying signal. This is usually the case, unless the delay between consecutive signals is greater than 50-100 milliseconds, in which cases echoes are perceived.

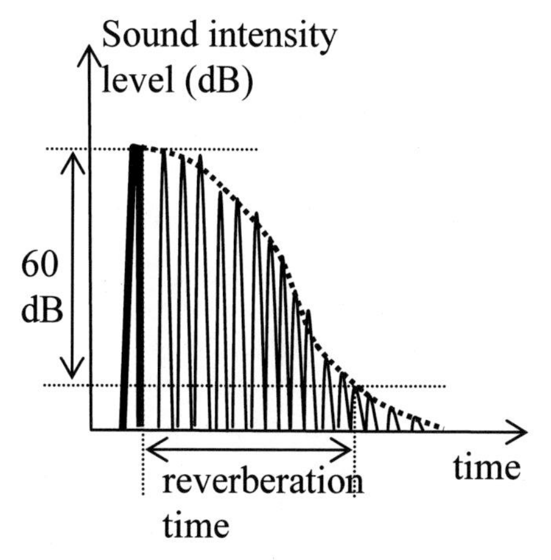

Figure 10.9: Reverberation time of a sound pulse. The original signal (thick line) is followed by a train of its reflections. The time interval between the end of the original signal and the arrival of a reflection that is 60 dB weaker is called the reverberation time. This train of reflections is perceived as one decaying sound (dotted curve).

In a room, signals are re-reflected many times by the walls, floor, ceiling, and other objects. In each reflection, the intensity of the signal decreases due to absorption by the reflecting surface and the geometric spreading of the waves. Eventually, the re-reflected signals are too weak to be recognized by the brain. The time interval between the arrival of the end of the original signal and the arrival of a re-reflection whose intensity level is 60dB below that of the original signal is called the reverberation time (RT) (This 60 dB is the same as intensity reduction by a factor of one million (10^6)). The intensity loss depends on the absorption coefficients of the reflecting surfaces and the volume and geometry of the space. If the absorption coefficients are large, the energy loss in each reflection is large, and it takes fewer reflections for the original signal to loose 60 dB. Thus, the reverberation time is shorter when the absorption is stronger and vice versa; if the surfaces are good sound reflectors, the reverberation time is longer. Another factor that affects the reverberation time is the mean time between reflections. The larger the mean distance that the sound travels between reflections, the longer is the travel time, and the longer is the reverberation time. In general, by increasing the size of the room without changing its reflection properties, the reverberation time increases. All those features have been combined together in a practical formula by Wallace C. Sabine (1868-1919). Sabine's formulas [10.5] provide an estimate of the reverberation time RT in a room, based on its volume V, the areas of its reflecting surfaces $A_1, A_2, \ldots, A_n$, and the absorption coefficients of those surfaces $\alpha_1, \alpha_2, \ldots, \alpha_n$.

$$RT = \frac{0.16V}{\alpha_1 \cdot A_1 + \alpha_2 \cdot A_2 +,\ldots,+ \alpha_n \cdot A_n} \qquad [10.5a]$$

where RT is in seconds, V in m^3 and A_i in m^2.

$$RT = \frac{0.05V}{\alpha_1 \cdot A_1 + \alpha_2 \cdot A_2 +,\ldots,+ \alpha_n \cdot A_n} \qquad [10.5b]$$

where RT is in seconds, V in ft^3, and A_i in ft^2.

Equations [10.5] do not include attenuation by the air inside the room. This effect is negligible in small rooms, but a correction term for it has to be included for large rooms.

The quantity $\alpha \cdot A$, which is called "sound absorption", provides the relative total absorption of a surface, as compared with the total absorption of a standard surface area. The unit of "sound-absorption" is the meter-sabin if A is expressed in m^2, and foot-sabin if A is expressed in ft^2. One foot-sabin is the absorption provided by a surface whose area is 1 ft^2 that completely absorbs sound (α=1), or by a surface whose area is 2 ft^2 and whose absorption coefficient is α=0.5, etc. A surface whose area is 20 m^2 with absorption coefficient 0.4 has sound absorption of 8 meter-sabin. An open window of area 1 ft^2 has sound absorption of 1 foot-sabin, because it does not reflect any sound. The denominators in equations [10.5] are expressed in sabins.

Example
Find the sound absorption (in meter-sabin and foot-sabin) of a 6mx2m brick wall for sound waves of frequency 2,000 Hz.

Solution
Sound absorption is defined as $\alpha \cdot A$, where α is the absorption coefficient and A is the area of the absorbing surface. According to table 10.3, α=0.03. The area of the wall is $A=12m^2=133ft^2$. The sound absorption of the wall is 0.36 meter-sabin=4 foot-sabin.

Sabine's formulas say nothing about the shape of the room or where the absorption is placed within it. These formulas are very useful as quick calculations, but if more information is needed for critical or otherwise unique spaces, computer modeling and even scale acoustic models can be used to get a better prediction of the performance of a design.

The optimal reverberation time for a room depends on its intended use. When speech comprehension is important, such as in classrooms, lecture halls, and theaters, the accepted reverberation time is 1 second. If the RT is much higher than this, then speech can become difficult to understand due to the prevalence of loud reflections. If it is much lower, then the room can feel eerie because the surfaces absorb almost all sound and give very little back as reflections. Reverberation times of 1.4±0.2 seconds are accepted for multipurpose auditoriums and contemporary churches, but churches with organs do well with much higher reverberation times because organ music was composed to be performed in very large reverberant cathedrals. This creates a substantial challenge since speech intelligibility remains very important in such spaces--something which requires a low reverberation time. Reverberation times of major opera houses are from 1 second to 1.8 seconds (e.g. La Scala, Milano 0.95 sec; Covent Garden, London 1.1 sec; National Theater, Taipei 1.4 sec; Metropolitan Opera House, New York 1.8 sec). Reverberation time of major concert halls are around 2 seconds (e.g. St. Andrew Hall, Glasgow, 2.2 sec; Concertgebouw, Amsterdam, 2.2 sec; Symphony Hall, Boston, 1.8 sec). Reverberation times of cathedrals may be as high as 3 second or more. Those are reverberation times for the 500-1,000 Hz range. The desired reverberation time can be achieved by selecting the sound absorption properties of surfaces in the room, according to formulas [10.5]. By placing or removing temporary absorption surfaces, e.g. like the adjustable absorptive banners at Strathmore, it is possible to adjust the reverberation time of a given room that is used for different purposes at different times, e.g. amplified and classical music, which require low and high reverberation times, respectively. When the room has to be used for both music and speech at the same time, for example in cathedrals, the use of line array loudspeakers, which have very tight dispersion on the vertical axis and therefore prevent sound from energizing reflective upper volumes, can help to resolve problems of unintelligible speech.

10.4.4 Transmission and attenuation of sound

When dealing with noise control in a room, one of the factors is outside noise transmitted through the walls to the room. Part of the sound that hits the wall from the outside is reflected back to the outside. The sound that enters the wall is the refracted part of the outside sound that hits the wall. The transmission coefficient τ is the ratio between the intensity of the sound that comes out of the wall and the intensity of the sound that entered the wall.

$$I_{transmitted} = \tau \cdot I_{entered} \quad [10.6a]$$

For example, if an intensity of $3x10^{-6}$ watt/m^2 entered a wall whose τ is 0.01, the intensity of the transmitted sound would be:
$I_{transmitted} = 0.01x(3x10^{-6}) = 3x10^{-8}$ watt/m^2.

In order to describe transmission by sound intensity levels (IL), which are expressed in dB, instead of by intensities (I), which are expressed in watt/m^2, we use the concept of "sound transmission loss" (STL). STL is the loss in dB to the entering sound intensity level. The sound transmission loss is also called the sound attenuated.

$$IL_{transmitted} = IL_{entered} - STL \quad [10.6b]$$

The relationship between the transmission coefficient τ and the sound transmission loss (STL) is given by

$$STL = 10 \cdot \log\left(\frac{1}{\tau}\right) \quad \text{or} \quad \tau = 10^{-0.1 \cdot STL} \quad [10.6c]$$

For example, when τ=0.01, STL=10·log(1/0.01)=10·log(100)=20dB. The entering wave looses 20dB as it passes through the obstruction, which corresponds to a reduction by a factor of a 100 of its original intensity. The higher the STL, the better the wall is at stopping sound from being transmitted.

It should be noted that α, ρ, and τ multiply intensities (watt/m^2) and not intensity levels (dB).

Sinusoidal sound waves, like sound waves of any other shape, can be reflected, absorbed, and transmitted by various mediums. For most materials, the coefficients of reflection, absorption, and transmission (ρ, α, and τ) depend on the frequency of the incident sinusoidal wave. Tables 10.3 and 10.4 provide values of those parameters for some materials for various sets of frequency bands.

In order to calculate the intensity of a reflected, absorbed, or transmitted sound wave of an arbitrary shape, the wave is first decomposed into its Fourier components, which are then grouped into bands. The outcomes of the process (reflection, absorption, or transmission) are evaluated separately on the sound intensity (watt/m^2) for each band, and then those affected intensities are added together.

Assembly\Frequency (Hz)	STC	125	250	500	1,000	2,000	4,000
½" gb.each side of 2-1/2"studs,24" oc,1.5"FG	45	22	38	51	57	54	46
Hollow 8" lightweight concrete block	44	32	36	42	47	47	50
6" solid concrete	56	43	49	53	56	62	65
8" solid concrete	58	43	48	55	58	63	67
¼" glass	31	25	28	31	34	30	37
Insulating glass unit ¼",lam.-3/8" air sp 3/16"	37	27	22	35	39	43	52

Table 10.4: Sound Transmission Loss of some materials. STC=sound transmission class, which represents the overall transmission loss.

Consider a sound beam, consisting of two bands, hitting a wall. Let the beam intensities be I_1 and I_2 and the transmission coefficients of the wall for those bands be τ_1 and τ_2. The total transmitted intensity (I) would be the sum of the individually transmitted bands: $I = \tau_1 \cdot I_1 + \tau_2 \cdot I_2$. Similar approach is used for calculating the intensities of the reflected and absorbed waves. It should be noted that sound intensities can be added up, but sound intensity levels, which are expressed in dB, are not added up directly and must be converted to sound intensities first, because of their logarithmic nature.

Example

A sound beam that hits a ¼" glass wall consists of 70 dB of 125 Hz and 60 dB of 1,000 Hz components. What are the total intensity levels of (a) the incident and (b) the transmitted beam? (Neglect sound reflection by the glass wall).

Solution

Following the notations of formulas [10.3], (or directly from table 10.1) we have: IL_1=70 dB, $I_1=10^{-5}$watt/m^2; IL_2=60 dB, $I_2=10^{-6}$watt/m^2.

(a) The total intensity of the incident beam $I_{incident}=10^{-5}+10^{-6}$ $=1.1x10^{-5}$watt/m^2, which is practically 70 dB.

(b) Based on table 10.4, STL_1=25 dB and STL_2=34 dB. Therefore, for the transmitted beams we have $IL_{1transmitted}$=70-25=45 dB and $IL_{2transmitted}$=60-34=26dB.

The intensities of the transmitted beams are, based on [10.3b]: $I_1=10^{-12}x10^{45/10}$= $3.2x10^{-8}$watt/m^2. $I_2=4x10^{-10}$ watt/m^2. The total intensity of the transmitted beam is approximately $3.2x10^{-8}$watt/m^2. This corresponds (based on [10.3a]) to intensity level of 45 dB of.

Paths of transmitted sound

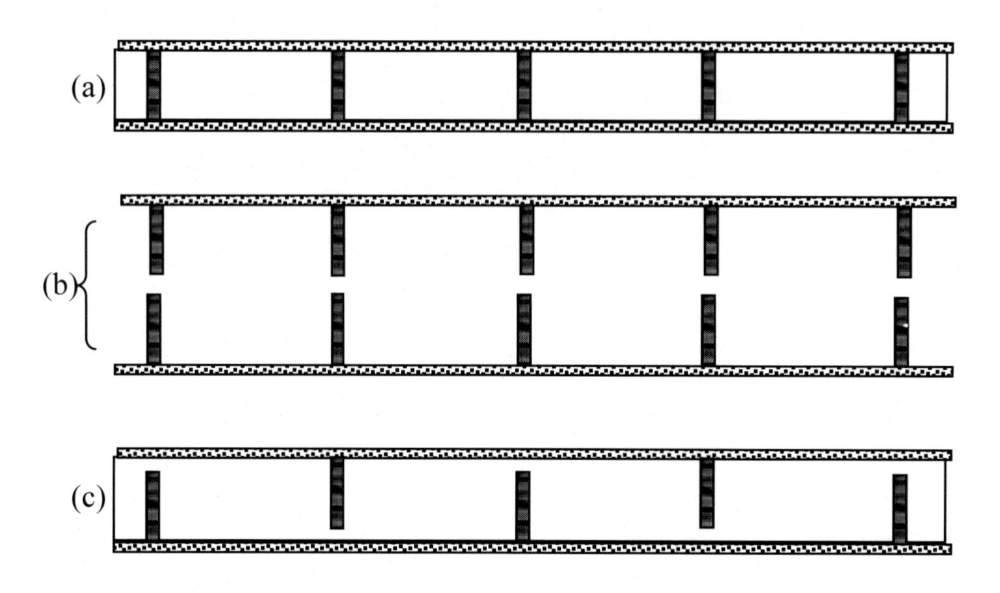

Figure 10.10: Isolation of rooms by walls. (a) Cross section of a wall consisting of two boards and a frame of studs. (b) Each board has its own frame, separated by a gap. (c) The boards are connected to alternate studs.

In many situations it is needed to isolate a room from outside noise. When sound waves in air hit a plate of any material, they cause vibrations in that plate. Those vibrations propagate to the other side of the plate and cause the air on the other side to vibrate, thus sound has been transmitted to the other side of the plate. If a wall consists of different elements (for example one layer of gypsum board connected to a frame of wood studs that are connected to another layer of gypsum board on the other side, figure 10.10a), sound may propagate through the wall using several pathways. Vibrations of the first layer of gypsum board will cause vibrations in the studs and in the air cavity between the boards. Both those vibrations will cause vibrations in the other layer of gypsum board. In order to reduce sound transmission through that wall, both pathways have to be attenuated. Transmission by the air cavity could be reduced by filling it with an insulator such as fiberglass. Transmission through the wood studs could be reduced by creating gaps between them and the layers of gypsum board. This could be accomplished in two ways: first (figure 10.10b), by having two detached wood frames, each supporting its own layer of gypsum board; second, by widening the space between the boards and alternating the studs between them (figure 10.10c). These arrangements will acoustically decouple the external surfaces of the wall. Because of the larger air gap in 10.10(b), it will perform better than 10.10(c), and both (b) and (c) will outperform (a), but both at a cost of lost floor area since the walls are thicker. Although (b) and (c) are stronger walls than (a), they are still too weak to stop all sound, especially low frequency sound which requires massive systems for adequate isolation. To further improve the performance of wall systems such as (b) and (c), multiple layers of gypsum board can be added, with staggered seams to evenly distribute weak points. Wall systems with as many as 5 layers of gypsum board on one side and 4 layers on the other are not uncommon when building a room which needs to be really well isolated. The dissimilar numbers of layers improves performance by staggering resonant weaknesses (the weakness of 5 layers is different from the weakness of 4 layers).

Although these de-couplings attenuate sound transmission through the wall, sound may still bypass the wall through the floor, the ceiling, open windows and doors, and air conditioning ducts. All possible paths that the sound may use have to be considered when isolating a room.

10. 5 Interference

10.5.1 Interference of waves

Imagine floating on an inner tube on a quiet lake. Then a jet-ski passes nearby. After a short while, waves from the wake of the jet-ski reach the tube, causing it to bob up and down at a certain amplitude. Now, a second jet-ski passes by, and waves from its wake reach the tube, as waves from the first jet-ski are still arriving. According to the superposition principle, the tube's oscillations would be affected by the two waves. If the waves reinforce each other, the amplitude of the tube's oscillation would be greater than each of the individual amplitudes separately. If the waves weaken each other, the combined amplitude would be smaller than the individual amplitudes.

Constructive interference is a term used when two waves of the same wavelength superimpose and reinforce each other throughout their entire wave cycle. Destructive interference is a term used when two superimposed waves of the same wavelength act against each other throughout their entire wave cycle. All other cases of superposition of waves of the same wavelength are called partial interference. In partial interference, the intensity of the superimposed wave is between that of constructive interference and destructive interference, depending on the portions of the cycle when the waves act with each other and against each other. The wavelength of any interference wave is the same as that of the waves that make it. Based on [10.1], waves that have the same wavelength also have the same frequency.

The phase difference between two waves of the same wavelength does not change as the waves keep passing through a common point (e.g. a float). Whether the interference between the waves at that point would be constructive, destructive, or partial depends on the phase difference of the individual waves at that point. As can be seen in figure 10.11a, if the phase difference between the two waves at their meeting point is zero, the waves will undergo constructive interference: They lift up the float together and pull it down together. If the phase difference between the two waves is half a wavelength (figure 10.11d), which is equivalent to a phase difference of 180^0, the waves will undergo destructive interference. The two waves will oppose each other throughout their entire cycle; when one wave tries to lift up the float the other tries to pull it down, and vice versa. If the two waves have the same amplitude, they will completely eliminate each other. Figures 10.11b and 10.11c illustrate two cases of partial interference. The amplitude of the superposition decreases as the phase difference increases from zero in 10.11a to 180^0 in 10.11d.

Constructive interference also occurs when the phase difference between the two waves is one wavelength, two wave lengths, and so on. Destructive interference also occurs when the phase difference between the two waves is one and one half wavelength, or two and one half wavelengths, and so on. This is the case because figures 10.11 also apply in those situations. In all other cases, the interference would be partial. Formulas [10.7] summarize the conditions for constructive and destructive interference of waves of the same wavelength λ,

$$\Delta L = \frac{1}{2}\lambda, \frac{3}{2}\lambda, \frac{5}{2}\lambda, \ldots \quad \text{destructive}$$
$$\Delta L = 0, \quad \lambda, \quad 2\lambda, \quad 3\lambda, \ldots \quad \text{constructive} \qquad [10.7]$$

where ΔL indicates the path difference of the two waves.

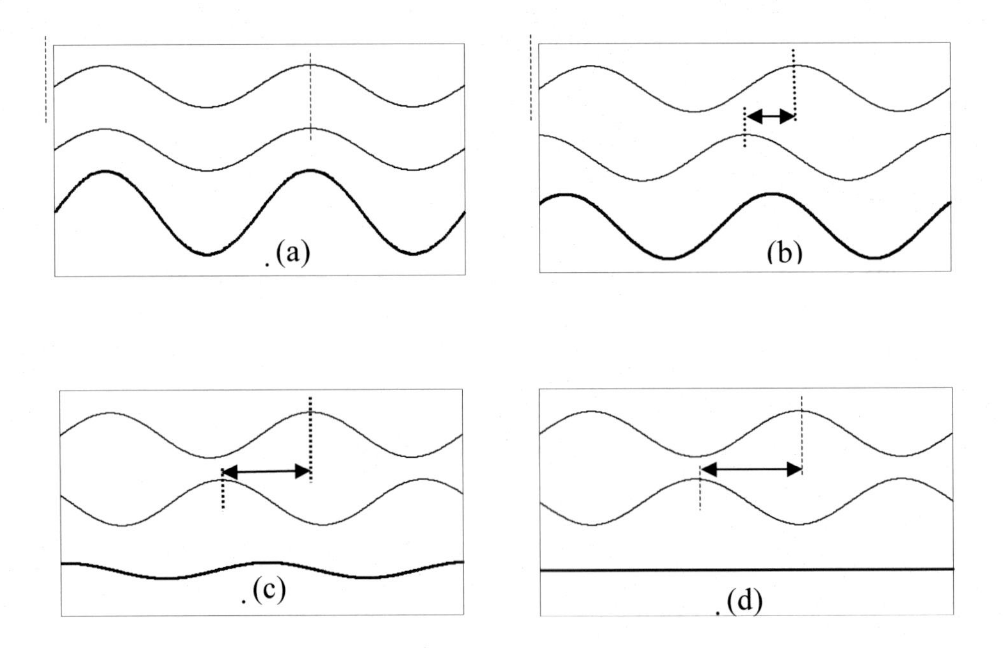

Figure 10.11: Interference types. Two waves of the same wavelength and amplitude (thin lines) interfere (thick lines). The type of interference depends on the phase difference between the waves (arrow). Same phase (a) results in constructive interference. Phase difference between zero and half a wavelength ((b) and (c)) result in partial interference. Phase difference of half a wavelength (d) results in destructive interference. The transition from (a) to (d) is gradual. The same patterns repeat themselves as the phase difference increases beyond one half of a wavelength. The two interfering waves and their superposition have the same wavelength.

If many inner tubes are floating on that lake, it is most likely that some would experience constructive interference, some destructive interference, and the rest would experience partial interference. It would all depend on the phase difference between the superimposed waves as they interfere at the location of each tube. Those phase differences would depend on when each wave has started and on the distance that it had to cover in order to reach the interference point.

Sustained destructive interference occurs only if the two waves act against each other throughout their entire presence. In practice, that may happen when a wave is split into two parts, and each part reaches the interference point with the appropriate delay of formula [10.7]. Such waves are called coherent. A radio that is connected to two speakers can send the same signals to both of them. Those signals can interfere destructively and constructively at different points of the room, depending on the phase differences, according to [10.7]. If speakers are connected with incorrect polarity (the + and -, or red and black, are switched) then signals from each channel will be out of phase with each other. This causes the left and right channels to move in opposition to each other, and the best seat from which to listen will become one of the worst.

In a superposition of waves of different wavelengths, the waves reinforce each other part of the time and weaken each other part of the time. Those situations, though, are not included in what is defined as interference.

10.5.2 Un-ringing a bell

It is said that "you cannot un-ring a bell". Well, now it is becoming possible to do so, at least in part. A sensor senses the pressure pattern of the sound wave that the bell (or any other source) is generating. An electronic circuit activates a vibrator at a point some distance away from the source, so that the pressure pattern that it creates in air is in opposite phase to the anticipated pattern of the bell's sound, once it arrives there. Each positive pressure of the bell is matched by an identical negative pressure, and each negative pressure of the bell is matched by a positive. The sound waves from the bell and from the vibrator interfere destructively, and both sounds are eliminated. It is possible to create this destructive interference, thus "un-ring the bell", only in some regions of the space; bell ringing will be heard in the other regions. This principle is already being implemented in a variety of active-noise-control application. They are used in reducing noise in air-conditioning ducts. A sensor senses the noise at one point in the duct, and an electronically controlled loud speaker emits the anti-phase signals further down the duct, so that the original noise and the matching sound from the loudspeaker destruct each other where they interfere. The same concept is applied to noise canceling headphones, which claim to block noise for the listener. These active approaches are most effective when used on steady state sounds such as the hum of a jet engine heard from within the cabin, and are not as successful with transient signals, such as the baby crying next to you. For general noise reduction, it is still hard to beat a pair of closed-backed headphones which surround the ear in a snug fit.

10.5.3 Standing waves and resonance

At the beach, the crests of the waves travel towards the shore, and eventually reach it. On the other hand, the crests of the waves in a guitar's string oscillate in one place, and never reach the ends of the string. Ocean waves are an example of traveling waves, and waves in a guitar's string are an example of standing waves. The entire shape of a traveling wave, including its crest, travels at the wave's velocity, while the entire shape of a standing wave, including its crest, stays in one place, just oscillating there. Figure 10.12 illustrates the oscillations of a standing wave. The amplitude of those oscillations changes from point to point. There are points on the wave that do not oscillate at all. These points are called nodes. The points where the amplitude of the oscillations is the highest –the crests of the standing wave – are called anti-nodes.

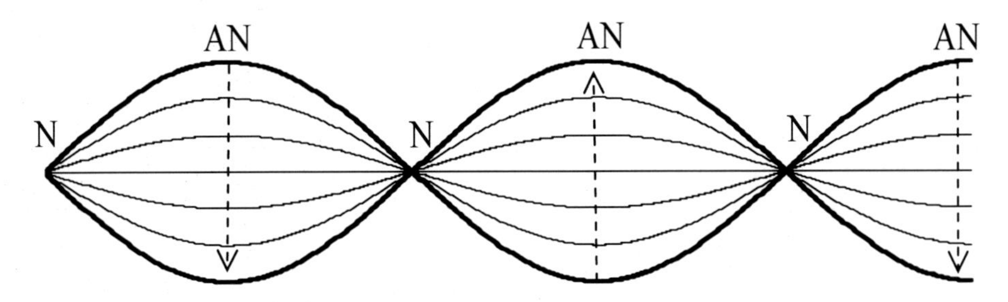

Figure 10.12: Oscillations of a standing wave. Nodes are indicated by N and antinodes by AN. Seven "snapshots" of a wave, taken during the first half period are shown. The directions of the arrows indicate the progression of the snapshots. The two end points of a standing wave may be two nodes, two anti-nodes, or a node and an anti-node.

When two traveling waves of the same wavelength and the same amplitude travel in the same place but in opposite directions, their superposition creates a standing wave. Some features of the standing wave are the same as those of the traveling waves, and some are different. The wavelength and the frequency of a standing wave are the same as those of the traveling waves that created it. The relationship between f, the frequency of the standing wave and λ, its wavelength, is given by $c=f\lambda$ (equation [10.1]), where c is the velocity of the traveling waves. The velocity of the standing wave itself is zero. The amplitude of a standing wave (the highest value of its crest) is twice the amplitude of the traveling waves that created it.

If two waves of different amplitudes but the same wavelength travel in opposite directions, the outcome of their superposition is a standing wave plus a secondary traveling wave. The secondary traveling wave is the "left-over" from the larger traveling wave, which is not used in the interference that created the standing wave.

When a sound wave in a rectangular room hits a wall perpendicularly, the reflected wave travels in a direction opposite to the incident wave and interferes with it. Since the two waves have the same wavelength and they are traveling in opposite directions, they create a standing wave. Reflected waves from one wall eventually reach the opposite wall and are reflected back from it, creating another standing wave. So, many standing waves are created in a rectangular room by the interference of waves that bounce back and forth between opposite walls. The same is true for the floor and ceiling path. If appropriate relations exist between the size of the room and the wavelength of the bouncing sound waves--a frequent occurrence for low frequencies in small rooms where the small room dimensions are on the same order as the wavelengths of sound--the multitude of standing waves interfere constructively with each other, and the amplitude of the resulting standing wave is much larger than the amplitude of the traveling waves that were emitted by the source. This is a resonance process and is of great concern in the design of recording studios and critical listening rooms such as home theaters. Uneven bass response can result, which means that some areas of the room have exaggerated low frequencies and other areas have too little low frequency energy. Mix engineers then under or over compensate for this room problem and recordings can sound uneven as a result. Oscillations of small amplitude (the original traveling waves) interact with a system that can oscillate at the same frequency (the air in the room that can oscillate as a standing wave). Energy is transferred from the small oscillations (the original sound) and builds up larger oscillations of the system (the standing waves).

This process is similar to the resonance of a vibrating spring or a pendulum, which were discussed in chapter 2. The main difference is that standing waves of a variety of wavelengths and frequencies can exist in the same room, while a pendulum or a spring-mass-system has only one frequency. The wavelengths and frequencies of standing sound waves that resonate between two parallel walls are given by:

$$\lambda = \frac{2L}{n}$$

or

$$f = \frac{c}{\lambda} = n\frac{c}{2L} \qquad [10.8]$$

where $n = 1,2,3,....$

c, is the speed of sound, and L is the distance between the walls.

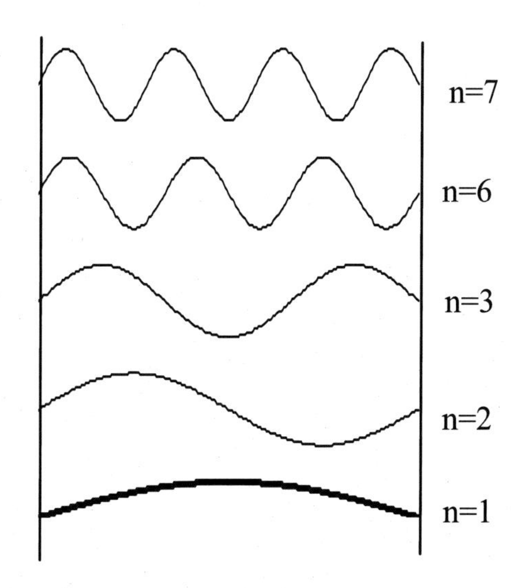

Figure 10.13: The fundamental mode (thick line at the bottom) and some of its harmonics.

The longest of those wavelengths is λ=2L. That means that half a wavelength fits between the two parallel walls. At the walls, the molecules of the air cannot vibrate. Therefore, when describing the sound wave by the velocity pattern of the air molecules, the nodes of that half wavelength will be at the walls, and the anti-node will be midway between the walls. This wave is called the fundamental mode, and its frequency (c/2L) is called the fundamental frequency. The waves that are characterized by higher n's in [10.8] are called harmonics of the fundamental wave. The frequencies of the harmonics are integer multiples of the fundamental frequency. Each harmonic consists of an integer number of half wavelengths of the fundamental standing wave (see figure 10.13).

Sound waves in a room can hit walls obliquely. They are then reflected and re-reflected by more than two walls. A variety of standing waves are generated by the interference of those waves. Each such wave is called a normal mode, and its frequency is called a normal frequency. The normal frequencies that may exist in a rectangular room whose dimensions are X by Y by Z are given by:

$$f = \frac{c}{2}\sqrt{\left(\frac{n_x}{X}\right)^2 + \left(\frac{n_y}{Y}\right)^2 + \left(\frac{n_z}{Z}\right)^2}$$

where [10.9]

$$n_x = 0,1,2,... \quad n_y = 0,1,2,... \quad n_z = 0,1,2,...$$

When two of the n's in [10.9] are equal to zero, equation [10.9] becomes [10.8].

Rooms of other shapes have their own sets of normal modes and normal frequencies. In general, if the walls of a room are not parallel, it will have a smaller number of possible standing waves in the audible range. Openings in the walls change the normal frequencies and the shape of the normal modes.

If some of the Fourier components of an indoor sound source have frequencies that match the normal frequencies of the room, they will resonate. For example, a subwoofer which produces very low frequencies in a typical sized living room. The intensity of the standing waves of the resonating frequencies will increase, while other frequencies will be unaffected. That will be most noticeable during the reverberation period. As a result, the original sound will be distorted. Such distortions may be pleasant or unwanted, just as we stated in a recording studio. As a positive example, the enclosure of a shower is made of sound reflecting surfaces and the dimensions of the enclosure are such that the lower frequencies of our voice resonate, while the higher frequencies do not resonate but are sustained by the reflective materials in the shower. This audible boost to the low frequencies makes our voice sound "fuller" and more self-enjoyable, and the reflective materials tend to make our voice sound stronger, so the entire experience is enjoyable enough that many people like to sing in the shower even though they do not like to sing in other room. However, what is enjoyable in a non-critical location such as a shower can be very detrimental to a critical location like a recording studio.

10.6 Diffraction of sound

10.6.1 Huygens' Principle

When a pebble is dropped into a pond, the disturbed water at the pebble's hitting point starts to oscillate up and down. These oscillations cause the adjacent water to oscillate up and down too. Thus, the water that was directly hit by the pebble became a wave source. That causes adjacent water points to oscillate, thus making them also sources of new waves, and so on. Huygens' Principle (Christian Huygens, 1629-1695) states that every point on a wave-front becomes a source to a new wavelet. In 3-D the wavelet is spherical and in 2-D it is circular. The envelope of all those new wavelets forms the new wave-front. Huygens' Principle applies to all kinds of waves, including sound waves. It provides a handy way of figuring out the shape of a wave front when the shapes of the wave source and the obstructions in the wave's path are known. Figure 10.14 illustrates this process in two types of waves in two dimensions: circular waves on the left, and linear waves on the right. At left, the wave starts at a point (point source), and the wave front that propagates from it has the shape of a circle. At the right, the wave starts as a line, and the segment of the wave-front that propagates away from that line has a straight-line shape. To the sides of the straight line, the wave-front is curved. Such waves can be observed in water. A circular wave is formed by a pebble and a linear wave by a straight stick thrown into the water. In three dimensions, the wave fronts of point sources (such as the

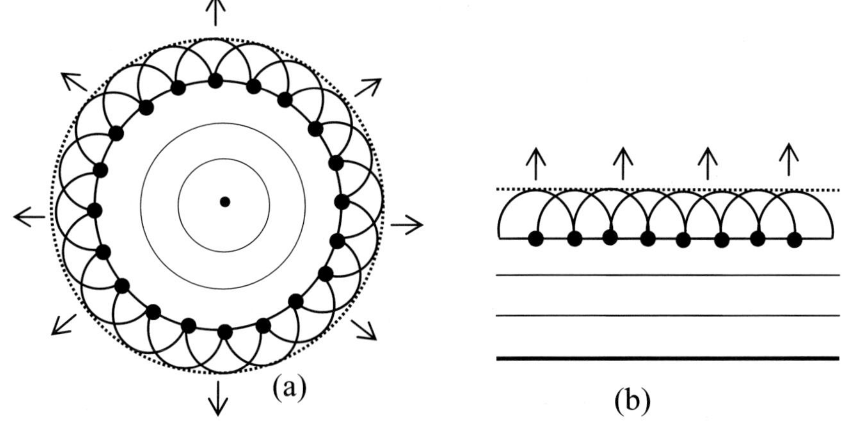

Figure 10.14: Huygens' Principle. Points on the current wave front (black dots) act as sources of circular wavelets in two dimensions (spherical wavelets in three dimensions). The envelope of those wavelets (dashed line) is the new front of the wave. In (a), the original source of the wave is the small dot at the center, and a circular wave is formed. In (b), the wave starts at a line and a linear wavefront is formed. In both cases, two previous wave fronts are also shown (thin lines). The arrows show the direction of propagation of the wave. In three dimensions, (a) describes spherical waves for a point source and a cylindrical wave for a line source perpendicular to the paper. (b) describes a cylindrical wave for a line source, and a plane wave for a plane source, (perpendicular to the paper).

sound of an exploding fire-cracker in mid air) are spherical. The wave fronts of line sources are cylindrical (such as the noise coming from a busy straight highway). Small segments of spherical and cylindrical wave fronts can be approximated as plane waves. At great distances from their sources, small portions of spherical and cylindrical wave fronts become practically plane waves. In all those cases, the direction of the wave propagation is perpendicular to the wave front.

10.6.2 The diffraction process

One basic property of waves is their ability to go around obstructions. This is called **diffraction**. For example, the sounds of a radio spread through an open window all over the street, even to points from where the view of the radio is blocked by the walls. Huygens' Principle provides a general explanation to the mechanism of diffraction. Figure 10.15 illustrates how a plane wave diffracts through an opening in an obstruction. As the plane wave hits the obstruction, the point on the wave front that reaches the opening acts as a source of a spherical wave, which diffracts through the opening, and propagates in all directions behind the obstruction.

Because of diffraction, a sound beam expands after it has passes through an opening (e.g. sounds from a radio that passes through a window). Diffraction also affects reflected sound beams. Sound waves that are reflected from surface elements expand and can reach regions beyond those prescribed by the θ_R of the law of reflection (figure 10.7). The extent of the diffraction depends on the wavelength of the sound. For a given obstruction, the shorter the wavelength the smaller is the extent of the diffraction. Figure 10.15 illustrates sound diffraction

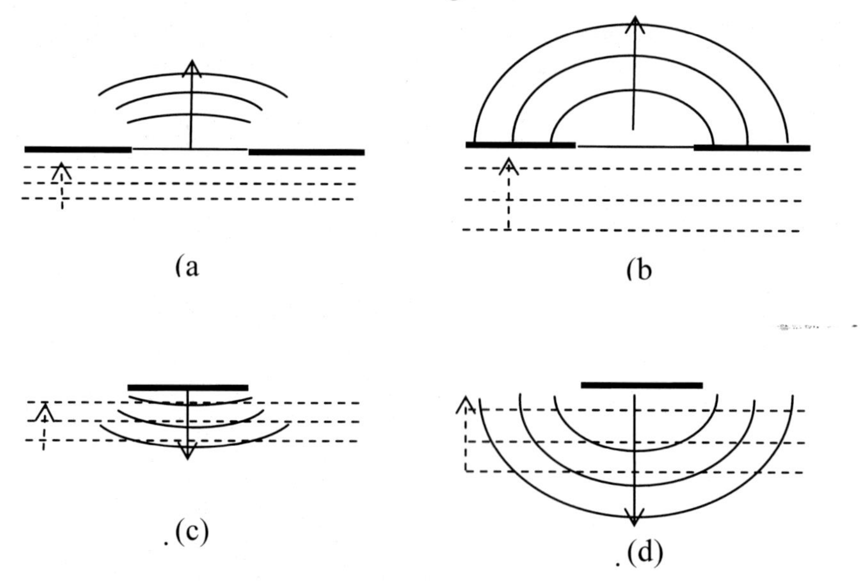

Figure 10.15: Diffraction of sound through openings ((a) and (b)) and in reflection ((c) and (d)). Dotted lines indicate wave fronts before the diffraction; arrows indicate direction of sound. Short wavelength: (a) and (c), long wavelength: (b) and (d).

when two waves of two wavelengths pass through openings and when they are reflected by an obstruction. Everyone has experienced the sound of music from outside a building where it is being produced. It is usually characterized by low frequency sound, which makes it sound muffled. This is primarily the result of two things: the first is the relative ease at which high frequency sounds are isolated as compared to low frequency sounds, and the second is the strength of diffraction of low frequency versus high frequency sound. Both of these make it much easier for low frequencies to find their way out of a room and to the ear of a passerby.

According to Huygens' Principle, each point on a wave front acts as a source of a spherical wavelet. The envelope of those wavelets forms the new wave front. This mechanism provides qualitative explanations for diffraction processes. However, it does not provide a quantitative explanation for the intensity of the diffracted waves. In order to figure out that intensity at any given point, it is necessary to consider the superposition of all the diffracted wavelets that arrive there. The intensity at a point would depend on whether the interference there is constructive, destructive, or partial. Calculating the intensity of a diffracted sound at different points may get complicated. The following is the simplest diffraction situation. It gives us a glimpse into various aspects of the process, and can also serve as a starting point for more complex diffraction cases.

Diffraction in a double slit

Consider a plane wave that arrives at two slots in a wall. The wall is parallel to the wave front. Figure 10.16 illustrates what happens in a cross sectional plane (the slots, the wave fronts, and the wall are perpendicular to the plane of the paper). The two wavelets that start from the slots have the same wavelength, frequency, and phase as the original wave front that caused them. At any point on the center line between the slots, the two wavelets interfere constructively, because they arrive there with the same phase. The further to the side that we move from the center line, the greater the phase difference between the two wavelets, because of the greater difference between the distances that the wavelets had to travel to get there. As a result, the two wavelets interfere there partially. Their resultant intensity decreases gradually as we move away from the center line. As can be seen in figure 10.16, eventually we will reach a point where the wavelets interfere destructively. As we keep moving, the wavelets interfere partially. This same thing happens on both sides of the center line. The lines where the two waves interfere destructively are called the first nodal lines (figure 10.16). (If we consider the three dimensional situations, those lines become planes perpendicular to the paper). The two wavelets will also interfere constructively a second time, at lines that are called the first anti-nodal lines.

Figure 10.16 illustrates the situation where the wavelength λ is equal to the distance d between the slots (λ=d). Will the same interference-patterns occur for other relationships between λ and d? It is clear that at the center line the two wavelets will always interfere constructively, regardless of the relationship between λ and d, because of the equal distances traveled by the left and right rays. Will there always be a nodal line? For the first nodal line to occur, the path difference between the two wavelets that intersect at the nodal line should be half a wavelength ($\lambda/2$). There is a theorem in geometry stating that the difference between the lengths of two sides of a triangle is shorter than the third side. Applying this theorem to the double slit configuration implies that the difference between the distances covered by the two intersecting wavelets is always shorter than the distance d between the slits. Therefore, if $\lambda/2>d$, two intersecting wavelets will never interfere destructively; sound levels everywhere behind the two slits will be higher than zero.

If now the entire segment between the two slits is cut out, each point on that opening becomes a source of a wavelet that propagates to the other side. Those wavelets will propagate to the entire space behind the opening, without forming destructive interference at any point there. This is because any two of those wavelets will satisfy the condition $\lambda/2>d$, which means no destructive interference between the wavelets. This is something that we can notice on a daily basis. The wavelength range of human voice is roughly from 6 cm to 17 m. Voice that passes through small opening fans out, and we can clearly hear it, even if we are not on a straight line to the source.

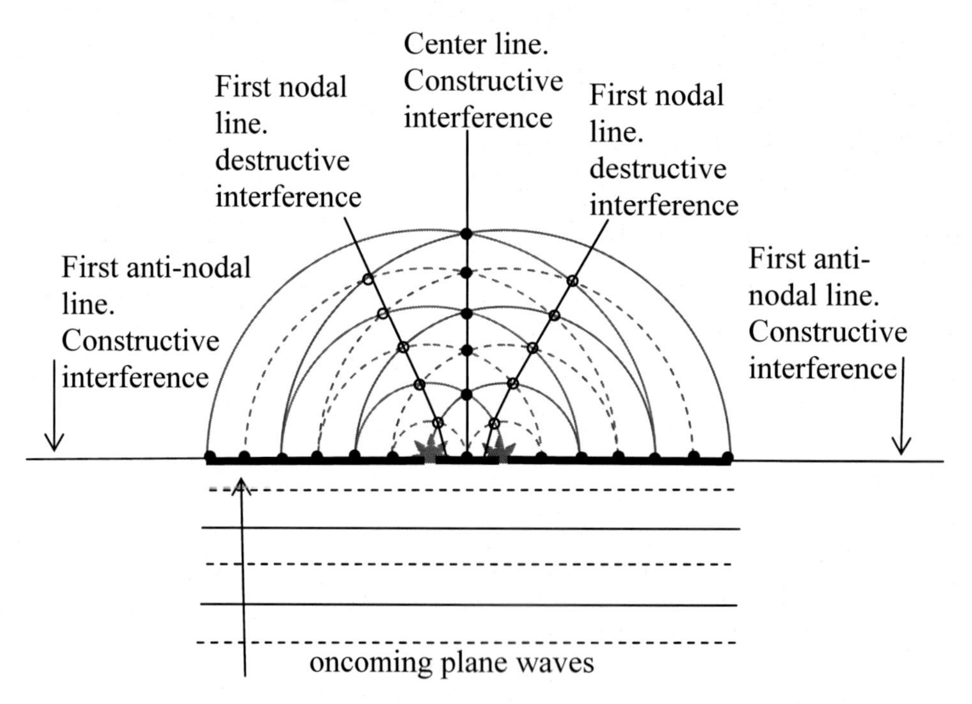

Figure 10.16: Interference pattern of a double slit (λ=d). Solid line-crests; dashed lines-troughs.

Similarly, a reflecting surface whose width is smaller than $\lambda/2$ will reflect an incident parallel sound beam in all directions, and not according to the reflection law as illustrated in figure 10.7. In general, any opening and any reflecting surface has a typical dimension D. When sound passes through an opening or is reflected from a surface, and the sound's wavelength satisfies $\lambda>D$, there will be significant diffraction, as illustrated in figures 10.15. If $\lambda<D$, there could be

destructive interference between the various wavelets. It was found that for practical architectural applications, if $\lambda<D/5$, that destructive interference will limit the fanning out of the diffracted sound. In those cases, reflected sound will obey the reflection rule (incidence angle equals reflected angle, figure 10.7). Sound rays that pass through an opening will continue in straight lines, without significant diffraction. The smaller the λ, the less significant the fanning out of the sound due to diffraction.

If the sound beam is a superposition of sinusoidal waves of different wavelengths (Fourier components), each such component will have its own wavelets. Since the extent of the diffraction depends on wavelength, different Fourier components will diffract to different extents. In general, long wavelengths diffract more than short wavelengths.

10.6.3 Diffraction in architectural contexts

As was just said, in most architectural applications, it may be assumed that a sound beam practically does not diffract if $\lambda/D \leq 0.2$. In other words, such sound beams that hit an obstruction of dimension D are reflected according to the Law of Reflection (figure 10.7), and such sound beams that pass through an opening of dimension D continue to propagate on their original straight lines. On the other hand, when λ is much greater then D, the extent of the diffraction is large. In those cases the phase difference between the interfering wavelets at any point is small, and the interference is almost constructive everywhere. Those distinctions have to be considered in applications.

Diffraction and Reflection

In a number of architectural applications surfaces are used to redirect sound beams by reflecting them to selected regions. For example, panels are hung from the ceiling of auditoriums and concert halls to direct the sound to audience members and to musicians onstage for communication purposes (it is important for musicians to be able to hear each other, and themselves). Usually, the length and width of such reflectors are at least five times the longest wavelength that has to be reflected. Under such conditions, the sound would be reflected according to the law of reflection (figure 10.7), with negligible diffraction.

Diffraction and Absorption

There are many architectural applications in which it is desirable for sound that reaches a wall to be absorbed by it. In some cases, the purpose of the absorption is to reduce the level of the reflected sound. In other cases, the purpose is to attenuate the sound that passes through the wall. In most cases, some combination of both of these types of absorption is desired, where the reduction of level of reflected sound is intended to control reflections within the space, and a reduction in the level of transmitted sound is meant to protect surrounding spaces. Luckily, the two requirements are not mutually exclusive. In the first case--the reduction of reflected level--a common solution consists of panels of porous materials, such as fiberglass or recycled cotton, covered by some sort of acoustically transparent protective material such as carefully chosen fabric or perforated metal. In many cases where reverberation control is desired, but the potential distortion caused by the treatment to the sound's Fourier components is not critical, such as the case in a banquet hall, covering material which is not completely transparent can be used. This allows designers to choose perforated wood or lattice materials which may be more visually appealing than fabric or perforated metal. Be aware that not all covering materials transmit sound equally, and the more transparent a covering material is, the more sound will have access to the absorptive material behind. A general rule states that the smaller, more numerous, and more closely spaced the holes are, the more transparent the material will be. Dictating percent open alone is not enough. This can be imagined by considering a material of 25% open made of 4 large holes, versus a material of 25% open made of 10,000 closely spaced smaller holes. In the former case, the bridges of material between the four holes are of significant size and will not be transparent to high frequency sound, whereas the latter case minimizes the bridge size and therefore be much more transparent. When choosing a covering material to be placed in front of an absorptive surface, it is helpful to choose as transparent a material as possible, but it is not always necessary and there exists a lot of flexibility to meet both acoustic and visual criteria in many cases..

The combination of a lattice of some material, such as wood, a porous material and a solid panel (often the wall itself), can increase the effectiveness of an absorber of sound. The incident sound passes the lattice and the porous materials on its way to the solid panel behind. It is then reflected or passes through, according to the particular case. By selecting a lattice of the appropriate openings and an appropriate porous material, and by placing them in the appropriate distance from the solid panel, it is possible to exploit diffraction effects to improve the overall absorption properties of the entire ensemble. The openings of the lattice determine the amount of diffraction of the sound that passes through them. The porous material attenuates some of the sound passing through it. The placement of the different layers determines the amount of destructive interference between incident and reflected waves, which will further reduce the intensity of the sound.

The lattice may be shaped as a pegboard, or as a grid of long narrow slots, or a variety of other custom configurations which are now possible with the proliferation of computer numerically controlled (CNC) routing techniques and services. One way to characterize this type of covering is by its visual transparency, which is the ratio between the total area of its openings and its total area. For example, visual transparency of 0.1 means that the area of the openings is 10% of the area of the screen. The visual transparency of the lattice can help to determine what portion of the energy of the incident sound beam passes through if the scale of the lattice and openings are considered carefully relative to the frequencies and wavelengths under consideration. The shape of the waves that have passed through will depend on the size of each opening and the spacing to surrounding openings. Consider the case when the incident sound hits a pegboard in the normal direction. If the size of each opening is much greater than the wavelength, most of the beam that passes straight through the openings will not expand by diffraction. If, on the other hand, the size of the openings is smaller than the wavelength, the original beam will turn into many spherical wavelets. The sound transmission loss in the porous absorber depends on the distance traveled in the absorber (its effective thickness). If the openings of the pegboard are wide, the passing sound travels through the porous material in the shortest path to the wall behind. If the openings of the pegboard are small, the spherical wave fronts of the passing sound have a longer path to transverse till they reach the wall, and therefore they lose more of their intensity. So, a lattice with the appropriate openings size increases the efficiency of the absorbing material behind it for some frequencies by causing the sound to travel longer distances in the absorber. If, in addition, the lattice itself is thick, such as a thick wooden pegboard, there will be additional losses at some frequencies due to air friction with the walls of the openings.

After the sound passes through the lattice and the absorbing material, it reaches the solid panel, and some of it is reflected. In order to maximize the absorption of the reflected sound, the absorbing layer should be placed at that distance from the solid panel where it would be the most effective. In order to optimize this effect we have to consider some details of the absorption process.

In general, sound absorbing materials convert the energy of the incident sound into other forms of energy. Those energies eventually are transformed into heat. Since the energy of a sound wave is small, that heat is unnoticeable. As sound propagates through porous materials, such as fiberglass, kinetic energy associated with the acoustic vibrations of the air molecules is converted into heat due to air friction with the fibers. This conversion is most efficient in regions where the acoustic kinetic energy of the molecules is large. When sound is reflected from a wall, the air molecules that are in contact with the wall do not vibrate. Consequently, the reflected sound wave has a node at the wall. The first velocity anti node of the reflected sound is at a distance of one quarter of a wavelength from the wall. At that distance, the kinetic acoustic energy of the air molecules is the largest. Therefore, if an absorbing panel is placed at a distance of $\lambda/4$ from the wall, its effectiveness in absorbing a reflected sound of wavelength λ will be maximal.

In critical acoustic applications such as concert halls and recording studios, absorptive coverings must be very carefully chosen, because unless they are completely transparent to all frequencies, they cause some reflections of sound energy back into a space. For example, a covering of pegboard consists of a lot of solid surface surrounding the holes. For most frequencies, the sound wave will pass right through, but for some very high frequencies the solid portions of the pegboard become effective reflectors. Imagine a musician playing a violin near a pegboard surface which covers absorption. Middle and low frequencies will pass through the pegboard and be absorbed, while high frequencies are reflected by the solid portions of the surface. This causes an unpleasant and distracting distorted sound to return to the musician. Designers must always be aware of the context in which acoustic materials are being applied.

Another common mechanism of reducing reflected sound is by resonance cavities. A resonance cavity is a chamber that has a small opening in one of its walls. When a sound wave of a certain wavelength, which depends on the dimensions of the chamber, hits the opening, the air inside the chamber resonates. That takes away energy from the incident sound. Those chambers are called resonance cavities or Helmholtz cavities (Hermann von Helmholtz, 1821-1894). In building applications, special cement blocks have narrow slots is one of their faces. When walls are built with those blocks, such that all the slots face the inside of a hall, that wall behaves like a wall made of resonance cavities. It absorbs more sound than a smooth wall. The cavities of the blocks may be filled with porous material to increase the absorption of sound. Some cavities have a metal divider to allow resonances of several wavelengths of sound.

Diffusion by diffraction

In designing auditoriums, it is often desirable to diffuse the sound that is reflected from a wall. This is an effective way to keep sound energy within a room (as opposed to absorption) while lowering its overall level and creating an immersive experience for the listener. The original sound reaches the wall from a well defined direction. In diffusion, that sound is reflected in all directions, thus losing its original directionality. One way of diffusing sound is by devices called Quadratic Residue Diffusers, which are based on diffraction. This diffuser, which is available commercially and relatively easy to construct in a custom manner, consists of arrays of slots (or wells) of the same width and different depths. The slots are separated by rigid walls made of wood or similar materials. The width and depths of the slots, their arrangement, and their number are determined by the range of the sound frequencies that need to be diffused. The width of each slot, d_{slot}, is determined by the highest frequency f_{max} that has to be diffused: $d_{slit} = 0.5\frac{c}{f_{max}} = 0.5\lambda_{max}$. The width of the entire diffuser is determined by the lowest frequency f_0 that has to be diffused. The width of the entire diffuser is at least twice the longest wavelength that has to be diffused. Those two conditions are reminiscent of the condition for the onset of diffraction in the case of a double slot. (It should be noted that the λ_{max} is not the longest wavelength that has to be diffused. Rather, it is the wavelength of the highest frequency). Other formulas determine the optimal number of slots and their depths. These formulas are outside the scope of this book, and they could be found in the professional literature (e.g. Architectural Acoustics; Principles and Design, Appendix C, by Metha et al. Prentice Hall, 1999) and in industrial specs.

Sound barriers for highways

It is now common to build walls or berms along highways in order to reduce the noise caused by automobile traffic to adjoining communities. Such barriers reduce the noise level, but they never eliminate it completely, because of sound diffraction over the top of the barrier. The amount by which the noise is reduced will depend on the topography of the area, the type of the soil and ground cover, the placement of the barrier, the local traffic conditions, and the properties of the barrier. Barrier attenuation refers to the reduction in the noise level due to the insertion of the barrier in free field conditions. There are formulas for barrier attenuation in idealized situations. It is assumed that the barrier is a very long rectangle (semi infinite barrier), and that it blocks all the sound that tries to pass through it. One formula deals with a point sound source and another with a straight long sound source, parallel to the barrier (such as a highway). The formulas are:

$$\begin{aligned} A &= 10\log(20N+3) \quad \text{(point source)} \\ A &= 10\log(20N+3) - (20N)^{0.3} \quad \text{(line source)} \end{aligned} \qquad [10.10]$$

where A is the barrier attenuation in dB, N is Fresnel number, $N = \frac{2f \cdot d}{c}$, where f is the sound frequency, c is the speed of sound, and d is the difference between the distance traveled by the sound from the source to the top of the barrier then to the observer and the direct distance (through the barrier) between the source and the observer (Figure 10.17).

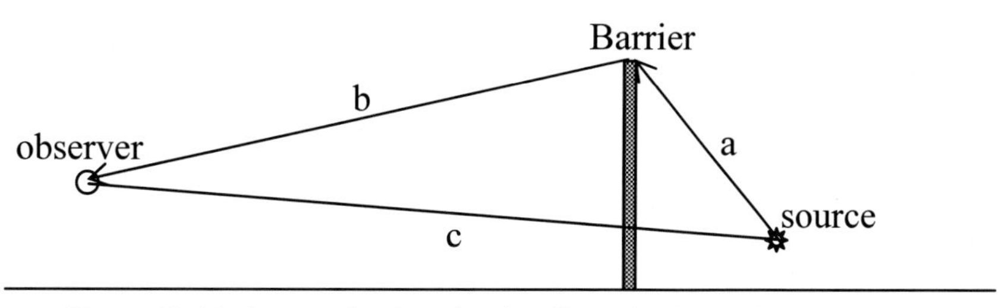

Figure 10.17: Attenuation by a barrier (formulas [10.10]) d=(a+b)-c

Formulas [10.10] tell us that the attenuation increases with N.
That means that high sound frequencies will be attenuated more than lower frequencies in any given source-wall-observer configuration.

N can also be written as $N = \frac{2d}{\lambda}$, where λ is the sound's wave length. When the source is close to the wall and the observer is far from it, N can be approximated by $N \approx \frac{2h}{\lambda}$, where h is the height of the wall. When the source and the observer are both very close to the wall, N can be approximated by $N \approx \frac{4h}{\lambda}$. By substituting these values of N in [10.10] we get bounds of what could be expected from a wall, where the most benefits are to observers close to the wall and the least are to observers farther away from it:

$$10\log(\frac{20h}{\lambda}+3) \le A \le 10\log(\frac{40h}{\lambda}+3) \quad \text{(point source)} \qquad [10.10a]$$

$$10\log(\frac{20h}{\lambda}+3)-(\frac{20h}{\lambda})^{0.3} \le A \le 10\log(\frac{40h}{\lambda}+3)-(\frac{40h}{\lambda})^{0.3} \quad \text{(line source)}$$

Usually, the total noise of a highway is measured in dBA. When the total noise of a source is measured in dB, all the frequency bands of that source have the same weight, so this is an absolute measurement of the noise of a source. Because of the lesser sensitivity of the ear to low frequencies, it is sometimes a more accurate representation of what we hear if the contributions of bands of lower frequencies are given less weight than bands of medium and high frequencies. In dBA, the weight given to the contribution of low frequency bands is approximated according to the threshold-of-hearing line in figure 10.5. When the typical spectrum of a highway is incorporated into [10.10], the attenuation of a barrier can be found directly from figure 10.18.

Trees at the top of the berm and trees that protrude above the barrier often help to reduce the level of sound for some frequencies, but may also diffract sound into the acoustical shade of the barrier and diminish the barrier's effectiveness. Formulas [10.10] are approximation to idealized circumstances, and they only provide estimates to real situations. They are not valid for values of c of figure 10.17 that are greater than 300m, due to sound reflection by the atmosphere.

Example

A sound barrier, 10 m high, is erected 25 m from a highway. (a) Find the attenuation of the highway's noise at a point 200 m from the barrier and 3 m higher than the highway. (b) Find the attenuation of the 50 Hz component of the highway's noise.

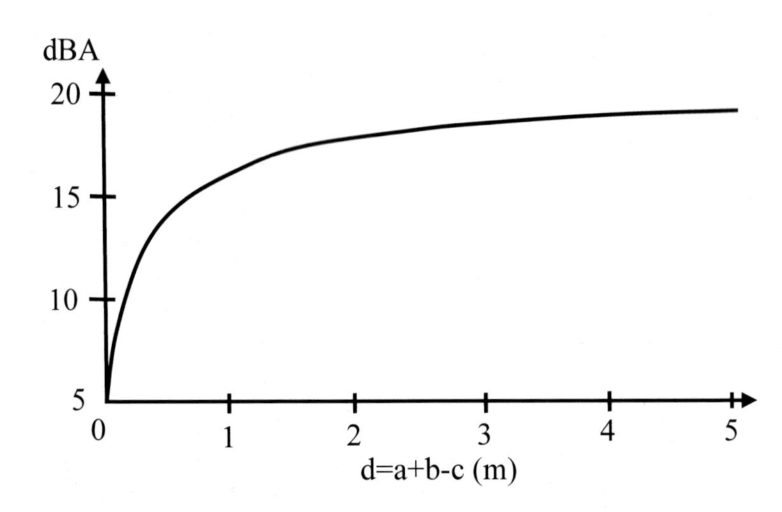

Figure 10.18: Attenuation (in dBA) of a standard highway noise by a semi-infinite barrier, as a function of d (figure 10.17).

Solution

(a) The following diagram (not to scale) illustrates the situation and the relevant variables, as used in figures 10.17 and 10.18.

In order to find the lengths of a, b, and c we use the Pythagorean Theorem for the applicable right angle triangles:

$$a = \sqrt{25^2 + 10^2} = 26.9\text{m}$$

$$b = \sqrt{200^2 + 7^2} = 200.1\text{m}$$

$$c = \sqrt{(200 + 25)^2 + 3^2} = 225.0\text{m}$$

d=a+b-c=2m

Based on figure 10.18, the attenuation would be approximately 18 dBA.

(b) The diagram that was used in part (a) is applicable also to the 50 Hz component, but now we have to use the second formula of [10.10]. Using the notation of this formula we get: f=50 Hz, c=340m/s, d=2m (from part (a)), N=2x50x2/340=0.59. Inserting in [10.10] we get: The attenuation is A=10xlog(20x0.59+3)-$(20\text{x}0.59)^{0.3}$=9.6 dB. So, the attenuation of the 50 Hz is smaller than the weighted attenuation of the entire spectrum. Longer wavelengths diffract more than shorter wavelengths, and therefore they are attenuated less.

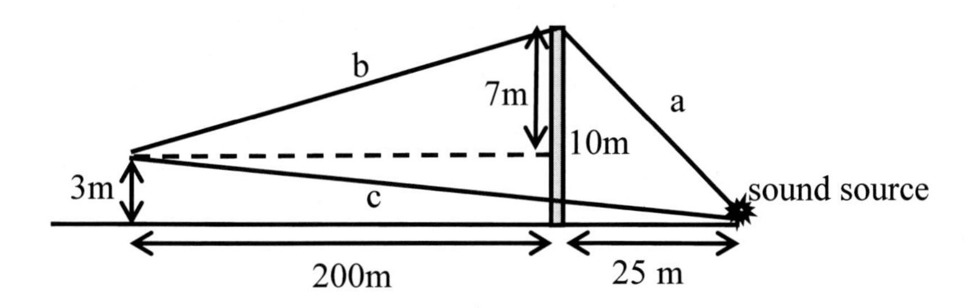

10.7 Case Study: The Music Center at Strathmore

By Zackery Belanger, Kirkegaard Associates, Chicago

The Music Center at Strathmore opened in 2005 in North Bethesda, Maryland. This new facility and second home to the Baltimore Symphony Orchestra houses a 1,976-seat concert hall and a full music and dance education center. Strathmore has presented a wide range of performing and visual arts since 1983, and the new Music Center expands their program to further enhance the lives of the public.

Figure 10.19. Strathmore Hall. Acrylic reflectors are aligned here horizontally in a canopy above the stage. The ceiling and portions of the walls are made of perforated metal, matching aesthetically with the design of the hall, while hiding sound reflectors and absorbers.

The concert hall is designed in the shoebox style for ample volume and desirable reflection structure. Several innovative acoustical systems were implemented in order to ensure high quality of sound throughout the hall for all types of musical performances.

One important acoustical system is the variable canopy of forty three independently-controlled clear acrylic reflectors. This array can be shaped and positioned to accommodate the full range of musical performances, from large symphonic and choral works to chamber and solo performances. The canopy primarily functions by providing musicians with important reflections for communication with each other onstage. A low ceiling would also provide reflections for communication, but would result in a smaller volume and therefore undesirably reduced reverberation. The hard canopy array is a common and effective way of providing needed reflections from the solid portions while still giving sound access to the volume above through the spaces and perimeter for reverberation. The array also functions to provide early reflections to the front central portions of the audience where side wall reflections arrive too late in time relative to the direct sound. The variable capabilities of the array are needed to accommodate a variety of instrument and musician layouts. For instance, a canopy geometry which is projective (somewhat vertical) and directs energy into the house is appropriate for a solo performance, and a more retentive (somewhat horizontal) geometry is helpful for a large orchestra where communication is of great concern. There are infinitely many possible canopy configurations and it is important to note that the canopy is an integrated component of the acoustic character of the room, so exact predictions of right configurations are difficult. When a new canopy is in place, either in a new hall or a renovated one, a tuning process occurs where careful listening, experimentation with position, and consideration of musician feedback are all used to hone the right standard configurations. Even beyond the tuning process, the canopy configurations can be refined and even changed through the years to suit changes in management, program, and conductor. At Strathmore this daunting range of possibilities is easily manageable by computer control systems through which presets can be established, allowing predetermined complicated settings to be attained at the push of a button.

You may wonder how hard reflectors with gaps can provide even coverage onstage without producing similar gaps in reflections. The individual reflectors at Strathmore have a convexly curved surface of large radius which helps to spread sound gently to fill in the gaps in coverage. In addition, diffraction effects occur at the panel edges which provide further infill. For low frequency sound, the panels work together in an array effect to reflect sound which would not be reflected by individual panels alone. Successful canopy arrays such as the one at Strathmore require a balanced approach to panel size, spacing, curvature, and material choice, in addition to carefully chosen presets of array geometry. Too much overlap in coverage can easily lead to unpleasant interference effects, and monolithic hard canopies tend to block too much sound from reaching the volume above. Recent canopies designed for the Royal Festival Hall renovation in London, UK and the Rensselaer Polytechnic Institute Experimental Media and Performing Arts Center in Troy, NY have made use of a monolithic lightweight fabric which selectively passes low frequencies for reverberation and returns mids and highs for communication. Initial reactions to this new approach are positive, but the hard reflector array remains a mainstay of concert hall acoustics.

Although the canopy is the primary adjustable reflective acoustic system in the hall, most of the adjustable acoustic treatments are absorptive. Extensive areas of retractable absorption which are visually masked by acoustically transparent meshed and perforated metals can be deployed along the side, upstage, and rear walls of the hall. When deployed, these absorptive treatments lower reverberation time and control reflections. When retracted, the absorptive treatments are rendered ineffective and the reflective wall surfaces behind are exposed, raising reverberation time.{Pictures} Higher reverberation times are desirable for some types of performance, such as choral pieces, and low reverberation time is desirable for other types, such as amplified or spoken word performances. Strathmore's reverberation time (unoccupied) varies from about 1.7 seconds at 1000 Hz to 2.1 seconds at 1000 Hz depending on the absorption settings, and there are many configurations which give reverberation times between these two extremes which are utilized for a wide variety of performances. The acoustically transparent visual layer was carefully chosen and tested and gives the hall a consistent look across a variety of acoustic settings.

Figure 10.20 The acrylic canopy. Top, view from the main floor. Bottom Each panel of the canopy is individually aligned, according to the requirements of the upcoming event.

Audio and video systems include both a high-energy left-center-right sound reinforcement system and a smaller speech reinforcement system integrated with the architectural design. Both systems are designed as extensions of the natural acoustics of the hall. Halls such as Strathmore are ultimately used as multipurpose spaces, even though they are designed primarily for orchestral performances. Many non-orchestral performances require the use of an amplification system (a lecture or a film, for instance), so a house audio system is often utilized. Power and other electrical are provided for non-permanent touring amplified systems (such as for rock shows), but the house audio system is very convenient and is often used. During amplified events, two central canopy reflectors are tilted to provide room for the central loudspeaker array, which is lowered in from above. The remaining portions of the canopy are configured to direct rear-radiated energy from loudspeaker clusters toward absorption deployed in the attic. Adjustable absorption is all deployed to reduce reverberation time as much as possible. A general rule for the room acoustics for amplified music is to make the room as acoustically "dead" as possible. Rooms with lower reverberation times provide more intelligible speech and help to control the excessive sound energy often present with the use of amplification.

The remainder of the materials which make up the structure and interior of Strathmore are all carefully chosen and tested such that their acoustical properties contribute to the desired balance of spectrum, reverberation, and reflection structure. Airspaces are minimized within the wall, floor, and ceiling systems to prevent excess low frequency absorption. Lightweight systems which cover air cavities--most modern construction can be described this way--tend to vibrate easily when exposed to low frequencies. This vibration transfers energy into the air cavity where it is either lost to further transmission through the wall system, or damped out in the air cavity. This leaves a low-frequency delinquency in the hall. This can be avoided by providing very solid, massive materials, which will reflect low frequency energy back into the space as well as assist in keeping external noise out. Strathmore's walls are constructed of 12" thick concrete. All materials are considered, specified, and tested so their acoustic properties are beneficial to the spectrum of the concert hall. Most materials are kept as reflective as possible to allow the adjustable curtains and banners to provide the needed absorption and keep the room reflective in its most reverberant state. However, a fine porosity which results in very high frequency absorption is often helpful to reduce the brightness of a room in its reflective state. Seats are usually kept as reflective as possible in order to not exaggerate the audience absorption in an occupied configuration. However, some acousticians prefer to keep unoccupied seats absorptive to emulate the absorption of an audience during rehearsal conditions.

The overall geometry of the room is designed to sustain the right amount of reverberation without promoting distracting, unwanted reflections. Balcony fronts are shaped to prevent echoes to stage, and balcony undersides are shaped to provide infill reflections to under-balcony audiences.

Figure 10.21 Top: The perforated metal ceiling. Bottom, adjustable reflectors behind the ceiling.

Figure 10.22. The 1/16 scale model.

Although the classic shoebox design is known for providing exceptional acoustic performance, the form of the concert hall was sufficiently complex to warrant both scale and computer acoustic modeling. The primary benefits of computer modeling are low cost, ease of use, and full frequency analysis. Computer modeling is less expensive than scale modeling, but exhibits limitations based on geometry. At the time of this writing the two most commonly used acoustic modeling software programs are CATT and ODEON, both of which are very useful but limited to faceted geometries rather than curves, and neither model the diffraction component of sound. The classic scale model approach was pursued to explore these missing components of the computer model. Scale models inherently include curvature and diffraction since they employ built materials and real sound.

The Strathmore concert hall scale model was built at 1:16 scale to investigate several design concepts. This is quite large for a built model--the length of the interior of the full-scale hall is 177 feet (54 m), which scales to a model over 11 feet long. (Picture/s) This relatively large model is beneficial (even necessary), however, for acoustic modeling. When a hall's geometry is reduced for scale model testing, the wavelengths of sound used must be reduced by the same factor, because diffraction patterns depend on the ratio between the wavelength and the size of the reflecting element. This means an increase in frequency (Based on [10.1], $c=\lambda f$, a shorter wavelength has a higher frequency). For example, if you wish to test a hall for frequencies of 500 Hz to 2,000 Hz using a model built at 1:16 scale, then the frequencies of interest must be multiplied by 16 (to correspond to a division by 16 in wavelength), so the actual tested frequency range is 8000 Hz to 32,000 Hz. Much of this is in the ultrasonic range! Special equipment must be used for this type of testing since most audio equipment is only designed for the range of human hearing, which has an upper limit of 20,000 Hz. And there is another subtlety of this scaling which you may have noticed: even if we reduce the geometry of the hall, and the wavelengths of the sound used, we have no practical way to reduce the dimensions of the medium--air. The distances between air molecules remain the same no matter how we scale the model and wavelengths. Some researchers and acousticians have employed pressurized gas to decrease these distances, but for most of the industry this approach is not practical. When using regular air at atmospheric pressure, the air absorption of sound starts to become significant at high frequencies (and remember, we're in the ultrasonic range here!). A mathematical correction for this effect is used, and the models are built as large as possible (this explains the 1:16 scale of the Strathmore model) to keep the frequencies as low as they can be.

Testing is conducted within the scale model just as it is in a full size hall, by producing sound with a source, allowing the sound to fill the space, and recording the direct and reflected sound waves with microphones. Results are then analyzed to inform the decision making process during design. The physical limitations of scale model size, air absorption, and equipment limitations limit the frequency range which can be explored with a scale model. For Strathmore, a range of 250 Hz to 4000 Hz was possible. This leaves out much of the range of human hearing, but still proves valuable for geometric studies of reflection structure, and it happens that this range is of primary importance in concert hall design because many orchestral instruments fall within this range.

It should be noted that computer and scale model testing are used as complements to each other. Computer models are inexpensive and efficient, and significant changes are easy to make, but they currently lack true geometric curves and diffraction and absorption coefficients tend to not translate well from lab tests. Scale models are much more expensive and time consuming, but model diffraction and curves well. A downside to scale model testing is the difficulty involved in scaling the properties of materials down such that they absorb and reflect ultrasonic frequencies in the same way as full scale materials absorb and reflect audible frequencies. Scale models should be considered geometric studies which can incorporate all-absorptive and all-reflective surfaces, but without the subtle differences present in the wide range of full scale materials which make up a concert hall. It is reasonable to believe that as better software programs are made available and computing power increases, acoustic scale modeling will become unnecessary. For now, a combined scale and computer model approach remains the best approach for predictive testing of concert halls. However, experience, good judgment, and a receptive, inventive design team are invaluable in the design of a successful concert hall.

Figure 10.23: Acoustic plan of Strathmore Music Hall. Shaded areas indicate the original and the reflected sound patterns from the various reflectors at the top of the hall.

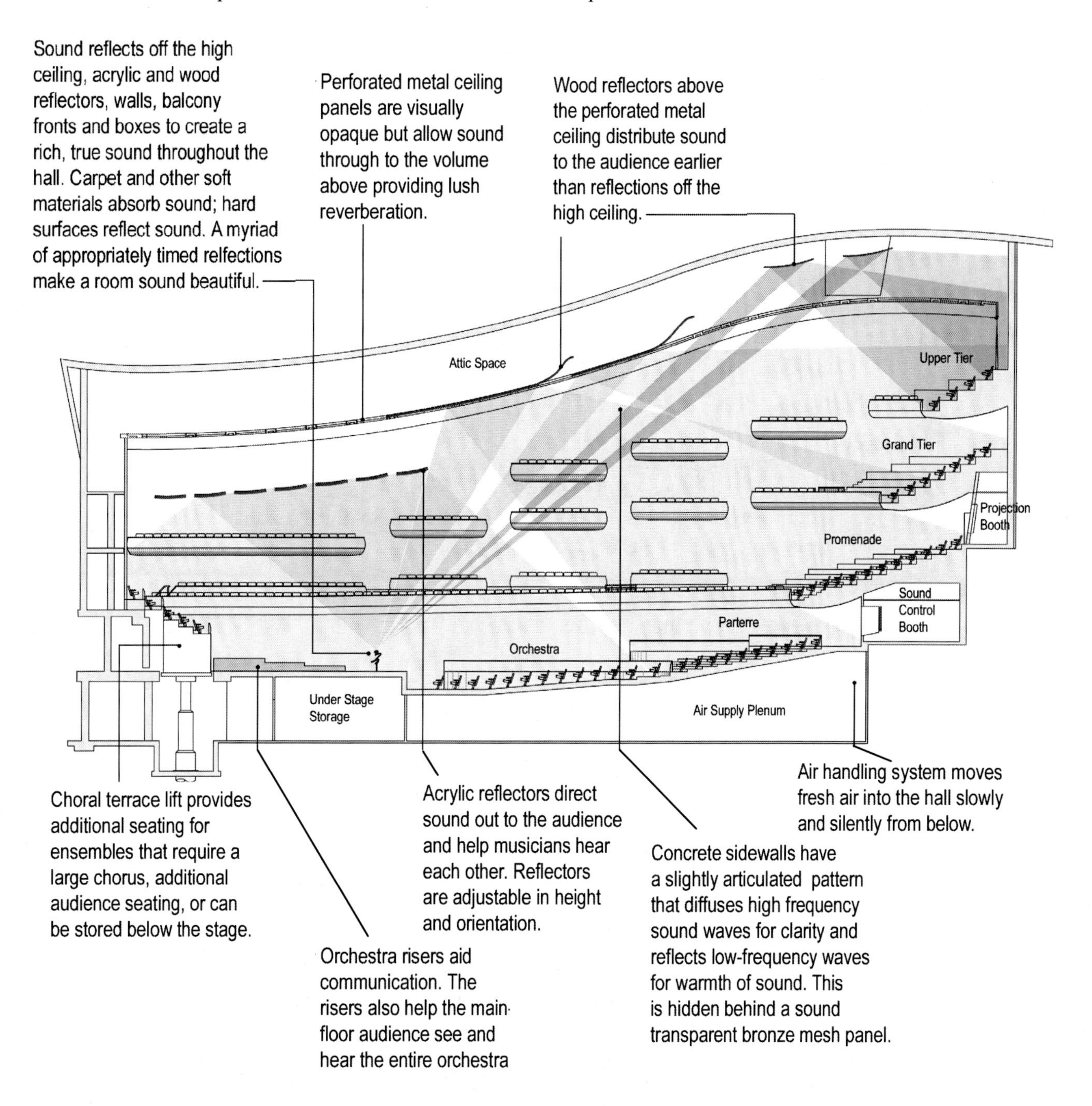

The design team for Strathmore was led by design architects William Rawn Associates, architect of record Grimm + Parker, and included theater consultant Theatre Projects Consultants of South Norwalk, Connecticut and acousticians Kirkegaard Associates of Chicago.

PROBLEMS

Sections 10.1, 10.2

1. A gun at the starting line of a 100 meter dash is used to start the race. The referee at the finish line is instructed to start the stop watch as she sees the flash of the shot. By how much would her reading be different had she started the stop watch when she hears the sound of the shot?

2. (a) What is the wavelength in air of a sound wave whose frequency is 60 Hz? (b) What would be the wavelength of that sound wave in water?

3. What are the lowest and highest frequencies of the octave band whose center frequency is 4,000 Hz?

Section 10.3

4. A point sound source emits a sinusoidal wave at a frequency of 2,000 Hz. The power of the sound at the source is 10^{-5}watt. What are: the sound intensity, the sound intensity level and the loudness of that sound at (a) a distance of 1m from the source? (b) at a distance of 2m from the source?

5. The sound from a busy highway contains a 500 Hz Fourier component whose power at the highway is 1.3×10^{-6} watt/m. What are: the sound intensity, the sound intensity level and the loudness of that Fourier component at (a) a distance of 50 m from the highway? (b) at a distance of 100m from the highway? (Ignore sound interactions with the ground).

6. The sound intensity level of a violin playing in a concert hall, as measured in the center of the hall, is 50dB. What would be the sound intensity level at the same point if three identical violins would play the same music from the same area on the stage?

7. When a 1.5'x3' window is open, the sound-intensity-level of the street's noise inside the room is 60 dB. When the window is closed, it blocks all the street's noise. (a) What would be the noise-intensity-level inside the room when the window is left slightly open, so that the open area is 18 square inch? (b) What would be the street's noise-intensity-level when the window is open to 18 square inches, if when the window is completely closed the street noise-intensity-level inside the room is 20 dB?

Section 10.4

8. A sound reflecting panel is placed on the ceiling of a lecture hall, as shown in the figure. Using the image method find by graphing: (a) the area on the floor

onto which the sound will be reflected by the reflecting panel. (b) The sound intensity level of the reflected sound reflected by the panel at the listener's location, if the intensity of the source is $6x10^{-5}$watt, and the reflective coefficient of the reflecting panel is 0.95. (Measure the needed distancing from your graph, and use the scale to find the real distances.) (c) What is the sound intensity level at the listener's due to the sound reflected from the ceiling, if the ceiling's reflecting coefficient is 0.8?

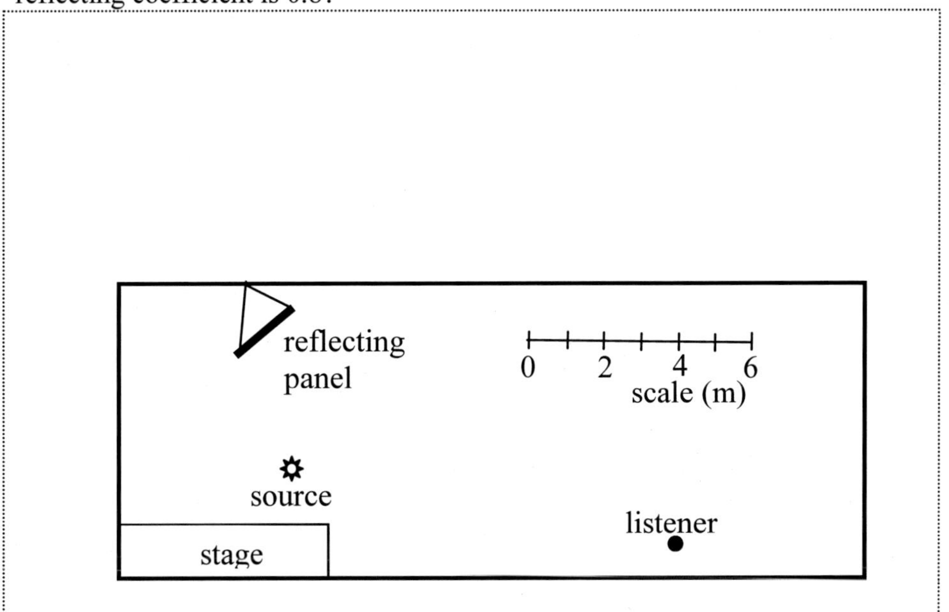

(Note: This drawing is of the center cut of the hall. The source and the listener are found on that plane, and the sound reflecting panel crosses it and is perpendicular to it. In general, sound reflection problems are three dimensional. The same principles apply also to the 3-D cases; however 3-D cases are usually solved with computers).

9. A sound point source, situated 2.0 m above a brick floor, is emitting sound at $4x10^{-5}$ watts. (a) Find the sound intensity level due to that source at a listening point 3.5 m above the floor and 15 m away. (b) Find the sound intensity level at that point when the reflection from the floor is taken into account.

10. Find the sound absorption (in meter-sabin and in foot-sabin) of a 2ft x 3ft glass window for sound waves of frequency 125 Hz and 4,000 Hz.

11. What is the sound transmission loss of a wall whose transmission coefficient is 0.02? What would be the sound intensity level of a sound after passing that wall, if its intensity level entering the wall was 70 dB?

12. The floor of a 20m x 18m x 8m chapel is covered by a wall-to-wall heavy carpet. The windows take 20% of the area of the walls, which are typical frame structure (1/2" gypsum board, 2"x4" nominal studs, 1/2 "gypsum board). The ceiling is made of ½" gypsum board. What is the reverberation time for sound of frequency 2,000 Hz? How can you bring the reverberation time to 1.4 seconds by

selecting other materials for the chapel? (Which materials will you choose, and how much?)

13. What is the transmission coefficient of a wall whose sound transmission loss is 30 dB? What would be the sound intensity of a sound after passing that wall, if its intensity entering the wall was $2x10^{-4}$ watt/m^2?

14. Analysis of the acoustics of Roman amphitheaters. In Roman amphitheaters, the seats were arranged in semi-circular rows. Most amphitheaters had a low wall behind the stage, or no wall at all. Very few, such as the one at Aspendos, Turkey had a high back-wall behind the stage (figure 10.24). Drawings A and B are of the plan at the level of the lower rows, and drawing C is the center vertical cross section. The dots denote the center of the circles of the seats and the stars indicate arbitrary sound sources on the stage.

Using ray-tracing, analyze the following issues: (a) Do direct sounds from those sources that are reflected by some seats reach other seats on the same row? If yes, which seats reflect to which? (Use drawings A and B) (b) What is the lowest back-wall that is needed for reflecting the sounds of those two sources to all the seats? (Use drawing C.)

15. The spectrum of a $4x10^{-6}$watt/m^2 sound, entering an 8" wall made of hollow light concrete blocks, is : 25% of the energy is in the 125 Hz band, 35% of the energy is in the 500 Hz band, and 40% of the energy is in the 4,000 Hz band. What would be the total intensity of that sound upon leaving the wall?

16. Two separate point sources A and B emit sinusoidal sound waves of wavelength 0.12m. Their waves have the same phase at the sources. After a while, those waves interfere. Find the kind of interference (constructive, destructive, or partial) at the following points: (a) 0.82m from A and 0.76m from B. (b) 1.14m from A and 1.26m from B. (c) 1.12m from A and 1.12m from B. (d) 1.14m from A and 1,16m from B. (e) 2.62m from A and 2.98m from B. (f) 4.77m from A and 4.47m from B.

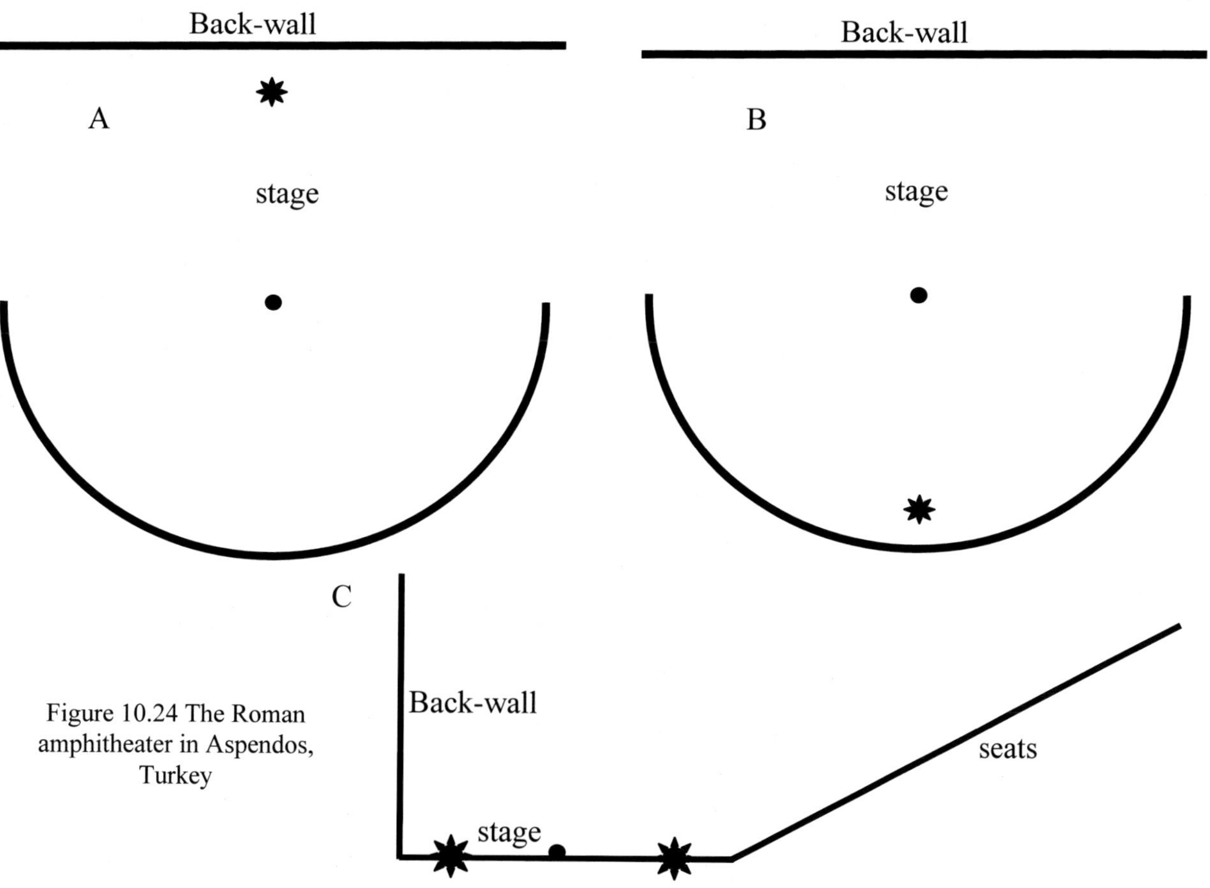

Figure 10.24 The Roman amphitheater in Aspendos, Turkey

Sections 10.5, 10.6

17. What are the wavelengths of the fundamental mode and its first and second harmonics of standing sound waves between two parallel walls, 2.5m apart from each other?

18. What is the frequency and wavelength of the standing wave with the longest wavelength that can happen in a room 3.5m x 4.2m x2.2 m? What is the frequency of the next standing wave in that room?

19. Sound waves of various wavelengths are passing through two narrow slots separated by 0.8m. What is the shortest wavelength that would pass those slots without undergoing total destructive interference?

20. An absorbing panel has to be placed in front of a sound reflecting wall. At what distance from the wall should the panel be placed so that it would be most effective in absorbing sound whose frequency is 250 Hz?

21. A window of a house is 4 m above the ground. The house itself is 110m from a highway. A 8 m high sound barrier is built 12 m from the highway. The base of the house and the highway are at the same elevation. (a) What is the sound attenuation provided by the barrier to that window? (b) What is the attenuation of the 100 Hz band? (c) What is the attenuation of the 2,000 Hz band?

Conversion Factors

Length
1 m=39.3701 in.=3.28084 ft=1.0936 yd
1 km=0.62137 mi
1 in.=2.54 cm
1 yd= 0.91440 m
1 mi=5280 ft=1,609 m= 1760 yd

Area
1 m^2=1.196 yd^2=10.76 ft^2=1,550 in^2.
1 in^2=6.452x$10^{-4}$$m^2$=6.452 cm^2=1/144 ft^2
1 ft^2=9.290x$10^{-2}$$m^2$=144 in^2=1/9 yd^2
1 km^2=0.3861 mi^2=1.076x$10^7$$ft^2$

Volume
1 m^3=1,000 L=35.31 ft^2=61,020 in^3=
=264.2 gal (US fluid)

Time
1 s=1/60 min= 1/3,600 h= 1.157x10^{-5} d=
=3.168x10^{-8} yr
1 min=60 s
1 h=3,600 s
1 d=8.640x10^4 s=1440 min=24 h
1 yr=3.156x10^7 s= 365.24 d

Speed
1 m/s=3.600 km/h=3.281 ft/s= 2.237 mi/h

Acceleration
1 ft/s^2= 0.30480 m/s^2
Standard g=9.80665 m/s^2=32.174 ft/s^2

Mass
1 kg= 6.582x10^{-2} slug
1 slug=14.59 kg

Force
1 N=0.2248 lb

Mass-Weight relations
1 kg mass weighs 2.205 lb at standard g.
A weight of 1 ounce (oz) =1/16 pound at standard g has a mass of 2.835x10^{-2} kg.
A weight of 1 pound (lb) at standard g has a mass of 0.4536 kg.

Pressure
1 pascal (Pa)=1 N/m^2= 9.869x10^{-6}atm=
=1.450x10^{-4} lb/in^2
1 atm=1.013x10^5Pa=760 mm of mercury (at 0^0C)= 29.92 inches of mercury=
=406.8 inches of water (at 0^0C)=

Work and Energy
1 J=0.7376 ft·lb= 0.2388 cal=
=9.478 Btu=2.778x10^{-7}kW·h=
=3.735x10^{-7}hp·h=6.242x10^{18} eV
1 ft·lb=1.356 j=0.3239 cal=
=1.285 x10^{-3}Btu
1 cal=4.186 J
1 Btu=1055 j=252.0 cal=
=2.930x10^{-4} ^{7}kW·h
1 eV=1.602x10^{-19}J

Power
1 W=1 J/s=0.7376 ft·lb/s=
=1.341x10^{-3} hp=0.2389 cal/s=3.413 Btu/h
1 hp=550 ft·lb/s=745.7 W

Plane Angle
1 rad=57.30^0 =0.1592 rev
1 rev=2π rad

Solid Angle
1 sphere=4π steradians (sr)

Solutions of Selected Problems

Chapter 1

Algebra

1. d=7w 3. c=100d 5. O=16p 7. d=7w+30.5m (approximate) 9. S=(x+75+y+98)/4
11. 211 13. -13 15. 50 17. 6 19. 3.333 21. 2.333 23. 4.529 25. 7 27. 1.125
29. 26 31. -3.1667 33. 0.95 35. -8 37. any x 3 9. t=S/v, v=S/t
41. a=2(X-x0-v0t)/t2, v0=(X-x0-0.5at2)/t 43. k=HL/(A(t2-t1)), t2=HL/(kA)+t1
45. S=(2f1L)2m, m=S/(2f1L)2 47. (3,4) 49. (3.5, -5.5)
51. (1.8, 1.8) 53. (-1, -1.143) 55. (0.5, 2) 57. (1,1) 59. True, False 61. nine times

Geometry

1. 110cf 3. a 40; b 44; c 52; d 145; e 34

Trigonometry

3. 6.61m, 0.82676 5. 20.53 m; 223.9 cm; 0.80f 7. 45.33, 25.93, 19.53, 117.5^0
9. (a) 15.6 feet.(b) 20.1 feet.

Units

1 a) 3,153,600 s b) 3.11×10^{-8} year c) 18f c) 2 yard d) 432 I^2 e) $2\times10^6 cm^2$
f) 7.29 m^3 g) 9.33 cents/min h) 200 cm/sec i) 6.11 cm/sec

3. $\frac{4.17\ USD}{gallon}$, which is 2.86 times higher than in the US.

Chapter 2

1. (a) 4.5m (b) 11.3m (c) 11.3m (d) 4.5m 3. (a) 60mph (b) 95.54 km/hr
(c) 26.82 m/s. 5. 23.09mph.

7.

Time (sec)	X=3t	X=-3t	$X=4+3t-0.5t^2$
0	0	0	4
1	3	-3	6.5
2	6	-6	8
3	9	-9	8.5
4	12	-12	8

9.
Car a: 55 km east of A, moving east at 30 km/hr.
Car b: 55 km east of A, moving west at 30 km/hr.
Car c: 15 km east of A, moving east at 30 km/hr.
Car d: 15 km east of A, moving west at 30 km/hr.

11. d=0.2m; t=0.01s; v=? v=0.2/0.01=20 m/s.
13. 83.22m 15. 234.4m 17. 1.04 m/s. 19. 3.85 m/s, down
21. 12,500N. 23. 0.33slugs. 25. 735N. 27. $D=w/V=mg/V=\rho Vg/V=\rho g$.
29. 6874.7N. 31. 16.3N. 33. 0.314m. 35. a) $\mu=0.33$. b) 2.15 f/s^2. c) 109 lb.
37. The situation in a chain is the same as in the cable described in Figure 2.11. The only difference is that a cable is made of segments, and a chain is made of links.
39. (a) $-5m/s^2$. The acceleration is in the same direction of the force that causes it.

(b) 1.6m. c) 0.8s. (d) 0.51.
41. 8.2×10^{-7}Hz. 43. F=-kx (a) 367.5N/m (b) 0.02m; c) 0.57s (d) 7.83cm. (e) 0.22m/s. (f) -0.219m/s moving up.
45 (a) 2Hz. (b) The boxer jumps at this same frequency, so there is a risk of resonance. To avoid it, change the mass and/or the wire that supports the chandelier, so that k/m changes.

Chapter 3.

1,2. (a) (60m,103.92m) (b) (-60m,103.92m) (c) 78.10m; 50.19^0 with respect to the positive x direction. (d) 78.10m; 50.19^0 with respect to the x axis in the third quadrant=230.19^0 with respect to the positive direction of the x-axis.
3. 45^0 with respect to the positive x direction; 45^0 north of east.
(b) 21.8^0; 201.8^0 with respect to the positive x direction; 21.8^0 south of west.
(c) 565.69m; 538.52m; 1,081.67m;
5, (1,a) 55m; (1,b) 0^0; (1,c) 5.5m/s; (1,d) 0^0.
(2,a) 5m; (2,b) 180^0; (2,c) 0.5m/s; (2,d) 180^0.
(3,a) 39.1m; (3,b) 50.2^0 north of east; (3,c) 3.91m/s;
(3,d) t50.2^0.
7. first contestant wins.
9.(a) The new axis is in the direction of the force, 30^0 with respect to the x-axis. In the new axis: $a=20m/s^2$ in the direction of the new axis, which is 30^0 with respect to the x-axis. (b) In the old axis: $a_x=17.32m/s^2$. (c) $F_y=40N$; $a_y=10m/s^2$..
11. 55N, 12.22 m/s^2.
13. F=150N; the torque of F has to be CW.
15. (a) (a) 0 (b) -3.9 N·m.

Chapter 4

1. 30N m 50.28^0 3. 40 lb. 30 lbs.
5 (a) 861N; 340N (b) 1,525N; 1,168N (c) 2,102N ; 2,319N (d) F_3=1,546N; F_5=326.7 N; F_2=0.9063 $F_3=F_4$=1,401N (e) F_3=497.6N; $F_2=F_4$=1,433N. (f) F_3=331.7N
F_4=1,277N. This is the compression at the bottom of the beam, all the way up to the point of attachment of the weight. At the top of the beam, in the beam's direction: F_2=311.7N. This is the compression at the top of the beam, all the way down to the point of attachment of the weight.
7. 42.5 lbs.
9. R^1_y=3,675N. R^1_x=1337.6N.
11. 6.17 lbs 35 lbs 130.6 lbs 0.088; 0.5; 1.87 respectively.
13. Each segment exerts 30,000/2=15,000 lbs on each post. Therefore, each of the outer posts supports 15,000 lbs, and the center post supports 30,000 lbs.
15. $T_0=7.35\times10^6$N. $T_6=9.41\times10^6$N.
17

i	Alfa i [4.21]	gama i [4.27]	delta i [4.27]
0	0.00	49.11	49.11
1	60.00	81.05	70.89
2	73.90	86.90	85.05
3	79.11	88.47	87.89

19. (a) $R_1=R_3=833.3N$ $R_2=8,333N$. (b) $R_1=R_3=4,762N$ $R_2=476.2N$ (c) $R_1=R_3=3,333N$ $R_2=3,333N$
21. 67,620 Pa.
23. 40,000N.
25. (a) $A=1.78x10^{-4}m^2$. diameter=0.015m. (b) diameter=0.017m.
27. (a) 34,300Pa (b) 269.3N
29. (a) 10,662Pa. (b) 101,292Pa. (c) 10,000Pa. (d) 7,840Pa.
30. 7,683.2Pa.

Chapter 5

1. 322.1Pa. 173.9N. Its weight is 86.73 N.
3. 75,760N
5. Junction of first and second stories: 18,163N. Junction of second and third floors: 9,081N.
7. The horizontal bending of the top was 0.015 m. The building survived.
9. T= 334 lbs.

Chapter 6

1. (a) 4.59J; 0J. (b) 4.59J (c) 230N.
3. (a) 9,600J (b) 7,056J (c) 2,544J. (d) 10.63 m/s. (e) 9,600J.
5. (a) 1.98 m/s. (b) 1.85 m/s.
7. $6.42.
9. 2,200m/s.
11. 46,800 $u_X=14.18m/s=u_Y$. 66,177kg m/s 46,794 kg m/s 46,794 kg m/s 31.2m/s

Chapter 7

1. 135 ^0C. 3. 36/1.8=20^0C. 5. 0.044f.
7. When heated the water will spill, and when cooled it will contract below the rim.
9. 36.8 PSI. 11. 3.62 m^3. 13. 1.84 kcal. 15. 160.3 kcal. 17. 26.8 ^0C.
19. 0.002 lb moisture/lb dry air
21. 740 gram. 23. 9.56 BTU (b) 0.008lbs.
.25. The plywood wall is a better insulator.
27. R=1.07 for the winter and R=1.31 for the summer.
29. (a) 0.12 (b) 1,327 BTU hr. 31. 70.2 BTU. 33. $110. 34 217 BTU.

Chapter 8

1. (a) $3.375x10^{-3}N$ in the positive x direction.
(b) $5.625x10^{-3}N$ in the negative x direction.
(c) $-2.25x10^{-3}N$ (in the negative x direction).
3. (a) 4N, (b) 1.78N.
5. 12μJ.
7. (a) 2,420W. (b) 7.33A, (c) 9.6W, won't heat (d) 1,500W. too hot.
9. 2.51mm.

11. (a) 0.048A, (b) 0.12A, (c) 1.2A.
13. 2,000W.
15. Based on [8.10-11]. V=120V, I_1=1.2A, I_2=0.6A. P_1=144W, P_2=72W. P=216W.
17. 2.5×10^{-4}A, V_2=425V, $R_2=1.7\times10^{6}\Omega$.
19. (b) The current in each wire would be 10A. The current that should trigger the circuit breaker is 30A. Any wire in the entire system, including wires in the room, should be able to carry at least 30A.

Chapter 9

1. 0.13 s. 3. 3x1010Hz. Microwave. 5. 3.54×10^{18}, 2.93×10^{18}
7. yellow, saturation=62.5%; cyan, saturation=33.3%; magenta, saturation=57.1%.
9. 4.91×10^{18}. 11 1,005 lumen. One bulb of 1,200 lumen.
13. 105lux. 15. 0.539=54%. 17. (a) 2.25×10^{8}m/s; (b) 1.24×10^{8}m/s.
19. 250. 21. 33.40

Chapter 10

1. By 0.29 seconds shorter.
3. 2,828Hz , 5,657Hz
5. 36.2dB, 37phon (approximately), (b) (b) 34phon (approximately).
7. (a) 44dB. (b) 44dB.
9. (a) 1.34×10^{-8}W/m^2, 41dB. (b) (assuming 100% reflection) 2.65×10^{-8}W/m^2, 44dB.
11. 53dB.
13. τ=0.001, 2×10^{-7}W/m^2.
15. 7.35×10^{-8}W/m^2.
17. n=1, λ=5; n=2, λ=2.5; n=3, λ=1.67m.
19. 1.6m
21. (a) approximately 17dBA. (b) 12,24dB. (c) 20.85dB.

Credits

Unless otherwise noted, each table and graph was prepared by the author from several sources, and each figure was made by the author. Abbreviations have been used in the text to note other copyright holders of figures and tables as follows:

ASHRAE:
© 1993, 1981 American Society of Heating, Refrigerating and Air-Conditioning Engineers, Inc. ASHRAE Fundamentals Handbook, I-P edition. Used by permission. Figure 5.1, Tables 7.4a, 7.4b, 7.4c, 7.5a, 7.5b, 7.5c.

MSCUA:
© Manuscripts, Special Collections, University Archives, University of Washington Libraries. Used by permission: Images UW21413, UW21422, UW20731, FAR028. Page 57.

NIST:
Courtesy of the National Institute of Science and Technology. Page 39.

USGS:
Courtesy of the United States Geological Survey. Figures 5.6, 5.7, 5.8, 5.9, 5.11, 5.15, 5.16, 5,16a.

Lechner:
Norbert Lechner "Heating, Cooling, Lighting: Design Mrthods for Architects. Second Edition." ©2001 by John Wiley & Sons, Inc. Used by Permission. Figures 7.6, 7.10. Table 7.5c.

Cover: National Park Service, Jefferson National Expansion Memorial (arch). and Bev Sykes Background).

© Fig 1.6 left, Allen Luke, Fig 1.6 right Roberto Caucino, Fig 4.12 National Park Service, Fig 4.19 Jeffery Bochert, Fig 4.20, National Park Service, Fig 5.13 FEMA, Fig 5.22 left Chris Pritchard, Fig 10.9 left Selahattin Bayram, Fig 10.19 right DieBuche.

INDEX

M

N

O

P

Q

R

S

T

U

V

W

X

Y

Made in United States
North Haven, CT
03 February 2024